Recent Advances and Future Trends in Pavement Engineering

Recent Advances and Future Trends in Pavement Engineering

Special Issue Editor

Patricia Kara De Maeijer

MDPI • Basel • Beijing • Wuhan • Barcelona • Belgrade • Manchester • Tokyo • Cluj • Tianjin

Special Issue Editor
Patricia Kara De Maeijer
University of Antwerp
Belgium

Editorial Office
MDPI
St. Alban-Anlage 66
4052 Basel, Switzerland

This is a reprint of articles from the Special Issue published online in the open access journal *Infrastructures* (ISSN 2412-3811) (available at: https://www.mdpi.com/journal/infrastructures/special_issues/pavement_engineering).

For citation purposes, cite each article independently as indicated on the article page online and as indicated below:

LastName, A.A.; LastName, B.B.; LastName, C.C. Article Title. *Journal Name* **Year**, *Article Number*, Page Range.

ISBN 978-3-03936-316-2 (Hbk)
ISBN 978-3-03936-317-9 (PDF)

Cover image courtesy of Patricia Kara De Maeijer.

Contents

About the Special Issue Editor

Patricia Kara De Maeijer is a civil engineer, and a researcher in EMIB research group at the Faculty of Applied Engineering at the University of Antwerp (Belgium). Patricia's main research areas are the design and installation of fiber Bragg grating (FBG) sensors to monitor pavement behavior in real-time (prototypes CyPaTs-UAntwerp and BVDZ-Port of Antwerp), heavy-duty pavement design; concrete technology; asphalt and bitumen; and the recycling of industrial wastes and by-products. Patricia is an active member of RILEM (International Union of Laboratories and Experts in Construction Materials, Systems and Structures) and a member of several RILEM Technical Committees. Patricia has authored and coauthored more than 50 publications in international scientific journals and conference proceedings.

infrastructures

MDPI

Editorial

Special Issue: Recent Advances and Future Trends in Pavement Engineering

Patricia Kara De Maeijer

RERS, EMIB, Faculty of Applied Engineering, University of Antwerp, 2020 Antwerp, Belgium; patricija.karademaeijer@uantwerpen.be

Received: 31 March 2020; Accepted: 3 April 2020; Published: 5 April 2020

Abstract: This Special Issue "Recent Advances and Future Trends in Pavement Engineering" has been proposed and organized to present recent developments in the field of innovative pavement materials and engineering. For this reason, the articles and state-of-the-art reviews highlighted in this editorial relate to different aspects of pavement engineering, from recycled asphalt pavements to alkali-activated materials, from hot mix asphalt concrete to porous asphalt concrete, from interface bonding to modal analysis, from destructive testing to non-destructive pavement monitoring by using fiber optics sensors.

Keywords: interface bonding; moisture damage; alkali-activated materials; RAP gradation; hot mix asphalt dynamic modulus; porous asphalt concrete; FOS; FBG; flow number; rutting; modal analysis

---

The twelve articles and state-of-the-art reviews of this Special Issue, "Recent Advances and Future Trends in Pavement Engineering", partly provided an overview of current innovative pavement engineering ideas, which have the potential to be implemented in industry in the future, covering some recent developments.

The interface bond between layers plays an important role in the behavior of pavement structure, especially in asphalt pavements. However, this aspect has not yet been adequately considered in the pavement analysis process due to the lack of advanced characterizations of actual conditions. Recently, it became one of the most important research topics in the field of pavements. RILEM TC 206-ATB, TC 237-SIB, TC 241-MCD and TC 272-PIM are among the most important international research activities considering this topic. Le et al. [1] suggested an interesting methodology for considering the interaction between pavement layers represented by a horizontal shear reaction modulus. Using this methodology, the field condition of the interface bond between the asphalt layers of experimental pavements in a full-scale test can be assessed using back-calculation from non-destructive testing. This study is a very good example providing a better understanding of the structural behavior of asphalt pavements and can contribute to a better evaluation of their long-term performance.

Moisture susceptibility is still one of the primary causes of distress in flexible pavements, reducing the pavements' durability. A very large number of tests are available to evaluate the susceptibility of a binder aggregate combination. Tests can be conducted on the asphalt mixture, either in a loose or compacted form, or on the individual components of an asphalt pavement. Apart from various mechanisms and models, fundamental concepts have been proposed to calculate the thermodynamic tendency of a binder aggregate combination to adhere and/or debond under wet conditions. Soenen et al. [2] summarized literature findings on the applied test methods, the obtained results, and the validation or predictability of these fundamental approaches.

The need to differentiate the pavements according to the final intended use has created different paving solutions, in terms of construction technology and materials. From the traditional bituminous pavements, the new design solutions encompass the application of special asphalt concretes (porous or colored asphalt mixtures), paving blocks, cobble stone pavements or a special ultra-thin surface layer. Paving blocks represent a suitable alternative to cobblestone or bituminous sidewalks, bike or pedestrian lanes and to historic pavements, especially in old city centers. These are commonly employed as paving solutions due to the relatively low production and laying costs. Tataranni [3] suggested the use of a waste basalt powder to produce alternative paving blocks through the alkali-activation process. The production of paving blocks through the alkali-activation of waste basalt powder seemed to be a viable alternative for interlocking modular elements.

Utilizing recycled asphalt pavements (RAP) in pavement construction is known as a sustainable approach with significant economic and environmental benefits. RILEM TC 264-RAP experts conduct scholarly international research and knowledge dissemination with a focus on asphalt material recycling. While studying the effect of high RAP contents on the performance of hot mix asphalt (HMA) mixtures has been the focus of several research projects, limited work has been done on studying the effect of RAP fraction and particle size on the overall performance of high RAP mixtures produced solely with either coarse or fine RAP particles. It was concluded by Saliani et al. [4] that the RAP particle size has a considerable effect on its contribution to the total binder content, the aggregate skeleton of the mixture, ultimately the performance of the mixture and that the black curve gradation assumption is not representative of the actual RAP particles contribution in a high RAP mixture.

The dynamic modulus of hot mix asphalt (HMA) is a fundamental material property that defines the stress-strain relationship based on viscoelastic principles and is a function of HMA properties, loading rate, and temperature. Because of the large number of efficacious predictors (factors) and their nonlinear interrelationships, developing predictive models for dynamic modulus can be a challenging task. In the research of Ghasemi et al. [5], results obtained from a series of laboratory tests including mixture dynamic modulus, aggregate gradation, dynamic shear rheometer (on asphalt binder), and mixture volumetric were used to create a database which was used to develop a model for estimating the dynamic modulus.

One of the most remarkable effects on the urban environment is the increase in impermeable surfaces which leads to problems related to water infiltration into the ground and the increase in wash-off volumes. The use of permeable and porous layers in urban applications for cycle lanes, footpaths and parking areas is growing in interest, considering their potential to tackle issues such as the urban runoff and the urban heat island effect. Tataranni and Sangiorgi [6] suggested the production of a low impact semi-porous concrete with a transparent polymeric binder and pale limestone aggregates. The application of synthetic aggregates seems to be a viable solution for the production of innovative and eco-friendly mixtures, allowing the recycling of waste materials.

Pavement design is essentially and usually a structural long-term evaluation process. It is very hard to devise an efficient method to determine realistic in situ mechanical properties of pavements, where the determination of strain at the bottom of asphalt pavement layers through non-destructive tests is of great interest. As it is known, fiber Bragg grating (FBG) sensors are the most promising candidates to effectively replace conventional strain gauges for a long-term monitoring application in a harsh environment. Kara De Maeijer et al. [7] summarized an overview of the recent developments worldwide in the application of fiber optics sensors (FOS) in asphalt pavement monitoring systems; to find out if those systems provide repeatable and suitable results for a long-term monitoring; if there are certain solutions to validate an inverse modelling approach based on the results of a falling weight deflectometer and FOS.

In the design of pavement infrastructure, the flow number is used to determine the suitability of a hot-mix asphalt mixture (HMA) to resist permanent deformation when used in a flexible pavement. Islam et al. [8] investigated the sensitivity of the flow numbers to the mix factors of eleven categories of HMAs used in flexible pavements. The flow number increased with increasing effective binder content, air voids, voids in mineral aggregates, voids filled with asphalt, and asphalt content.

Rutting is one of the most common distresses in asphalt pavements. The problem is particularly prevalent at intersections, bus stops, railway crossings, police check points, climbing lanes and other heavily loaded sections, where there is deceleration, slow moving or static loading. The most widely used methods to identify the source of rutting among flexible pavement layers are destructive methods; field trenching and coring methods. Chilukwa and Lungu [9] used the Transverse Profile Analysis method (TPAM), which is a non-destructive method to determine the layers of pavement responsible for rutting on sections. It was established that the TPAM was a simpler, faster and less costly method of determining the source of rutting failure compared to the traditional methods used in Zambia.

Rutting resistance can be improved by adding a small amount of RAP in asphalt mixes without significantly changing properties such as stiffness and low-temperature cracking resistance. However, there is no clear understanding of how RAP gradation and bitumen properties impact the mixture properties. Saliani et al. [10] indicated that the recovered bitumen from coarse RAP did not have the same characteristics as the fine RAP bitumen, and the interaction of RAP bitumen with virgin bitumen significantly depended on RAP particle size. The amount of active RAP bitumen in coarse RAP particles was higher than in fine RAP particles.

Kara De Maeijer et al. [11] investigated a feasibility of a natural peat fiber and finely ground peat powder as a modifier for bitumen. The rheological data showed stiffening effects of the powder fraction and the presence of a fiber network, which was strain-dependent and showed elastic effects. It was indicated that the fibers should improve the rutting resistance. The data revealed that the amount of added peat fibers and powder should be limited to avoid difficulties in the compaction of these asphalt mixtures.

Non-destructive testing (NDT) is an important part of optimizing any pavement management system. In the recent years, laser Doppler vibrometer (LDV) has been introduced to conduct non-contact measurements in road engineering. Hasheminejad et al. [12] investigated the quality of two types of commercially available LDV systems—helium–neon (He–Ne)-based vibrometers and recently developed infrared vibrometers. It was shown that the noise floor of the He–Ne LDV was higher when dealing with a non-cooperative dark surface, such as asphalt concrete, and it could be improved by improving the surface quality or by using an infrared LDV, which consequently improved the modal analysis experiments performed on pavement materials.

Acknowledgments: I would like to thank the authors who supported and contributed to my first Special Issue, the reviewers who dedicated their time to review the papers and MDPI Infrastructures Editorial Team.

Conflicts of Interest: The author declares no conflicts of interests.

References

1. Le, M.-T.; Nguyen, Q.-H.; Nguyen, M.L. Numerical and Experimental Investigations of Asphalt Pavement Behaviour, Taking into Account Interface Bonding Conditions. *Infrastructures* **2020**, *5*, 21. [CrossRef]
2. Soenen, H.; Vansteenkiste, S.; Kara De Maeijer, P. Fundamental Approaches to Predict Moisture Damage in Asphalt Mixtures: State-of-the-Art Review. *Infrastructures* **2020**, *5*, 20. [CrossRef]
3. Tataranni, P. Recycled Waste Powders for Alkali-Activated Paving Blocks for Urban Pavements: A Full Laboratory Characterization. *Infrastructures* **2019**, *4*, 73. [CrossRef]
4. Saliani, S.S.; Carter, A.; Baaj, H.; Tavassoti, P. Characterization of Asphalt Mixtures Produced with Coarse and Fine Recycled Asphalt Particles. *Infrastructures* **2019**, *4*, 67. [CrossRef]
5. Ghasemi, P.; Aslani, M.; Rollins, D.K.; Williams, R.C. Principal Component Neural Networks for Modeling, Prediction, and Optimization of Hot Mix Asphalt Dynamics Modulus. *Infrastructures* **2019**, *4*, 53. [CrossRef]
6. Tataranni, P.; Sangiorgi, C. Synthetic Aggregates for the Production of Innovative Low Impact Porous Layers for Urban Pavements. *Infrastructures* **2019**, *4*, 48. [CrossRef]
7. Kara De Maeijer, P.; Luyckx, G.; Vuye, C.; Voet, E.; Van den bergh, W.; Vanlanduit, S.; Braspenninckx, J.; Stevens, N.; De Wolf, J. Fiber Optics Sensors in Asphalt Pavement: State-of-the-Art Review. *Infrastructures* **2019**, *4*, 36. [CrossRef]
8. Islam, M.R.; Kalevela, S.A.; Nesselhauf, S.K. Sensitivity of the Flow Number to Mix Factors of Hot-Mix Asphalt. *Infrastructures* **2019**, *4*, 34. [CrossRef]
9. Chilukwa, N.; Lungu, R. Determination of Layers Responsible for Rutting Failure in a Pavement Structure. *Infrastructures* **2019**, *4*, 29. [CrossRef]

10. Saliani, S.S.; Carter, A.; Baaj, H.; Mikhailenko, P. Characterization of Recovered Bitumen from Coarse and Fine Reclaimed Asphalt Pavement Particles. *Infrastructures* **2019**, *4*, 24. [CrossRef]
11. Kara De Maeijer, P.; Soenen, H.; Van den bergh, W.; Blom, J.; Jacobs, G.; Stoop, J. Peat Fibers and Finely Ground Peat Powder for Application in Asphalt. *Infrastructures* **2019**, *4*, 3. [CrossRef]
12. Hasheminejad, N.; Vuye, C.; Van den bergh, W.; Dirckx, J.; Vanlanduit, S. A Comparative Study of Laser Doppler Vibrometers for Vibration Measurements on Pavement Materials. *Infrastructures* **2018**, *3*, 47. [CrossRef]

infrastructures

Article

Numerical and Experimental Investigations of Asphalt Pavement Behaviour, Taking into Account Interface Bonding Conditions

Minh-Tu Le [1], Quang-Huy Nguyen [1] and Mai Lan Nguyen [2,*]

1 LGCGM—Structural Engineering Research Group, National Institute of Applied Sciences (INSA), 35700 Rennes, France; Minh-Tu.Le@insa-rennes.fr (M.-T.L.); Quang-Huy.Nguyen@insa-rennes.fr (Q.-H.N.)
2 MAST-LAMES, Gustave Eiffel University, Nantes Campus, F-44344 Bouguenais, France
* Correspondence: mai-lan.nguyen@univ-eiffel.fr

Received: 17 January 2020; Accepted: 18 February 2020; Published: 21 February 2020

Abstract: The interface bond between layers plays an important role in the behavior of pavement structure. However, this aspect has not yet been adequately considered in the pavement analysis process due to the lack of advanced characterizations of actual condition. In many pavement design procedures, only completely bonded or unbounded interfaces between the layers are considered. For the purpose of the better evaluation of the asphalt pavement behavior, this work focused on its investigation taking into account the actual interface bonding condition between the asphalt layers. Based on the layered theory developed by Burmister (1943), the actual interaction between pavement layers was taken into account by introducing a horizontal shear reaction modulus which represents the interface bonding condition for a given state. The analytical solution was then implemented in a numerical program before doing forward calculations for sensitivity analysis which highlights the influence of the interface bonding conditions on the structural behaviors of asphalt pavement under a static load. Furthermore, the numerical program was applied through an original experimental case study where falling weight deflectometer (FWD) tests were carried out on two full-scale pavement structures with or without a geogrid at the interface between the asphalt layers. Backcalculations of the FWD measurements allowed determining field condition of the interface bond between the asphalt layers. The obtained values of the interface shear modulus in pavement structure with a geogrid are smaller than the ones in pavement structure without geogrid. Moreover, all of these values representing field performance are at the same order of magnitude as those from dynamic interlayer shear testing.

Keywords: asphalt pavement; interface bonding; shear reaction modulus; numerical analysis; falling weight deflectometer

1. Introduction

Asphalt pavement is generally considered as being a multilayered structure comprising of successive material layers. The kinematics of the disorders in this type of structure are related to the nature of the materials used, to the conditions of the construction and more particularly to the layers properties as well as the bonding conditions between layers. Among these conditions, a good interface bond between the asphalt layers ensures the estimated performance of the designed pavement structure. Moreover, the majority of current works for the rehabilitation of existing road network as well as for new pavement structures use thinner and thinner overlayers, which require an effective bonding. However, conventional design methods consider that the interface between two pavement layers is perfectly bonded, or unbonded, depending on the nature of the layers involved. In situ inspections revealed that lack of bonding or damage to the bonding layer (interface) leads to rapid and considerable structural damage. The principle of dimensioning is based on the fact that the layers deformed by bending depend on their own characteristics (thickness, Young modulus and Poisson ratio), but also on the other layers on which they are glued. When there is an absence or failure of bonding at the interfaces (on the top or at the bottom of the layers), each layer works independently. Deformations and constraints on both sides of the interface are then more important than when the layers are glued.

Burmister [1] first derived the analytical solutions for a two-layered elastic system and subsequently extended them to a three-layered system [2–4]. Over the years, the theory has been extended to an arbitrary number of layers [5]. However, the interface bonding condition still has not been well considered in most of the modelling processes. Since the 1970s, many experimental methods have been applied to assess the capability of tack coats as well as the internal cohesion of the two involved pavement layers. Experimental methods can be divided into two main groups according to the situation of testing, in laboratory or in situ. In laboratory, direct shear tests with or without normal stress are most commonly used in the assessment of adhesion properties between two asphalt layers. Shear tests with normal stress allow the consideration of the presence of a wheel load on the road by not only its horizontal force but also its vertical influence [6,7]. However, the application of normal stress makes the experiment much more complicated. Therefore, the direct shear test method without normal stress is the most utilized one [8–11]. Most of these tests are inspired from the Leutner shear test [12]. With monotonic loading, they allow us to rapidly evaluate the influence of different factors on bond strength at the interfaces between pavement layers [13,14]. In parallel to these quasi-static tests, several dynamic shear tests developed recently [15,16] should lead to more reliable field performance characteristics. In field evaluation, until now there have been very few methods. Some pull-off test methods can be found in the literature, but are rare or only in development. In France, the destructive ovalization test has been developed since 1970s, aiming to evaluate bond conditions at the interface between pavement layers under moving wheel loads [17,18]. However, it is not often used due to the complex interpretation of the measurements. Recently, the non-destructive method of using a Falling Weight Deflectometer (FWD) [19] device has been applied quite commonly for pavement assessment through measured pavement surface deflections. Several researches using this method were performed with the same objective of investigating pavement layers interface bonding, but without relating the measured pavement deflections with interface bonding characteristics.

This present paper focuses on numerical and experimental investigations of asphalt pavement behaviour taking into account actual bonding condition at the interface between the asphalt layers. For that purpose, a theoretical background on the analytical solution of multilayered pavement structure is firstly presented. It is then improved by introducing a shear reaction modulus to take into account the bonding condition of the interface between the pavement layers. Next, the improved solution is implemented in a numerical program, which is used to perform a parametric study to investigate the sensitivity of pavement responses to the interface bonding conditions. Finally, the developed solution is applied through an original experimental case study where falling weight deflectometer (FWD) tests were carried out on two full-scale pavement structures to investigate field condition of the interface bond between the asphalt layers.

This paper is an expanded version of the conference paper [20] from the same authors. All parts of the work have been developed with more completed and self-supported elements, in particular, the analytical solution and the experimental case study. New elements have also been added in this expanded version to support both the model developed in the analytical solution and the result obtained in the original experimental study. They are the sensitivity analysis part and the characteristics of materials and structures of a full-scale pavement in the experimental part.

2. Analytical Solution Background and Improvement

2.1. Analytical Solution Background

Asphalt pavement is typically modelled using a multilayered structure based on the layered theory of Burmister. Each layer is considered as linear elastic isotropic (having an elastic modulus and a Poisson ratio) and infinite in the horizontal plan. The thickness of each layer is finite, except the bottom layer which is infinite. The interface bonding conditions between the layers are only bonded or unbonded. Figure 1 presents the multilayered pavement structure in cylindrical coordinates with r and z are the coordinates in the radial and vertical directions, respectively. The load applied on the surface of the pavement is a uniform vertical pressure of magnitude q and has a circular form of radius a. The analytical results to the problem described above are the stress, strain and displacement fields in the pavement structure. As discussed in the objectives of the work, for further improvement purpose in the paper and especially with the numerical implementation developed by the authors, the main steps and equations of the analytical solution to the problem described above, to which improvements will be made in the next paragraph, are presented here. Other details for this analytical solution can be found in the literature [5].

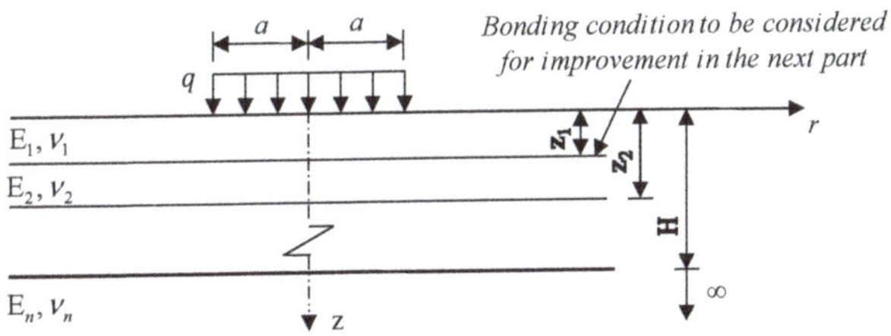

Figure 1. Multi-layered pavement structure.

Equation (1) presents the axisymmetric layered elastic responses (stresses and displacements) under a concentrated load.

$$\begin{bmatrix} (\sigma^*_{zz})_i \\ (\tau^*_{rz})_i \\ (u^*)_i \\ (w^*)_i \end{bmatrix} = \begin{bmatrix} -mJ_0(m\rho)\{1 & 1 & -(1-2v_i-m\lambda) & (1-2v_i+m\lambda)\} \\ mJ_1(m\rho)\{1 & -1 & (2v_i+m\lambda) & (2v_i-m\lambda)\} \\ \frac{1+v_i}{E}J_1(m\rho)H\{1 & -1 & (1+m\lambda) & (1-m\lambda)\} \\ -\frac{1+v_i}{E_i}J_0(m\rho)H\{1 & 1 & -(2-4v_i-m\lambda) & (2-4v_i+m\lambda)\} \end{bmatrix} \begin{bmatrix} e^{-m(\lambda_i-\lambda)}A_i \\ e^{-m(\lambda-\lambda_{i-1})}B_i \\ e^{-m(\lambda_i-\lambda)}C_i \\ e^{-m(\lambda-\lambda_{i-1})}D_i \end{bmatrix} \tag{1}$$

where $(\sigma^*_{zz})_i$ and $(\tau^*_{rz})_i$ are the vertical and shear stresses, $(u^*)_i$ and $(w^*)_i$ are the horizontal and vertical displacements of layer i; H is the distance from the pavement surface to the upper boundary of the bottom layer $\rho = r/H$ and $\lambda = z/H$; J_0 and J_1 are Bessel functions of the first kind and order 0 and 1 respectively; A_i, B_i, C_i and D_i are constants of integration to be determined from boundary and continuity conditions; m is a parameter. The superscript i varies from 1 to n and refers to the quantities corresponding to the ith layer. A star super is placed on these stresses and displacement due to a concentrated vertical load $-mJ_0(m\rho)$, not the actual stresses and displacements due to a uniform pressure q distributed over a circular are of radius a.

The stresses and displacements as a result of the uniform pressure q distributed over the circular load of radius a are obtained by using the Hankel transform (Equation (2)):

$$R = q\alpha \int_0^\infty \frac{R^*}{m} J_1(m\alpha)dm \tag{2}$$

where $\alpha = a/H$; R^* is the stress or displacement as a result of concentrated load $-mJ_0(m\rho)$; R is the stress or displacement as a result of load uniform q. So, the boundary and continuity of the multilayered pavement structure by the load $-mJ_0(m\rho)$ and uniform q distributed are the same.

2.1.1. At the Surface, $z = 0$

At this position, $i = 1$ and $\lambda = z/H = 0$, the surface stresses conditions are:

$$(\sigma^*_{zz})_1 = -mJ_0(m\rho) \text{ with } 0 \leq r \leq a \tag{3}$$

$$(\sigma^*_{zz})_1 = 0 \text{ with } r > a \tag{4}$$

$$\tau^*_{rz} = 0 \tag{5}$$

2.1.2. Between the Layers i and $i + 1$, $0 < z < H$

(a) Fully bonded

The layers are fully bonded with the same vertical stress, shear stress, vertical displacement and radial displacement at every point along the interface. Therefore $\lambda = \lambda_i$. The continuity conditions are:

$$(\sigma^*_{zz})_i = (\sigma^*_{zz})_{i+1} \tag{6}$$

$$(\tau^*_{rz})_i = (\tau^*_{rz})_{i+1} \tag{7}$$

$$(u^*)_i = (u^*)_{i+1} \tag{8}$$

$$(w^*)_i = (w^*)_{i+1} \tag{9}$$

(b) Unbonded

At the interface, the vertical stress and vertical displacement remain the same, but the shear stresses are equal to zero on both sides of the interface. Equation (7) is replaced by:

$$(\tau^*_{rz})_{i+1} = (\tau^*_{rz})_i = 0 \tag{10}$$

2.1.3. At the Lowest Layer, $i = n$, $z \geq H$

The bottom layer is semi-infinite ($z \to \infty$) and all responses (stresses, displacements) approach zero as z approaches ∞, so λ approaches infinity. From Equation (1) for the lowest layer with $i = n$ and λ approaches infinity, we have ($e^{-m(\lambda_n-\lambda)} \to \infty$) and ($e^{-m(\lambda-\lambda_{n-1})} \to 0$), to all responses (stresses, displacements and strains) approach zero, coefficients A_n and C_n will become zero.

2.2. *Improvement Taking into Account Actual Bonding Condition*

In a general case, the layers interface bonding condition can be considered as partially bonded. The layers interface behavior can be described according to Goodman's constitutive law [21] (Figure 2) in which the interface shear stress can be expressed as follows:

$$\tau = K_s \, \Delta u \tag{11}$$

where Δu is the relative horizontal displacement of the two layers at the interface; K_s is the horizontal shear reaction modulus at the interface.

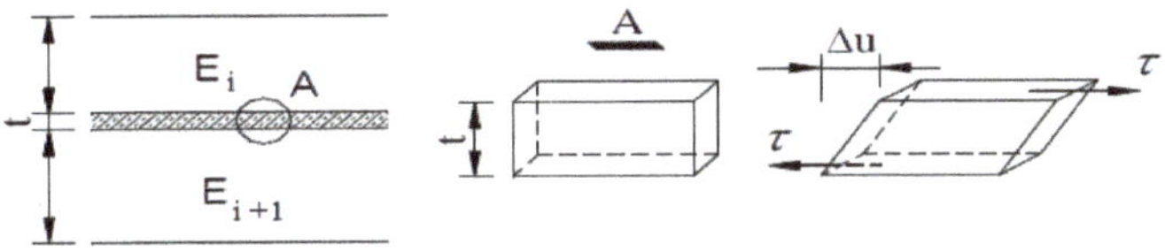

Figure 2. Modeling of the bonding between two faces at the interface.

The continuity conditions for this general case are:

$$(\sigma^*_{zz})_i = (\sigma^*_{zz})_{i+1} \tag{12}$$

$$(\tau^*_{rz})_i = (\tau^*_{rz})_{i+1} \tag{13}$$

$$(\tau^*_{rz})_i = K_s\left[(u^*)_{i+1} - (u^*)_i\right] \tag{14}$$

$$(w^*)_i = (w^*)_{i+1} \tag{15}$$

Substituting Equation (1) by these above conditions, one obtains:

$$\begin{bmatrix} 1 & F_i & -(1-2\nu_i-m\lambda_i) & (1-2\nu_i+m\lambda_i)F_i \\ 1 & -F_i & 2\nu_i+m\lambda_i & (2\nu_i-m\lambda_i)F_i \\ \frac{mE_i}{(1+\nu_i)K_s}+1 & \left(1-\frac{mE_i}{(1+\nu_i)K_s}\right)F_i & 1+m\lambda_i+\frac{(2\nu_i+m\lambda_i)mE_i}{(1+\nu_i)K_s} & \left(\frac{(2\nu_i-m\lambda_i)mE_i}{(1+\nu_i)K_s}-1+m\nu_i\right)F_i \\ 1 & -F_i & -(2-4\nu_i-m\lambda_i). & -(2-4\nu_i+m\lambda_i)F_i \end{bmatrix} \begin{bmatrix} A_i \\ B_i \\ C_i \\ D_i \end{bmatrix} = \\ = \begin{bmatrix} F_{i+1} & 1 & -(1-2\nu_{i+1}-m\lambda_i)F_{i+1} & 1-2\nu_{i+1}+m\lambda_i \\ F_{i+1} & -1 & (2\nu_{i+1}+m\lambda_i)F_{i+1} & 2\nu_{i+1}-m\lambda_i \\ R_iF_{i+1} & R_i & (1+m\lambda_i)R_iF_{i+1} & -(1-m\lambda_i)R_i \\ R_iF_{i+1} & -R_i & -(2-4\nu_{i+1}-m\lambda_i)R_iF_{i+1} & -(2-4\nu_{i+1}+m\lambda_i)R_i \end{bmatrix} \begin{bmatrix} A_{i+1} \\ B_{i+1} \\ C_{i+1} \\ D_{i+1} \end{bmatrix} \tag{16}$$

with $F_i = e^{-m(\lambda_i - \lambda_{i-1})}$; $R_i = \frac{E_i}{E_{i+1}} \frac{1+\nu_{i+1}}{1+\nu_i}$.

In Equation (2), the stress or displacement function for each layer has four coefficients of integration: A_i, B_i, C_i and D_i. All responses (stresses, displacements) can be calculated by these coefficients and integrations.

For n-layers system, the total number of unknown coefficients is 4n, which must be evaluated by the boundary and continuity conditions. With the lowest layer $A_n = C_n = 0$, there are only (4n-2) unknown coefficients.

All of these above conditions result in four equations for each of (n-1) interfaces and two equations at the surface, there are so (4n-2) independent equations. Thus, the (4n-2) unknown constants can be solved.

2.3. Numerical Implementation and Backcalculation Principle

The analytical solution including its improvement was implemented in a numerical program using Matlab [22]. This implementation is very important for research studies of the authors because it, with regard of specific or new features of pavement materials and structures, allows evaluating pavement responses under different loading configurations without depending on existing commercial software.

The developed numerical program can be used to determine pavement responses by forward calculation or to evaluate pavement properties by backcalculation. In forward calculation, based on given properties of pavement materials and structures, pavement responses in terms of stress, strain or deflection can be calculated directly. In backcalculation, which is frequently applied for FWD measurements, pavement properties can be evaluated by adjusting their seed values until getting the least squares differences between the calculated and measured pavement deflections. These investigations where the bonding condition at the interface of the asphalt layers were taken into account are presented in the following paragraphs 3 and 4, respectively.

3. Sensitivity Analysis

Sensitivity analysis using the developed numerical program is presented in this paragraph. The variation of some most important pavement responses under the loading of an FWD (with a circular plate of 0.3 m in diameter and a vertical static pressure of 0.92 MPa) in function of the interface bonding condition were evaluated. The main characteristics (with nominal values of the asphalt layers thickness) of the pavement structure used for this analysis are presented in Table 1.

Table 1. Pavement structure characteristics.

	Layer	E (MPa)	Poisson's Ratio	Nominal [1] Thickness (cm)	Actual [2] Thickness S-I (cm)	Actual [2] Thickness S-II (cm)
1	Asphalt surface	9000	0.35	6.5	6.6	6.3
	Interface	-	-	-	-	-
2	Asphalt base	9000	0.35	4.5	4.6	3.9
3	Subgrade	184	0.35	290	290	290
4	Concrete raft	55000	0.25	-	-	-

[1] Values used for sensitivity analysis; [2] Values measured in actual pavement structures in paragraph 4.

3.1. Strain Sensitivity to the Interface Bonding Conditions

In an asphalt pavement, the horizontal strain at the bottom of the asphalt layer is among the most important parameters because its magnitude will directly affect the pavement performance. Generally, the higher this magnitude is, the lower the pavement performance is. Figure 3 presents the horizontal strain at the bottom of each of the two asphalt layers of the investigated pavement structure in function of the bonding condition at the interface between the asphalt layers. As can be seen in this figure, when the bond modulus K_s decreases from infinite to nil, the horizontal strain at the bottom of the asphalt surface layer (EpsilonT_bottom_AC1) increases from 47 to 360 microstrains. The horizontal strain at the bottom of the asphalt base layer (EpsilonT_bottom_AC2) increases from 243 to a maximum value of 251 before decreasing down to 233 microstrains when K_s decreases from infinite to about 10 MPa/mm then continues to decrease to nil, respectively. Compared to the first strain, the shape of the second strain is different. This can be explained by the fact that in this case, the interface between the two asphalt layers is below their neutral axis. The position of the last one is a result from a combination of the pavement layers thicknesses and moduli. For the considered pavement structure, while the first strain is smaller than the second one when $K_s > 2$ MPa/mm, opposite result is obtained when $K_s < 2$ MPa/mm. The first strain is even much higher than the second one when K_s is close to nil, i.e., close to the unbonded condition of the interface. Based on these evaluations, it is possible to classify the interface bonding condition as follows:

- $K_s \leq 0.1$ MPa/mm: Poor bond to unbonded.
- 0.1 MPa/mm $< K_s <$ 100 MPa/mm: Partially bonded
- $K_s \geq 100$ MPa/mm: Good bond to fully bonded.

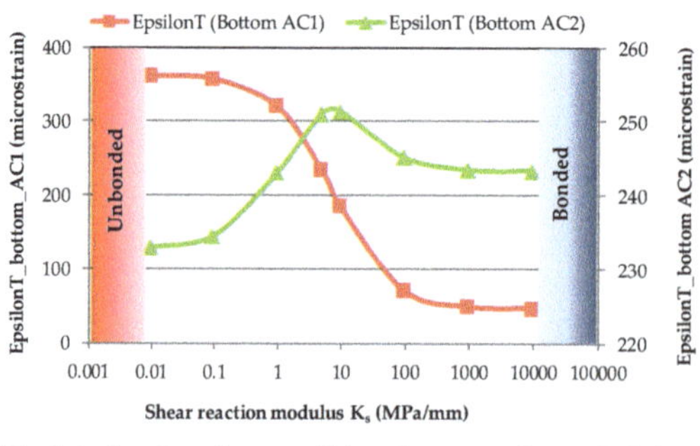

Figure 3. Impact of the interface bonding conditions between the asphalt layers on the horizontal strains at the bottom of the asphalt layers.

Moreover, the pavement responses are more sensible for K_s between 0.1 and 100 MPa/mm than when $K_s \geq 100$ MPa/mm or $K_s \leq 0.1$ MPa/mm. Among the two horizontal strains, the one at the bottom of the surface layer is more sensible with variation of K_s than the other one of the base layer. That means that the influence of the interface bonding condition is higher on the bottom of the surface layer than in the bottom of the base layer. This result can be explained by the fact that the interface is much closer to the bottom of the surface layer than the base layer.

3.2. Deflection Sensitivity to the Interface Bonding Conditions

In this parametrical study, five different deflection bowls of the pavement surface were calculated for five different bonding levels at the interface between the asphalt layers. The results are presented in Figure 4. It can be observed that when K_s = 100 MPa/mm, the pavement response is very close to the one where the interface is fully bonded. Similarly, when K_s = 0.1 MPa/mm, the pavement response is very close to that where the interface is fully debonded. For K_s = 5 MPa/mm, this bonding level gives a deflection bowl near to the middle position between the two previous cases. These observations affirm once more the classification in the previous paragraph.

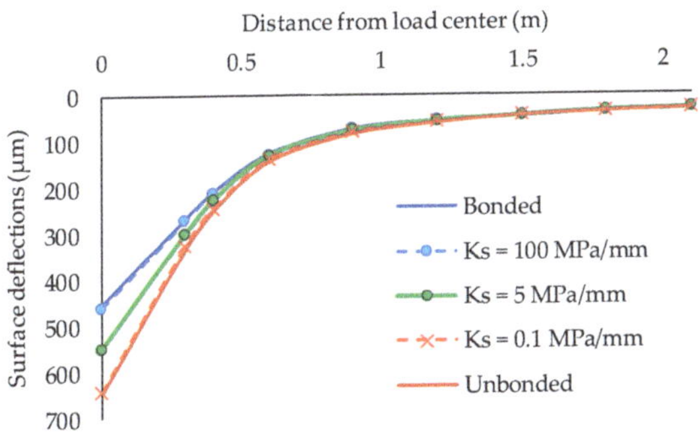

Figure 4. Deflections surface with varying values of interface bonding condition.

4. Evaluation of Pavement Interface Bonding Condition in an Experimental Case Study

The developed solution is applied in this part to evaluate field conditions of the interface bond between the asphalt layers of full-scale pavement structures in an experimental case study.

4.1. Pavement Structures and Materials Characteristics

In order to evaluate the field interface bonding conditions, two specific full-scale pavement structures at the accelerated pavement testing (APT) facility of IFSTTAR were chosen. They have the same design, which is composed of two asphalt concrete layers built on a homogenous and well-controlled subgrade of 2.9-m-thick unbound granular material and sand. The subgrade has a mean value of stiffness modulus of 184 MPa. All pavement layers were built above a concrete raft inside a watertight concrete lining. The same asphalt concrete material was used for both asphalt layers in both structures. The asphalt material is a hot mix whose formulation is a standard semi-coarse asphalt concrete of class 3 (according to the standard EN 13108-1). The unique difference between the two structures is the bonding condition at the interface between the asphalt layers. In the first structure, noted S-I, the asphalt surface layer was laid directly above the asphalt base layer. In the second one, noted S-II, there is a geogrid at the interface between the asphalt layers. One can notice that the surface layer is thicker than the base layer. The reason is that in order to get advantage of geogrid-based reinforcement in new pavement, the geogrid must be installed below the apparent neutral axis of the asphalt layers. For rehabilitated pavement, the overlay above the geogrid is often thinner than the existing base layer. A same tack coat material made of a classical cationic rapid setting bitumen emulsion (classified as C69B3 according to EN 13808) was applied at the interface between the asphalt layers with an application rate of 350 g/m^2 and 700 g/m^2 in the case without and with geogrid, respectively.

Asphalt concrete material was extracted during the construction of the full-scale pavement. The loose mix was then used for fabrication in the laboratory by a roller compacter of slab with the same air voids content as targeted in the field. The complex modulus of the obtained asphalt material was measured using two points bending test (according to EN 12697-26). The results obtained at five different frequencies (3, 6, 10, 25 and 40 Hz) and six different temperatures (−10, 0, 10, 15, 20 and 30 °C) are plotted in Figure 5 in isotherm curves.

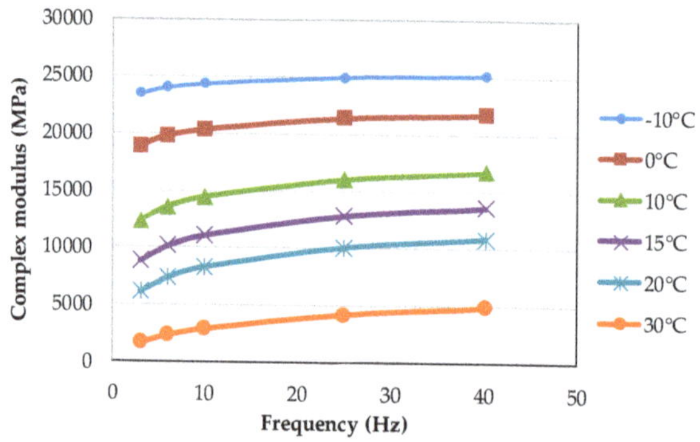

Figure 5. Isotherms of complex modulus of the tested asphalt concrete material.

4.2. *Evaluation of Bonding Condition at the Interface of the Asphalt Layers*

For this evaluation, a dedicated FWD tests campaign was carried out. Measurements were performed at three different locations on each pavement structure with the same load level of 65 kN. The circular load plate of the FWD used for these measurements has 0.3 m in diameter. The distances of the geophone sensors are 0, 0.3, 0.45, 0.6, 0.9, 1.2, 1.5, 1.8, 2.1 m from the load plate, respectively. The temperature measured by thermocouple sensors in the middle depth of the asphalt surface and base layers during these FWD measurements were close to 23 °C and 21.5 °C, respectively.

The actual thicknesses (Table 1) of the pavement layers were obtained from levelling measurement during the construction. The stiffness modulus of each asphalt layer (the same as in Table 1) was taken from the complex modulus measured in the laboratory. They were determined taking into account the temperature and frequency variations in function of the asphalt layer depth according to [23]. The Poisson's ratio of each pavement layer material was assumed to be equal to 0.35 for asphalt and unbound granular materials and 0.25 for concrete raft.

The backcalculation process was applied here to determine the shear reaction modulus K_s at the interface between the asphalt layers. In this case, all the pavement layers moduli were known, only the interface bonding condition was the unknown parameter.

Figure 6 presents the measured and calculated deflections associated with a value of shear reaction modulus for each point of FWD measurement. Good results of calculated deflections can be observed. They fit well with the measured values. These obtained values of K_s are in accordance with the initial assumption of the interface bonding condition between the asphalt layers of the two investigated pavement structures: structure S-I has good interface bond condition at points 1, 2, 3 with K_s equal to 531, 109 and 131 MPa/mm (>100 MPa/mm), respectively; intermediate interface bond conditions were obtained in structure S-II at points 4, 5, 6 with K_s equal to 74, 76 and 69 MPa/mm (0.01 MPa/mm < K_s < 100 MPa/mm), respectively.

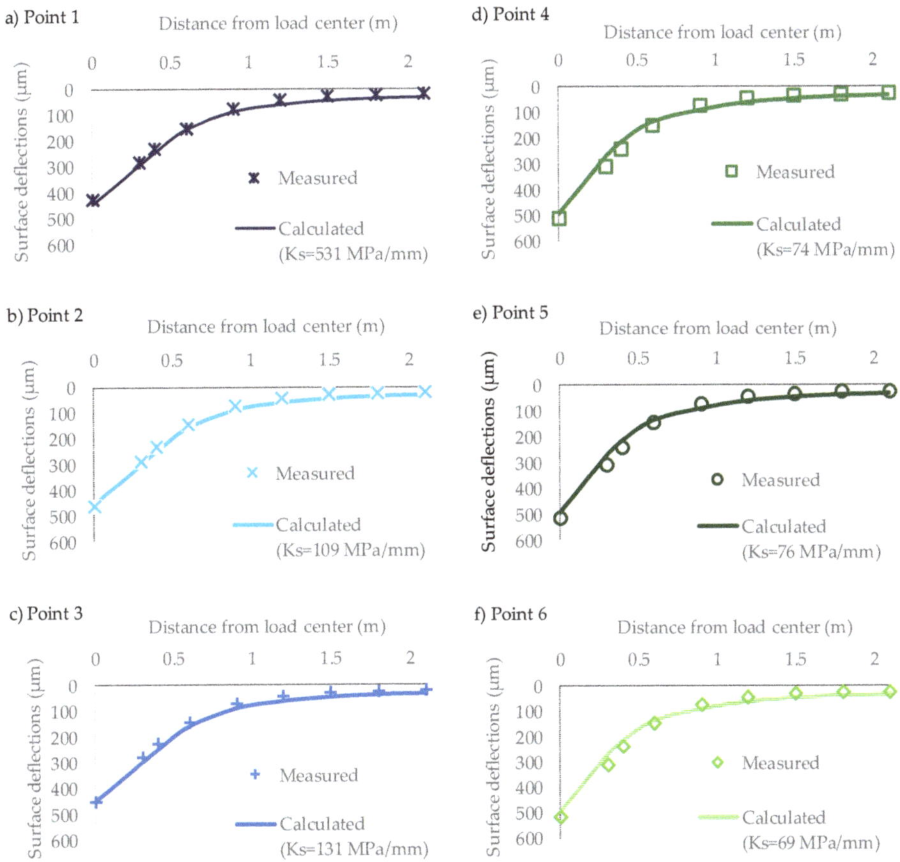

Figure 6. Measured and calculated deflections in structures S-I (points 1, 2, 3) and S-II (points 4, 5, 6) and the associated interface shear reaction moduli.

One can note some differences in the K_s values obtained for structure S-I, which vary between 109 and 531 MPa/mm. However, as analyzed in paragraph 3, when K_s is higher than 100 MPa/mm (good bond), pavement responses (strains and deflections) are much closer to the case with fully bonded condition. In that case, even though the difference in terms of K_s value is high, the difference in terms of pavement deflection is little. This experimental result confirms those observed in paragraph 3.2 of the sensitivity analysis. For structure S-II, the three K_s values are very similar, which means that the interface bonding condition is quite homogeneous, at least within the investigated pavement section, and is at the same intermediate bonding level. Moreover, K_s values in structure S-II with geogrid at the interface between the asphalt layers are smaller than the ones in structure S-I without geogrid. It confirms the literature review made in [24] that the use of a geogrid reduces the interlayer bond and hence reduces the instantaneous structural response of the pavement. However, as the geogrid could delay the reflective cracking, if properly installed, it can contribute to the long-term performance of the pavement. Furthermore, one can note that the experimental K_s values obtained for both pavement structures in this case study are at the same order of magnitude as those from dynamic shear tests [15,16] than from quasi-static shear tests [6,14]. This result confirms the position, as stated in [25] that dynamic tests represent better the field condition of interface bonding than static tests and hence are more suitable for characterization, modelling and design studies of the structural behaviors of pavements. It joints also the point of view of the Task Group 3 of the actual RILEM Technical Committee 272-PIM [26] working on dynamic interlayer shear testing.

5. Conclusions

The work presented in this paper focused on a better evaluation of structural behavior of asphalt pavement. The analytical solution based on the layered theory was improved by introducing a shear reaction modulus (K_s) to take into account the interface bonding condition between the asphalt layers. It was implemented in a numerical program using Matlab and then applied in the following parts of the research study:

- The numerical sensitivity analysis showed clearly the influence of interface bonding condition on pavement responses under the loading of an FWD. It allows classifying the interface bonding condition depending on the shear reaction modulus: poor bond to unbonded for $K_s \leq 0.1$ MPa/mm; partially bonded for 0.1 MPa/mm $< K_s <$ 100 MPa/mm; good bond to fully bonded for $K_s \geq 100$ MPa/mm.
- In the experimental case study on two full-scale pavement structures, the presented original procedure made it possible to determine an actual value of K_s for each evaluated pavement position and to differentiate the interface bonding level of the two investigated pavement structures.

With the procedure presented in this paper, the field condition of the interface bonding between asphalt layers can be assessed for better evaluation of pavement behaviors and for further performance assessment. Future works will focus on improving this procedure without possessing pavement layers modulus as among input parameters. For the experimental full-scale pavement structures, the interface bonding condition between the asphalt layers of the investigated pavement structures can be evaluated at different temperatures under different load levels together with the evolution of pavement damage during the accelerated test.

Author Contributions: All the authors contributed the conceptualization and methodology of this work; the theoretical development and numerical implementation together with the sensitivity analysis as well as the experimental case study were contributed by M.-T.L. and M.L.N.; the original draft was preprared by M.-T.L.; its review was performed by Q.-H.N. and M.L.N.; the editing of the draft was finalized by all the authors. All authors have read and agreed to the published version of the manuscript.

Funding: This research received no external funding.

Acknowledgments: The authors acknowledged the managers and staffs of the IFSTTAR APT facility for providing support to perform experimental study on full-scale pavement.

Conflicts of Interest: The authors hereby declare no conflict of interest regarding the publication of this article.

References

1. Burmister, D.M. The theory of stress and displacements in layered systems and applications to the design of airport runways. *Highw. Res. Board Proc.* **1943**, *23*, 126–144.
2. Burmister, D.M. The general theory of stress and displacements in layered soil systems. I. *J. Appl. Phys.* **1945**, *16*, 89–94. [CrossRef]
3. Burmister, D.M. The general theory of stress and displacements in layered soil systems. II. *J. Appl. Phys.* **1945**, *16*, 126–127. [CrossRef]
4. Burmister, D.M. The general theory of stress and displacements in layered soil systems. III. *J. Appl. Phys.* **1945**, *16*, 296–302. [CrossRef]
5. Huang, Y. *Pavement Analysis and Design*, 2nd ed.; Prentice Hall: Englewood Cliffs, NJ, USA, 2003.
6. Romanoschi, S.; Metcalf, J. Characterization of asphalt concrete layer interfaces. *Transp. Res. Rec. J. Transp. Res. Board* **2001**, *1778*, 132–139. [CrossRef]
7. Zofka, A.; Maliszewski, M.; Bernier, A.; Josen, R.; Vaitkus, A.; Kleizienė, R. Advanced shear tester for evaluation of asphalt concrete under constant normal stiffness conditions. *Road Mater. Pavement Des.* **2015**, *16* (Suppl. 1), 187–210. [CrossRef]

8. Canestrari, F.; Ferrotti, G.; Parti, M.N.; Santagata, E. Advanced testing and characterization of interlayer shear resistance. *Transp. Res. Rec. J. Transp. Res. Board* **2005**, *1929*, 69–78. [CrossRef]
9. Canestrari, F.; Ferrotti, G.; Lu, G.; Millien, A.; Partl, M.; Petit, C. Mechanical testing of interlayer bonding in asphalt pavements. In *Advances in Interlaboratory Testing and Evaluation of Bituminous Materials*; Partl, M., Bahia, H.U., Canestrari, F., de la Roche, C., Di Benedetto, H., Piber, H., Sybilski, D., Eds.; Springer: Berlin/Heidelberg, Germany, 2013; pp. 303–360. [CrossRef]
10. West, R.C.; Zhang, J.; Moore, J. *Evaluation of Bond Strength between Pavement Layers*; NCAT Report 05–08; National Center for Asphalt Technology: Auburn, AL, USA, 2005.
11. Raab, C.; Partl, M.N.; Abd El Halim, A.O. Evaluation of interlayer shear bond devices for asphalt pavements. *Balt. J. Road Bridge Eng.* **2009**, *4*, 176–195. [CrossRef]
12. Leutner, R. Untersuchung des Schichtenverbundes beim bituminösen Oberbau. *Bitumen* **1979**, *41*, 84–91.
13. Destrée, A.; De Visscher, J. Impact of tack coat application conditions on the interlayer bond strength. *Eur. J. Environ. Civ. Eng.* **2017**, *21*, 3–13. [CrossRef]
14. Petit, C.; Chabot, A.; Destree, A.; Raab, C. Recommendation of RILEM TC 241-MCD on Interface Debonding Testing in Pavements. *Mater. Struct.* **2018**, *51*, 96. [CrossRef]
15. Diakhate, M.; Millien, A.; Petit, C.; Phelipot-Mardelé, A.; Pouteau, B. Experimental investigation of tack coat fatigue performance: Towards an improved lifetime assessment of pavement structure interfaces. *Constr. Build. Mater.* **2011**, *25*, 1123–1133. [CrossRef]
16. Freire, R.; Di Benedetto, H.; Sauzéat, C.; Pouget, S.; Lesueur, D. Linear Viscoelastic Behaviour of Geogrids Interface within Bituminous Mixtures. *KSCE J. Civ. Eng.* **2018**, *22*, 2082–2088. [CrossRef]
17. Kobisch, R.; Peyronne, C. L'ovalisation: Une nouvelle méthode de mesure des déformations élastiques des chaussées. *Bull Liaison Lab Ponts Chauss* **1979**, *102*, 59–71.
18. Goacolou, H.; Keryell, P.; Kobisch, R.; Poilane, J.P. Utilisation de l'ovalisation en auscultation des chaussées. *Bull Liaison Lab Ponts Chauss* **1983**, *128*, 65–75.
19. Alavi, S.; Lecates, J.F.; Tavares, M.P. *Falling Weight Deflectometer Usage*; The National Academies Press: Washington, DC, USA, 2007. [CrossRef]
20. Le, M.T.; Nguyen, Q.H.; Nguyen, M.L. Numerical analysis of double-layered asphalt pavement behaviour taking into account interface bonding conditions. In Proceedings of the CIGOS 2019-Innovation for Sustainable Infrastructure, Hanoi, Vietnam, 31 October–1 November 2019. [CrossRef]
21. Goodman, R.E.; Taylor, R.L.; Brekke, T.L. A model for the Mechanics of Jointed Rocks. *J. Soil Mech. Found. Div.* **1968**, *94*, 637–659.
22. The MathWorks. *MATLAB and Statistics Toolbox Release 2012b*; The MathWorks, Inc.: Natick, MA, USA, 2012.
23. Le, V.P.; Lee, H.J.; Flores, J.M.; Kim, W.J.; Baek, J. New approach to construct master curve of damaged asphalt concrete based on falling weight deflectometer back-calculated moduli. *J. Transp. Eng.* **2016**, 142. [CrossRef]
24. Nguyen, M.L.; Blanc, J.; Kerzreho, J.P.; Hornych, P. Review of glass fibre grid use for pavement reinforcement and APT experiments at IFSTTAR. *Road Mater. Pavement Des.* **2013**, *14*, 287–308. [CrossRef]
25. Nguyen, N.L.; Dao, V.D.; Nguyen, M.L.; Pham, D.H. Investigation of Bond between Asphalt Layers in Flexible Pavement. In Proceedings of the 8th RILEM International Conference on Mechanisms of Cracking and Debonding in Pavements, Nantes, France, 7–9 June 2016. [CrossRef]
26. RILEM TC 272-PIM. Available online: https://www.rilem.net/groupe/272-pim-phase-and-interphase-behaviour-of-bituminous-materials-359 (accessed on 15 January 2020).

infrastructures

MDPI

Review

Fundamental Approaches to Predict Moisture Damage in Asphalt Mixtures: State-of-the-Art Review

Hilde Soenen [1,*], Stefan Vansteenkiste [2] and Patricia Kara De Maeijer [3]

[1] Nynas NV, Bitumen Research, 2020 Antwerp, Belgium
[2] Belgian Road Research Center, 1200 Brussels, Belgium; s.vansteenkiste@brrc.be
[3] RERS, EMIB, Faculty of Applied Engineering, University of Antwerp, 2020 Antwerp, Belgium; patricija.karademaeijer@uantwerpen.be
* Correspondence: hilde.soenen@nynas.com

Received: 31 January 2020; Accepted: 19 February 2020; Published: 21 February 2020

Abstract: Moisture susceptibility is still one of the primary causes of distress in flexible pavements, reducing the pavements' durability. A very large number of tests are available to evaluate the susceptibility of a binder aggregate combination. Tests can be conducted on the asphalt mixture, either in a loose or compacted form, or on the individual components of an asphalt pavement. Apart from various mechanisms and models, fundamental concepts have been proposed to calculate the thermodynamic tendency of a binder aggregate combination to adhere and/or debond under wet conditions. The aim of this review is to summarize literature findings and conclusions, regarding these concepts as carried out in the CEDR project FunDBits. The applied test methods, the obtained results, and the validation or predictability of these fundamental approaches are discussed.

Keywords: moisture damage; surface free energy components; cohesion; binder–aggregate adhesion

1. Introduction

Moisture in asphalt pavement structures can lead to phenomena such as stripping, raveling, and pothole formation, limiting the lifetime and durability of the pavement. Moisture sensitivity has been studied extensively in the literature, resulting in an enormous number of possible test procedures, which have been classified into various levels [1]. The first level consists of tests, conducted on the individual components, traditionally comprising of the binder, the aggregate, and possible additives. Nowadays this level will also include renewable, as well as secondary (waste) materials [2–4]. The next level includes tests conducted on loose asphalt mixture, while subsequent levels consider tests involving compacted asphalt mixtures, and finally compacted mixture in a pavement under field conditions. It is obvious that the number of parameters and the test complexity increase as tests move from the individual components to the pavement level.

Commonly, in the literature, the occurrence of moisture damage is divided into adhesive or cohesive failure. Adhesive failure being a loss of adhesion between the aggregate and the binder interface. This will result in clean aggregate surfaces after failure [5,6]. Cohesive failure occurs if moisture weakens the binder or mastic phase leading to a failure inside this binder or mastic film. In this case, the aggregates will still be covered with bitumen after failure. In addition to this, a third possibility has been identified where damage is caused by the fracture of aggregates, particularly when the mixture is subjected to freezing [5,7,8]. The actions by which damage occurs can be further divided into at least five different mechanisms, detachment, displacement, spontaneous emulsification, pore pressure, and hydraulic scour. An overview is given by Little et al. [9].

Moreover, also, various mechanisms have been proposed to explain adhesion between bitumen and aggregates [6,10]; including a chemical reaction [11], a thermodynamic approach based on surface energies [8], molecular orientation [12], molecular dynamics [13], and mechanical adhesion [14].

In asphalt pavements, moisture damage is most likely related to a combination of mechanisms, which depend on the pavement materials, the mix design, the traffic loading, and the climatic conditions. Due to this complexity, a large number of possible parameters have been identified, but it is not clear which of these are decisive and determine the behavior [15].

In the literature, two fundamental concepts have been proposed to calculate thermodynamic work of adhesion and debonding in the presence of moisture between a binder and an aggregate. The first one is the surface energy component concept which was first applied by Texas A&M researchers to bitumen and aggregates and who developed also the methodologies for the measurement of surface energies for bitumen, respectively aggregates [8]. Another concept is based on the Hamaker equation, which was developed for materials having only Lifshitz–Vander Waals interactions [16]. This concept has also been applied to bitumen aggregate adhesion.

The aim of this paper is to summarize literature findings and conclusions, regarding the test methods, results, and the validation or predictability of these fundamental approaches. The current paper is based on activities conducted for the Project FunDBitS (Functional Durability-related Bitumen Specification, CEDR Transnational Road Research programme Call 2013: Energy Efficiency—Materials and Technology) [17], updated with recent literature.

2. Concepts to Calculate the Bitumen–Aggregate Adhesion

2.1. Calculation of Adhesive Bond Strength and Debonding by Water from Surface Energy Components

Researchers at Texas A&M University have applied the methodology of measuring surface energy components as a base to calculate the adhesion of bitumen to an aggregate surface [8]. They also developed and evaluated the most suitable test methods to determine surface energy components for bitumen and for aggregates. In this concept, surface energy components of bitumen and aggregates are derived separately, and the data allow calculating the interfacial work of adhesion in dry, as well as in wet conditions. The concept is based on the Van Oss–Chaudhury–Good (VCG) theory of wettability and is very well explained in for example, [16,18–23].

The surface free energy (SFE) of a material is defined as the amount of work required to create a unit area of a new surface of that specific material in a vacuum [20,24–26]. This surface energy can be divided into different parts (Equation (1)); a first part, relating to Liftshitz–van der Waals interactions and referred to as γ^{LW} and a second part referring to asymmetrical polar interactions, described as acid–base interactions γ^{AB} or electron acceptor, respectively donor parts. The Lifshitz–van der Waals component of the surface energy comprises the following interactions: Keesom (dipole–dipole interactions), Debye (dipole-induced–dipole interactions), and London dispersion forces (induced dipole–induced dipole interactions). In literature, it was shown later that the LW part should only include the London dispersive interactions, while Keesom and Debye interactions should be included in the acid–base part [27]. In this paper, the notation LW part is kept, although it refers to the dispersive part only.

$$\gamma = \gamma^{LW} + \gamma^{AB} = \gamma^{LW} + 2\sqrt{\gamma^{+}\gamma^{-}} \quad (1)$$

γ—total surface energy;
γ^{LW}—dispersive part of the surface energy;
γ^{AB}—acid base part of the surface energy;
γ^{+}—Lewis acid component or electron acceptor of surface energy;
γ^{-}—Lewis base or electron donor component of surface energy.

The interaction of two materials in vacuum or the free energy change of adhesion (ΔG_{12}) between two materials 1 and 2 can be formulated as a function of their respective surface energy components as shown in Equation (2). The free energy change is equal in magnitude but has the opposite sign as the work of adhesion, W_{12}.

$$\Delta G_{12} = -W_{12} = -\left(2\sqrt{\gamma_1^{LW}\gamma_2^{LW}} + 2\sqrt{\gamma_1^{+}\gamma_2^{-}} + 2\sqrt{\gamma_1^{-}\gamma_2^{+}}\right) \tag{2}$$

The subscripts 1 and 2 refer to the respective surface energy components of the two substances 1 and 2. Equation (2) shows that the interaction of two materials in vacuum is always negative, meaning there is always an attraction. Equation (2) cannot be zero since for all materials γ^{LW} is a finite and positive number. Based on Equation (2), it is possible to calculate the surface energy components for an unknown substance by measuring the surface energy of the unknown versus at least three probe compounds of known surface energy components. From these three liquids at least two need to have (known) polar parts. Different options are available to test this experimentally and they will be discussed briefly in the experimental part.

Once the surface components for bitumen and aggregates are determined, their interfacial work of adhesion, the dry bond strength, can be calculated using Equation (2), where the subscripts 1 and 2 refer to the two substances tested, in this case, aggregate and bitumen. If in this equation, material 1 and 2 would be the same substance it becomes equal to two times the surface energy of this material (2γ in Equation (1)). Therefore, twice the surface energy of bitumen is related to the cohesive strength or bond energy of bitumen. The cohesive bond energy of a material is defined as the amount of work required to fracture the material to create two new surfaces of a unit area each, in a vacuum. Numerically this is equal to twice the total surface free energy of the material (Equation (3)). A higher magnitude of cohesive bond energy implies that more energy is required for a crack to propagate due to fracture.

$$\Delta G_{ii} = -2\gamma_i \tag{3}$$

Finally, consider a three-phase system comprising of bitumen, aggregate, and water represented by material 1 and 2 in medium 3, respectively (Equation (4)). If the medium water displaces bitumen from the bitumen–aggregate interface several processes occur. The interface bitumen–aggregate is lost, and this is associated with external work, $-\gamma_{12}$. At the same time, during this process, two new interfaces are created: between bitumen and water and between aggregate and water. The work needed for the formation of these two new interfaces is $\gamma_{13} + \gamma_{23}$. Therefore, the total work needed for water to displace bitumen from the surface of the aggregate is $\gamma_{13} + \gamma_{23} - \gamma_{12}$ (Equation (5)). In terms of free energy, the resulting free energy of adhesion of component 1 and 2 in medium 3 can be expressed using the same relations but with opposite signs.

$$W_{132} = \gamma_{13} + \gamma_{23} - \gamma_{12} \tag{4}$$

$$\Delta G^{a}{}_{132} = \gamma_{12} - \gamma_{13} - \gamma_{23} \tag{5}$$

In order to take both the LW and polar part into account, Equation 5 must be calculated as follows in Equation (6):

$$\Delta G^{a}{}_{132} = 2\left[\begin{array}{l}\sqrt{\gamma_1^{LW}\gamma_3^{LW}} + \sqrt{\gamma_2^{LW}\gamma_3^{LW}} - \sqrt{\gamma_1^{LW}\gamma_2^{LW}} - \gamma_3^{LW} + \\ \sqrt{\gamma_3^{+}}\left(\sqrt{\gamma_1^{-}} + \sqrt{\gamma_2^{-}} - \sqrt{\gamma_3^{-}}\right) + \sqrt{\gamma_3^{-}}\left(\sqrt{\gamma_1^{+}} + \sqrt{\gamma_2^{+}} - \sqrt{\gamma_3^{+}}\right) \\ -\sqrt{\gamma_1^{+}\gamma_2^{-}} - \sqrt{\gamma_1^{-}\gamma_2^{+}}\end{array}\right] \tag{6}$$

When $\Delta G^{a}{}_{132} < 0$ it indicates that there is an attraction between component 1 and 2 also when immersed in medium 3, and in this case, a displacement will not happen for a system in thermodynamic equilibrium. For $\Delta G^{a}{}_{132} > 0$ the interaction between 1 and 2 becomes repulsive. The magnitude of work of debonding can differ significantly depending on the surface energy components of bitumen and aggregates. Similarly, for describing the interaction between molecules or particles of material

1 suspended in liquid 3 one can write (Equation (7)). The latter is the driving force for phase separation of adhesives in aqueous media (Equation (8)). A large negative value would indicate a good resistance to debonding while a large positive value indicates easier stripping due to water, for systems in thermodynamic equilibrium. For practically all bitumen–aggregate systems the work of debonding W_{132} is negative or $\Delta G^{a}{}_{132}$ is positive indicating that debonding in the presence of water is thermodynamically favorable.

$$\Delta G^{a}{}_{132} = -2\gamma_{13} = -2\left(\sqrt{\gamma_1^{LW}} - \sqrt{\gamma_3^{LW}}\right)^2 - 4\left(\sqrt{\gamma_1^{+}\gamma_1^{-}} + \sqrt{\gamma_3^{+}\gamma_3^{-}} - \sqrt{\gamma_1^{+}\gamma_3^{-}} - \sqrt{\gamma_1^{-}\gamma_3^{+}}\right) \quad (7)$$

If the polar surface free energy components of a hydrophobic material (or two similar hydrophobic materials) are negligibly small, then the most important parameter in Equation (7) is $-4\sqrt{\gamma_3^{+}\gamma_3^{-}}$. This parameter represents the polar contribution to the cohesive energy of water. The value is -102 mJ/m^2 and is present in all types of interactions when immersed in water. In fact, this term is the main contributor to the interfacial attractions between nonpolar materials immersed in an H-bonding material such as water.

Based on three parameters: the dry adhesion, the cohesive strength of bitumen, and the free energy of adhesion in the water, two related energy ratios have been proposed: ER_1 and ER_2 according to Equations (8) and (9), respectively. In literature, the ratio between the adhesive bond energy values in the dry condition and in the presence of water, ER1, can be used to predict the moisture sensitivity of asphalt mixtures [16]. Another ratio ER_2 can be used; in this parameter the adhesive bond energy in the dry state is diminished with the bitumen cohesion, and this value is divided by the bitumen aggregate adhesion in the presence of water. In order to accommodate the effects of aggregate micro-texture on the bitumen–aggregate bond strength in the presence of moisture both bond parameters can also be multiplied by the specific surface area (SSA) of the aggregates. The procedure on how to calculate these parameters is very well described in the literature [9,19,20,28–30], but is it not fully clear which of these parameters is best suited to predict moisture damage. The term $\Delta G^{a}{}_{12}$ in Equations (8) and (9) refers to the interfacial free energy of adhesion between bitumen and aggregate in a vacuum (or air), while $\Delta G^{a}{}_{132}$ refers to the wet adhesion.

$$ER_1 = \frac{\Delta G^{a}{}_{12}}{\Delta G^{a}{}_{132}} \quad (8)$$

$$ER_2 = \left|\frac{\Delta G^{a}{}_{12} - \Delta G_{11}}{\Delta G^{a}{}_{132}}\right| \quad (9)$$

In the paper by Bhasin et al. [19], the free energy ratios calculated separately for the acid–base components, as shown in Equation (10), were used. The authors observed that the portion of the bond energy that results from the interaction of the acid component of asphalt and the base component of aggregate contributes the most to the total adhesive bond strength of the mixture [19]. Still, other combinations have been proposed by Hamedi and Moghadas Nejad [31].

$$R^{AB} = \left|\frac{\Delta G_{12}^{AB}}{\Delta G_{132}^{AB}}\right| \quad (10)$$

In addition to the VCG method, other methods to calculate adhesion are also often used, such as for example, the Owens–Wendt (OW) method, also known as the Kaelble method. In the OW method, SFE is a sum of two components: a dispersive (D) and a polar (P) part, where the dispersive part reflects only dispersive interactions, and the polar part is a sum of polar, hydrogen, inductive, and acid–base interactions. In the OW method, a minimum of two known solvents or media are needed to calculate the SFE components.

2.2. Calculation of Adhesive Bond Strengths in Various Media Based on the Hamaker Equation

Researchers at KTH (Royal Institute of Technology in Stockholm, Sweden) have used the Hamaker equation to estimate the interaction of bitumen and aggregate/mineral components having air or water as an intervening medium [32]. The Hamaker equation is used to estimate the van der Waals interaction, including dispersive, Keesom, and Debye interactions. The Hamaker's equation (Equation (11)) is composed of two parts: a first part describes the polar contribution and a second part—the dispersive contribution. In this Equation (11), subscripts 1 and 2 refer in this case to bitumen and aggregate while subscript 3 refers to the medium, either air or water. Calculations of Hamaker's polar part require accurate dielectric data, in particular, dielectric constants and for the dispersive part the refractive index of the interacting materials and the intervening medium.

$$A_{132} = \frac{3}{4}kT\left(\frac{\varepsilon_1 - \varepsilon_3}{\varepsilon_1 + \varepsilon_3}\right)\left(\frac{\varepsilon_2 - \varepsilon_3}{\varepsilon_2 + \varepsilon_3}\right) + \frac{3h\upsilon}{8\sqrt{2}}\frac{\left(n_1^2 - n_3^2\right)\left(n_2^2 - n_3^2\right)}{\sqrt{n_1^2 + n_3^2}\sqrt{n_2^2 + n_3^2}\left(\sqrt{n_1^2 + n_3^2} + \sqrt{n_2^2 + n_3^2}\right)} \quad (11)$$

ε_i—the static dielectric constant for material/medium I (in vacuum $\varepsilon_3 = 1$);
n_i—the refractive index of the material/medium I, in the visible region (in vacuum $n_3 = 1$);
h—Planck's constant (= 6.6261×10^{-34} Js);
k—Boltzmann constant (= 1.3807×10^{-23} J/K);
T—the absolute temperature;
υ—the main electronic absorption frequency (typically $\pm\, 3 \times 10^{15}$ s^{-1}).

If Hamaker's equation equals zero, there is no net force and the bodies are neither pulled together nor pushed apart. If the net force is positive then the bodies will adhere, if the net force is negative repulsion will occur. For most material combinations, the Hamaker equation is positive and the van der Waals force is attractive. The van der Waals force is always attractive between two surfaces of the same material and always attractive in a vacuum. The Hamaker's equation can be negative and repulsive for two different material surfaces interacting through a liquid medium (A_{123}). Relations between the Hamaker constant and the dispersive part of the surface energy have been proposed: For example, Israelachvili [33] has calculated the Hamaker constants of different liquids from their refractive indices and the surface tensions of these liquids using the following Equation (12):

$$A_{ii} = 24\pi r_{ii}^2 \gamma_i^{LW} \quad (12)$$

r_{ii}—the separation distance between interacting atoms or molecules;
γ^{LW}—the dispersive part of the surface energy.

Israelachvili [33] found a very good agreement between the calculated surface tension of saturated hydrocarbons and the corresponding experimental values using Equation (12) and r = 0.2054 nm. However, this was not true for polar substances. Israelachvili [33] concluded that Equation (12) may not be used to calculate the surface free energies of highly polar liquids, where short-range forces other than dispersion forces (e.g., hydrogen bonds) are involved. Later, the value for r was corrected (by the same author) to a value of 0.165 nm.

3. Summary of Experimental Studies Based on SFE Approach

3.1. Overview of Test Procedures to Determine Surface Energies

An overview of experimental methods, as was observed in the literature survey, to determine SFE or SFE components of bitumen and aggregates is presented in Table 1. The calculation method, if applicable is indicated. In the Owens–Wendt (OW) concept, SFE is a sum of two components: a dispersion (D) and a polar (P) part, while in the Van Oss–Chaudhury–Good (VCG) theory, SFE is calculated based on three components; a disperse, an acid, and a base component. The most popular test methods include the Wilhemy plate test (WP) and the universal sorption device (USD) for respectively, bitumen and aggregates. Due to its simplicity and because it can be used for both bitumen and aggregates, the sessile drop test is also used a lot. These three tests are briefly explained in the next paragraphs.

Table 1. Literature overview indicating the test methods used to determine surface energies of bitumen and aggregate.

Reference(s)	Method to determine surface energies of bituminous binders
[18,20,21,23,24,26,28–30,34–47]	Wilhelmy plate tests in probe liquids (VCG), ambient
[20,39,48–51]	Sessile drops of probe liquids on bitumen surface (VCG) ambient
[18,20,35]	Inverse gas chromatography (CVS)
[46,52]	Pending drop (100–140 °C) combined with sessile drop on PTFE (OW)
[53]	Sessile drops on a microtome-cut bitumen surface 20 °C (OW)
[39,54]	Sessile drops of probe liquids on bitumen surface (OW) ambient
[55]	Dynamic sessile drop measurements of probe liquids on a bitumen surface (VCG)
[56]	Pending drops of bitumen (100–130 °C) (γ total)
[53]	Pending drops at equiviscous temperatures (γ total)
[57]	Pending drops of bitumen at a fixed G* 209 Pa (γ total)
[39]	Wilhelmy plate tests in probe liquids (OW) ambient
[20]	Atomic force microscopy (dispersive component)
Reference(s)	**Method to determine surface energies of aggregates used in asphalt applications**
[18,19,21,23,24,26,28–30,35,36,38,40,41,45,46,58–60]	Universal sorption device (VCG)
[49–51,55]	Sessile drops of probe liquids on flat aggregate (VCG)
[53,57,61]	Sessile drops of probe liquids on flat aggregate (OW)
[20,28]	Micro calorimeter (VCG)
[20]	Inverse gas chromatography (VCG)

Legend: OW—Owens–Wendt theory resulting in two surface free energy (SFE) components: Dispersive and polar; VCG—Van Oss–Chaudhury–Good theory resulting in three SFE components: Dispersive, acid, and base.

The Wilhelmy plate method also referred to as plate method, was proposed by researchers from Texas A&M University to investigate bitumen [18,35]. Later, this test was applied by other researchers as referred to in Table 1. In this test, a thin bitumen coated glass plate is immersed or withdrawn from probe liquid at a slow and constant speed. The dynamic contact angles that develop between the bitumen film and the liquid are obtained, by comparing the weight of the sample slide in air to its weight in the liquid after correcting for buoyancy. The basic principle is schematically illustrated in Figure 1. The dynamic contact angle measured during the immersion process is called the advancing contact angle (a wetting process), while the dynamic contact angle measured during the withdrawal process is called the receding contact angle (a de-wetting process). Theoretically, for a surface that does not undergo any permanent change by meeting the probe liquid, the advancing and receding contact angles should be the same. However, in most cases, differences are observed [20] and they have been attributed to chemical and morphological heterogeneities of the surface, or also to roughness, swelling, rearrangement, inter-diffusion, and/or surface deformation. In the case of bitumen, it has been attributed to surface heterogeneities, and in bitumen, the advancing contact angles are used to determine surface energy components [20]. In the latter study, the effect of the choice of probe liquids on the precision of the SFE determination was investigated. In principle, for the OW method it would be enough to determine the contact angles with two known probe liquids and for VCG calculation with three probe liquids. However, if two or more of the probe liquids have similar surface free energy components, the calculated surface free energy components of bitumen will be very sensitive to small errors in the measurement of contact angles. It was advised by Texas A&M to use five probe liquids.

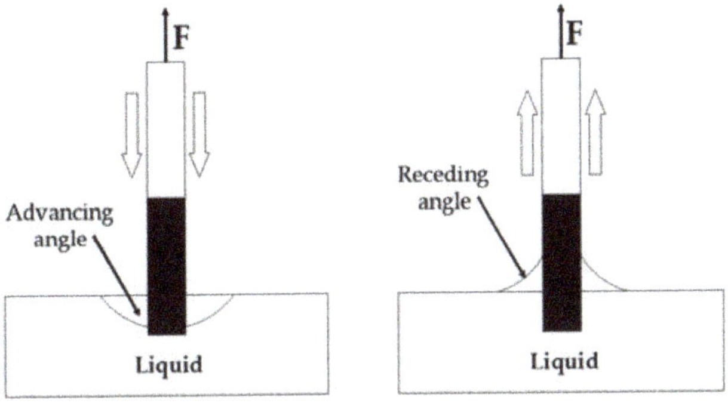

Figure 1. Schematic illustration of Wilhelmy plate technique [35].

The sessile drop method is based on a direct measurement of the contact angle of a known liquid on the surface of the material being tested [18]. This approach is very well described in other references such as [52]. While the plate technique measures a dynamic contact angle in a quasi-equilibrium state, the sessile drop approach usually measures a static contact angle although it can also be used in a dynamic mode. As in the case of Wilhelmy plate method, the three unknown surface energy components of a solid under investigation can be calculated once the contact angles of at least three known liquids on this substrate are measured. In this test, a drop (about 2 to 3 mm in diameter) of a probe liquid is dispensed on a horizontal, flat surface of the material being tested (this can be a bitumen film or a flat stone surface), and the contact angle is measured. In Figure 2, different contact angles are shown as a function of the degree of wetting. For a complete wetting, the contact angle is zero, while for two materials that have no wetting the contact angle is 180 °C. In that case, the drop will roll on the surface. Instead of working with known probe liquids, it is also possible to use solid substrates with known surface energy characteristics and use sessile drop measurements to determine the SFE components of an unknown liquid.

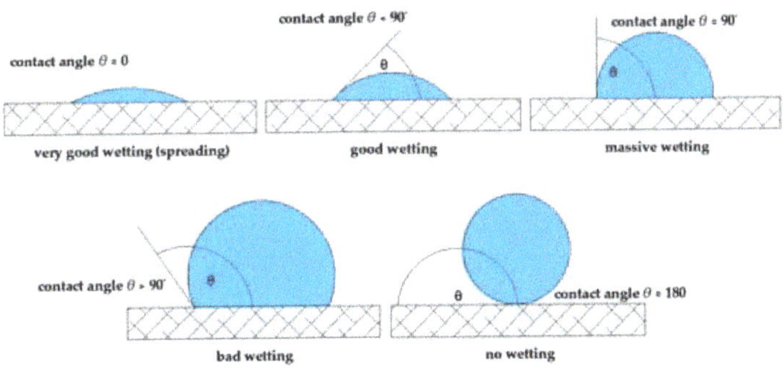

Figure 2. Overview of contact angles as a function of the wetting behavior [52].

Bitumen and aggregate can be tested using this method. In both cases, the sample needs to be flat and smooth, and aggregates are very often polished. Therefore, the surface does not represent a natural crushed state. In addition, since each measurement is made using a single drop that covers only an area of about 2 to 3 mm in diameter, the application of this test on heterogeneous aggregate surfaces will require a lot of repeats on different locations, in order to get a reasonable average representative of the whole surface area. Additionally, it has also been reported that nonpolar liquids cannot be used on high-energy aggregates surfaces since complete wetting occurs, resulting in a zero-contact angle. The quality of the drop pictures is also very important as stated in [56]. A summary of possible errors related with this technique is provided by Rudawska et al. [62].

The universal sorption device (USD) is a gravimetric sorption technique, used to measure the sorption characteristics of selected vapors on aggregate surfaces. According to literature it can be used in a static or dynamic way. Again, the sorption tests need to be conducted with at least three vapors with known surface energy components in order to be able to calculate the three SFE components of an unknown aggregate. Sorption methods are particularly suitable since they can accommodate different sample sizes, irregular shapes, mineralogy's, and surface textures associated with aggregates. In the method used by Texas A&M University, an aggregate fraction passing a 4.75 mm sieve and retained on a 2.36 mm sieve is tested. The chamber, including the suspended aggregate fraction, is vacuumed and the solute is injected into the system. A highly sensitive magnetic suspension balance is used to measure the amount of solute adsorbed on the surface of the aggregate at predetermined increasing levels of relative pressure. Little and Bhasin [20] noted that the specific surface area (SSA) of the aggregates is a required input to be able to compute SFE components. The methodology of using the universal sorption device is applicable only when the probe vapor molecules are adsorbed due to physical adsorption and not chemisorption [20]. Experiments by Texas A&M University have confirmed that the adsorption of the selected probe vapors on typical aggregate surfaces such as granite, limestone, and gravel is mostly due to physical adsorption. However, if aggregates are coated with chemically active materials and are used for testing, it must be ensured that the probe vapors do not react with the coating. Little and Bhasin [20] also show that clean aggregates are needed and physically adsorbed impurities such as water vapor, etc., need to be removed, as they will influence the SFE components. Water molecules and other impurities when adsorbed to the clean aggregate surface will lower its surface energy.

3.1.1. SFE Measurement Results on Bituminous Binders

In Table 2, values obtained for surface energy measurements on unmodified bitumen are shown. All these data are based on the Wilhelmy plate method combined with the VCG calculation procedure. Only data on unmodified binders are summarized, and the data are first grouped by the laboratory where they have been tested, and secondly by the reported publication date.

There seems to be a good agreement in the total surface energy levels found, although the levels found by Bahramian [39], and Zhang et al. [22] are slightly higher. Bahramian [39] also showed in his project slightly higher levels in the polar parts as compared to other references. In general, for bituminous binders, the polar parts are much smaller compared to the dispersive parts.

Standard deviations on the surface energy components have been reported for example in [18] and [39]. These standard deviations are small compared to the measured levels for the total and dispersive parts, but they are large for polar, acid, and base parts, indicating that these parts cannot be measured very accurately. Especially the small acid and base parts have sometimes very large standard deviations. In the paper by Luo et al. [42], the obtained levels are lower as compared to the other references because these authors added a correction to the Wilhelmy plate method. This correction takes into account the slight curvature of the liquid surface next to the plate. For some of the samples, this correction resulted in a change of more than 10% in the cohesive energy components.

Table 2. Overview of typical values of surface energies for unmodified bituminous binders, using the Wilhelmy plate test.

γ_{Total} mJ/m²	γ_{LW} mJ/m²	γ_{AB} mJ/m²	γ^+ mJ/m²	γ^- mJ/m²	Reference(s)	Notes
14–32	13–32	0–3	0–1.5	0–3	[18,20,24,28–30,35,36]	
12.1–12.5	10.6–11.2	1.3–1.5	0.7–0.8	0.6–0.7	[55]	
15.6	13.7	1.9	1.5	0.7	[21]	
19.1–30.6	18.8–30.6	0–0.4	0–0.34	0–5.08	[15,26,41,46]	
33.6–38.1	32.9–37.5	–	0.0–0.01	2.1–3.5	[22]	Two binders
14.0–15.6	12.5–13.5	1.5–1.9	1.4–1.5	0.4–0.7	[31]	Two binders
14.2–15.9	11.8–13.7	2.2–2.4	2.6–3.0	0.5–0.5	[45]	Two binders
10.45	6.85	3.60	3.47	0.93	[42]	Revised calculation method; one unmodified binder
20.2	19.3	0.9	1.2	0.2	[43]	One unmodified binder
26–39	18–33	6–8	4.9–6.7	1.7–2.4	[39]	Six pen grade 70/100 binders
20.5	17.8	2.5	1.5	1.2	[47]	One unmodified binder

γ_{Total}—total SFE; γ_{LW}—the nonpolar Lifshitz–van der Waals component; γ_{AB}—acid–base part of SFE; γ^+—the polar Lewis acid component; γ^-—the polar Lewis base component.

Comparisons of different tests methods, using the same sample set, were reported in [18,20,32,35,39]. For example, for the comparison between the sessile drop method and the Wilhelmy plate tests, Little and Bhasin [20] reported that both test methods produce results for the LW component with adequate precision. There was a trend between LW parts of both methods ($R^2 = 0.59$), but sessile drop measurements systematically resulted in larger dispersive parts, attributed to differences in the test methodologies. Differences were also reported in the acid components; a larger value was obtained using the Wilhelmy plate method. In the paper by Bahramian [39], a similar comparison was conducted, in this case, almost no differences were observed between both methods.

In the paper by Hirsch and Friemel-Göttlich [53], sessile drop measurements of probe liquids on a bitumen surface were used to determine SFE components of bitumen (OW calculation), in this study the author reported that especially for soft binders, very small polar components were obtained. This was related to separation effects at the bitumen surface. Normally bitumen is heated, covered on a glass plate, and afterward allowed to cool to room temperature. During this cooling, the surface will optimize its surface energy by selectively placing compounds with small surface energy at the surface. Especially, the accumulation of smaller more mobile paraffin at the surface can reduce the surface energy. The authors in [53] could avoid this phenomenon by using a freshly microtome-cut bitumen surface instead of an air-cooled surface. Unfortunately, in the paper the comparison of both surfaces is not included, only measured data on microtome-cut surfaces are reported. These indicate rather larger polar SFE components as compared to other references. The author also expected these effects in polymer modified binders.

Interesting results were also obtained by Khan et al. [54], he used the sessile drop method to determine the three SFE components. Additionally, for three binders, he compared several preparation methods of the bitumen surfaces. In method A, he first mixed the additives with the binders, this was either an adhesion improver or an emulsifier, and used the standard method to prepare a sample. This is heating a binder, preparing a hot surface, and leaving it in air to cool down. In addition to this, in method B, he prepared bitumen surfaces without additives in the standard way, and placed these surfaces underwater, in which the additives were mixed. After removing this surface from the water, the SFE components were determined. The data clearly showed that the effect of a surfactant or an emulsifier was only observed in method B. Only those surfaces that had been into contact with the water, had developed a clear polar part in their surface energy components.

The effects of testing time and temperature on SFE component determination is discussed in the papers [35,52,53,56]. A linear decrease in total surface energy with temperature has been reported, and all binders seem to vary in a very similar way, for 30 °C increase, the total surface energy decreases between 2 and 2.5 mN/m^2.

In many studies, different bituminous binders show differences in the total and dispersive parts of the surface energy and also small differences in acid or base parts as reported in the papers [18,20,35]. However, at this moment, it is not clear how these differences relate to other bitumen parameters, and/or to the bitumen chemistry. Korn [52] has reported that paraffinic binders have larger total surface energies compared to naphthenic bitumen, but paraffinic binders have a lower polar part. In the papers [53,57] only very little variation was observed in the total surface energies of various commercial binders, while in the paper [53], it was noted that more differences could be observed when comparing polar and dispersive parts, but the author also indicated that the errors bars on these parts are rather large. The influences of bitumen modification, antistrip additives and ageing on the surface energy, and its components are summarized in Table 3. In conclusion, it seems that the effect of additives such as antistrip agents is very hard to detect in SFE measurements. Moreover, polymer or wax modification does not result in large changes, and for aging different trends are reported.

Table 3. Overview of effect of bitumen type and bitumen modification including use of additives on surface energy and SFE components.

Effect Studied	Finding(s)—Conclusion(s)	Reference(s) *
Type of bitumen	• γ_{total} parafinic -> γ_{total} naphthenic; γ_{AB} napthenic -> γ_{AB} paraffinic; • γ_{AB} napthenic -> γ_{AB} paraffinic;	[52]
	• SHRP library binders, differences between the binders were observed but this did not relate to another bitumen parameter;	[18,20,28,34,35]
	• Very little difference between binders in γ_{total}.	[53,57]
Antistrip agent (0.4%–1% addition)	✓ Slight increase in γ_{total} but decrease in γ_{AB};	[52]
	✓ No clear trend on surface energy and SFE components;	[18,20,28,35]
	✓ γ_{total} and γ^- increase slightly, only for the harder binder;	[41]
	✓ γ_{total}, γ_{LW}, and γ^- increase slightly;	[48]
	✓ γ_{total}, γ_{LW}, and γ^- increase;	[31,45,60]
	✓ Method A (see text) $\gamma_{LW} \approx 40$, $\gamma_{AB} \approx 0$; ✓ Method B (see text) $\gamma_{LW} \approx 20$–25, $\gamma_{AB} \approx 10$–40.	[54]
Hydrated lime	• No effect.	[20,49]
Wax addition 3% wax addition	• Slight increase in γ_{total} and γ_{AB} due to wax;	[52]

Table 3. *Cont.*

Effect Studied	Finding(s)—Conclusion(s)	Reference(s) *
sasobit (2%–8%), paraffin wax (8%)	• Sasobit decreases γ_{total}, γ^+ is increased no effect on γ^- similar effect of paraffin wax, no effect of aspha-min;	[37,55]
sasobit (0.5%–8.0%), evotherm (0.5%–1.5%)	• γ_{total}, γ_{LW}, and γ^- increase slightly;	[48]
sasobit (3%), and RH (3%) (RH is a type of organic wax)	• For sasobit: for one binder γ_{total} decreases, for the second binder γ_{total} increases, basic part increases, acidic part decreases. For RH: For both binders γ_{total} decreases, basic part increases acidic part decreases.	[50]
Polymer modification [46,52,53]: SBS	✓ No effect on γ_{total};	[52]
	✓ γ_{total} increases;	[53]
[49]: SBS (3%), SBR (3%)	✓ SBS results in an increase in γ_{LW}, decrease in γ_{AB}, SBR almost no effect;	[49]
	✓ γ_{total} increases slightly for Wilhelmy plate tests at 23 °C, no effect for pendant drop tests at 120 °C.	[46]
Ageing (RTFOT + PAV)	• γ^- increases slightly;	[20]
	• No clear effect on γ_{total};	[18]
	• γ_{total} increases, γ_{LW} increases, γ^+ decreases, γ^- varies, γ_{AB} decreases;	[49]
	• Increase after RTFOR, reduction after PAV in γ_{total}, similar trend for LW and basic component, the acidic component increases after each aging step.	[47]
Nanomaterials carbon nanotubes 3%, nano-SiO_2 (6%), diatomite (6%)	✓ These additives have almost no influence.	[49]
WMA additive Aspha-min (1%–6%)	• No effect;	[37]
Cecabase RT® (0.2%–0.4%)	• γ_{total} increases;	[47]
Zycotherm (NZ)	• γ_{total} and all components increase.	[43]

* For each reference the respective test method can be found in Table 1.

3.1.2. SFE Measurement Results on Aggregates

Aggregates consist of an assemblage of one or more minerals, while minerals have a definite chemical composition and an ordered atomic arrangement as described in [18]. Consequently, it is expected that the surface of aggregates is rather heterogeneous. In the papers [52,53,63], aggregate types are divided into acid and basic types; acid aggregates consist mainly of quartz (silica dioxide or SiO_2) and are generally considered as water sensitive aggregates. The reason is that SiO_2 can form strong hydrogen bonds with water resulting in strong interactions. The author also reports that on these surfaces even after heating it is very likely that a monomolecular layer of water will remain. On the other hand, basic aggregates consist mainly of calcite and are considered as good adhering aggregates. Therefore, in general terms, bad adhering aggregates may include quartz, quartzite, hornblende, biotite, orthoclase, while good adhering aggregates comprise, e.g., basalt, augite, and olivine. Hirsch and Friemel-Göttlich [53] note that the chemical components at the stone surface with a high affinity for bitumen are in general elements such as Al, Fe, Mg, and Ca, while elements with a low affinity are Na and K. The reason given is that Ca^{2+} and Mg^{2+} ions form water-insoluble salts, while K^+ and Na^+ form water-soluble salts, with a negative effect on the bitumen stone adhesion. A similar observation was in [64], where it was shown that alkali feldspars induced a weak interface with bitumen. Lyne et al. [65,66] has given a literature overview on the effect of specific elements on the moisture sensitivity of bitumen—aggregate adhesion. Minerals containing alkali metals are prone to stripping and iron compounds are considered beneficial. Magnesium and calcium are also considered advantageous. The total picture of the effect of aluminum and silica is less clear. Hefer [18] has noted that carbonates may result in weak boundaries if the pH of the water drops below 6, since then, the carbonates may dissolve leading to failure.

Surface energy measurements on aggregates indicate much more variation between the surface energy components compared to the values observed for various bituminous binders. Little and Bhasin [20] have reported that the total surface energy is typically in the range of 50 up to 400 mJ/m^2 for aggregates. The magnitude of the LW component is smaller (30 to 60 mJ/m^2) compared to the magnitude of the base component (200 to 1000 mJ/m^2). Most aggregates have a small magnitude of the acid component ranging from 0 up to100 mJ/m^2, so the total polar part of the surface energy is not necessarily high and can still be small depending on the acid component. Hirsch and Friemel-Göttlich [52] noted that differences in the total surface energy and the distribution between dispersive and polar parts can be larger between aggregates belonging to the same class of aggregates compared to aggregates belonging to different classes. In the particular case in [53], when two quartzite aggregate types showed larger differences than what is observed between quartzite and granite types.

In the collected papers, large variations between aggregates were observed, even for aggregates that differ only slightly in their mineralogical composition. For example, in the papers [26,41], two limestone type of aggregates, consisting both for 96% of the mineral calcite, still showed a very large difference between the surface energy components. Regarding standard deviations, these are reported, for example, by Bhasin [28] and they are low compared to the measured values, except if an SFE component is very small.

Different test methods are available and have been used to evaluate the SFE of aggregates. From a comparison of different methods, the following conclusions were obtained:

- Inverse gas chromatography is not very successful for aggregates as stated by Hefer [18] as the retention time measured may correspond to only the high-energy functional groups present on the surface of the material and not to an average value as reported in the papers [20,28];
- The use of contact angle methods such as the sessile drop method are limited as they require a flat stone surface; and a large number of repeats are needed [53];
- Static and dynamic vapor sorption measurements are very suited for aggregates with irregular shapes as concluded in the papers [34,35];
- Atomic force microscopy (AFM) also requires a very flat surface and moreover, only a very small area is scanned at a given time as reported in the papers [34,35];
- Micro calorimetry characterizes only the enthalpy change and could only be used if entropy would turn out to be negligible as stated in the papers [34,35].

In several references the effects of polishing have been discussed. Hirsch and Friemel-Göttlich [53] have investigated the surface roughness after polishing and concluded that even after polishing differences in surface flatness still exists between different aggregates. In this case, slight variations in the surface hardness could, after polishing result in very small imperfections on the surface. The researchers tried to avoid the influence of surface irregularities by using the same sample preparation for all the aggregates, and by performing a larger number of contact angle measurements on different locations on the stone surface. The author also investigated the difference between cutting, scrubbing and polishing, when measuring contact angles with water. The polished stones gave the lowest contact angles. The author concluded that irregularities, which may be larger after cutting and scrubbing, hinder the wetting of liquids on the surface. This study also compared static and dynamic contact angles between probe liquids on stone surfaces. For some surface types differences between both contact angles were seen, while for others these were very similar. If differences were observed these could be attributed to the surface roughness. In this study, the author decided to determine the SFE components of the aggregates (polar and dispersive only) using dynamic contact angles. There is still a lot of discussion on the validity of methods that measure liquid contact angles for solids which are not easily prepared into a flat plane as stated by Miller [58].

In some references, the effect of using "freshly-cut" or "aged" aggregates is discussed. Korn [52] noted that after crushing, there is a reorganization of polar substances at the surface, and by adsorption of substances from the air (water molecules and dust) the surface energy of the aggregate decreases until after a few months a stable situation has been installed. Schellenberger [67] noted that freshly cut aggregates have a worse moisture sensitivity behavior compared to aggregates that have been stored for some time. Hefer [18] noted that possible reactions at the stone surface when water is absorbed are presented. Little and Bhasin [20] noted that for two aggregate types no differences due to aggregate storage could be observed. This was explained by the fact that crushed aggregates may within a few seconds after crushing accumulate surface contamination from the environment.

Miller et al. [58,59] evaluated the possibility to catalogue mineral properties, with the aim to relate the mineralogy to the surface energy components and also to check if it is possible to assign surface energy values to aggregates coming from a specific quarry. This last objective can only be possible if the surface energies of aggregates are not dependent upon the moment and at which point in the quarry they are taken. This last condition was however not met. Additionally, a simple relation between mineralogical composition and surface energy could not be developed because it was not always possible to draw clear correlations between surface chemistry and surface energy. The reasons for this observation are not so clear. Hirsch and Friemel-Göttlich [53] reported that determining the chemical and functional groups as active sites is useful to understand interfacial reactions but does not explain the magnitude or strength of these sites. Additionally, measured values of organic and inorganic coatings seemed to play an important role and also the influence of varying surface roughness was unclear.

Regarding aggregate modifications, in the paper by Arabani et al. [21] the influence of a polyethylene (PE) precoating of three types of aggregates was evaluated. The data clearly show that PE treatment decreases the polar components of the aggregates and increases the dispersive part. However, the base component was for all aggregates higher after modification. There was no explanation for this last observation. In the literature, numerous papers have evaluated the effect of hydrated lime. In the papers [18,53], it was postulated that hydrated lime ties up carboxylic acids and 2-quinolones in the bitumen, with the formation of insoluble calcium organic salts, which prevent these functionalities from reacting with a siliceous surface to form water-sensitive bonds. This leaves important active sites on the siliceous surface to form strong water-resistant bonds with nitrogen groups in bitumen (amines) resulting in a reduction of the water sensitivity of asphalt mixtures. In the paper by Moghadas Nejad et al. [40] the SFE components of the aggregates with and without hydrate lime treatment were determined, and the authors showed that hydrated lime treatment reduces the acid SFE and increases the base SFE of the two types of aggregates. However, Little and Bhasin [20] reported that the methodology of using the universal sorption device is applicable only when the probe vapor molecules are adsorbed due to physical adsorption and not due to chemisorption. It is unclear if this condition is met in the case of hydrated lime.

3.1.3. Dry and Wet Adhesion

Once the SFE components of bitumen and aggregate are determined, the dry and wet bond strengths can be calculated. In literature, normally dry bond strengths are positive and wet bond strengths become negative as stated by Hefer [18] indicating an attraction between bitumen and aggregate when dry that becomes a repulsion when wet. Miller et al. [59] investigated a large range of minerals and for some of these, free energies of adhesion between bitumen and these minerals were still negative even in the presence of water. These were typically minerals with a low acid and a low base component. Therefore, the latter minerals do not have a thermodynamically drive for an adhesive debonding. The fact that this was not observed for any of the aggregates studied was explained by the fact that aggregates are combinations of different minerals, resulting in average values.

Grenfell et al. [26] reported that a limestone aggregate, which had small acid and base components, also showed a negative free energy of adhesion with bitumen. A similar observation was made by Porot et al. [61]. In this case, all investigated aggregates showed negative free energies of adhesion in the presence of water, which may also be related to difficulties in the exact determination of acid and base components.

3.2. Sessile Drop Measurements in Dry and Wet Conditions

In some studies [49,52,53,56,57,67–70], the bitumen–aggregate affinity is derived from contact angle measurements, between bitumen drops in direct contact with a (flat) aggregate surface. In such tests, surface energy components are not determined, however, later tests allow a direct measurement of bitumen–aggregate interaction, which can be measured both in air, as well as in water. Therefore, a short summary of these types of tests is included.

Korn [52] has developed a procedure to measure contact angles of bitumen drops on substrates in air and in water. In this proposal, the temperatures for wetting and respectively, de-wetting by water are determined by the bitumen softening points. In the first step, a procedure to bring equally sized drops of bitumen on a substrate is described. For this purpose, the bitumen was heated, in this case, up to 150 °C, while the substrate was kept at room temperature. Several drops of bitumen were placed on the substrate using a needle. Afterwards, the covered substrates were heated to a temperature, 25 to 30 °C above the softening point. The contact angle data of bitumen to the substrate in air were recorded after 10 min at 70 °C. Subsequently, the substrate with the wetted bitumen drop was placed underwater at a temperature which is 5 °C below the R&B softening point, in this example at 40 °C. The drop contact angle was measured again after 2 h of storage underwater. The difference in contact angle before and after water storage is then calculated. Threshold values for the change after 2 h respectively 24 h water storage have been set and used as a measure of the resistance to stripping.

Several references have applied this type of tests as described in the papers [52,56,68], while Schellenberger [66] is using a similar approach, measuring contact angle underwater while adapting slightly the procedure. Korn [52] observed clear differences between stone types and between various bitumen types, especially in case of bitumen modified with antistrip additives, waxes, or polymers. However, it should be noted that for polymer and wax modified bitumen the wetting temperatures needed to be considerably higher, because of the increased viscosity. The authors also noted that differences in surface texture influenced the contact angles, and finally, that there is no relation between the total surface energy of a binder and the wetting of this binder on a particular stone.

Schellenberger [67] mentioned that the procedure was adapted, and, in this test, all unmodified binders gave rather low contact angles, so a good wetting was not achieved. The test could clearly show the effect of additives, which was also aggregate type dependent. Furthermore, the author noted that this test cannot be used to investigate PmB's, as it does not allow preparing small drops when the bitumen is polymer modified. In the paper, the effect of additives was confirmed in rolling bottle tests and in boiling water tests. Nehrings [68] used the same procedure as Korn [52] to evaluate the effect of adhesion improvers, however, no effect was observed when measuring contact angles in the dry situation, only the water conditioning step allowed to distinguish the binders. Wistuba [56] conducted very similar tests as Korn [52], the authors report the need for minimum standards for image quality, in order to improve the repeatability and reproducibility of the test method.

In the papers [49,52,57,69,70], contact angles were only evaluated in the dry situation. Some studies described in the papers [49,53,69] indicate that effects of additives such as adhesion promotors can be observed even when testing dry contact angles, while other papers [57,70] indicate that these tests cannot differentiate binders with and without adhesion promotors.

In summary, it seems that contact angle measurements in the dry and wet state may be a successful way to estimate the water sensitivity. However, at this moment the validation is mainly limited to comparing binders with and without adhesion improvers, or other modifications. There are still a lot of open questions with regard to this test such as: should binders be compared at an equal viscosity state or at an equal temperature, what is the effect of using a polished or a flat aggregate surface and what about the accuracy, how large is the effect of a minor unevenness or a scratch on the spreading of bitumen drops.

3.3. Correlations between SFE Measurements and Water Sensitivity Observations

3.3.1. Correlations between Calculated Bond Strengths and Laboratory Tests Indicative of Moisture Damage

Table 4, is a selection of research papers, investigating the validation of surface energy predictions based on laboratory tests. The number of test samples from each study, as well as the water conditioning and test procedures are included in Table 4, while the findings are summarized in the next paragraphs.

Apeagyei et al. [71] evaluated six combinations and determined compressive strength ratios, between wet and dry conditioned tests. In this paper, possible correlations between predicted and tested ratios were not presented. Instead, from the surface energies of binder and aggregate, and from the measured compressive strength ratios, a new parameter, the percentage of the surface area of aggregate exposed to water was calculated. According to the authors, this percentage is a significant index to quantify the level of adhesive fracture. In this study, the water conditioning took place prior to testing the asphalt mixture samples, but the authors concluded that the introduction of moisture in either a liquid or vapor state during cyclic loading may well be more damaging than simply moisture conditioning an asphalt concrete sample prior to testing. The authors also noted a filler effect: This essentially occurs when a filler is intimately mixed or dispersed with the binder. In fact, hydrated lime intimately mixed with the bitumen may provide a filler effect that not only stiffens the mixture at high temperatures but also provides a toughening effect that leads to increased fatigue life and increased low-temperature fracture resistance.

In the papers [20,28,30], 12 binder–aggregate combinations were investigated, the authors came to the following conclusions: the SSA, the specific surface are of the aggregates, is a parameter that should be added to the surface energy parameters to predict the moisture sensitivity. When including SSA, correlations improved, and residual errors decreased. Moreover, in general, the correlation between the SFE parameters and the ratio of dry versus wet fatigue life was better than the correlation with the ratio of dynamic modulus in tension. The best correlations of 0.84, 0.83 were obtained between the fatigue ratio (wet/dry) versus log (ER1*SSA) and versus log (ER2*SSA), respectively.

Howson et al. [24] investigated six binder–aggregate combinations, based on uniaxial pull-off tests using a DSR. For each test, the total work of fracture was derived. The latter was higher for limestone as compared to andesite (for each binder and at each conditioning level). The change in the total work of fracture compared to the dry sample was calculated (in limestone this tended to increase with conditioning time, in andesite this tended to decrease). The average of this change related well to the values given by the energy parameter (ER2), R2 = 0.89. A lot more conclusions were obtained in this paper: for example, the authors noted an effect of the aggregate type, even if the failure was purely cohesive. They also noted changes in the patterns of (cohesively) fractured surfaces after water conditioning, and proposed a mechanism based on an increased ability of the binders to flow, as well as cavitation and fibril formation due to the action of water. The authors also noted that an increase in film thickness resulted in increases in total work of fracture, as it allowed more energy to be dissipated in the bulk of the viscoelastic asphalt binder.

Hirsch and Friemel-Göttlich [53] compared surface energies and rolling bottle tests on 25 binder–aggregate. Two criteria for the assessment were used; stone coverage should be larger if γ stone is large and γ binder is low. Additionally, the bond between aggregate and binder is stronger if the distribution of polar to dispersive parts is more equal. In this study, these assumptions could not be confirmed, and no correlations were obtained. The reasons for this are not clear but the fact that SSA was not included in the predictions was stated as a reason, a second reason could be related to the possibility that the binder forms a good adhering, but brittle film which could break during the rolling bottle test; or a strong cohesive binder film is formed, which is not well adhering, but not breaking during rolling action.

Arabani et al. [21] used three aggregate–binder combinations and each aggregate was tested in three states; without treatment, after coating it with low density poly(ethylene), and again after coating with high density poly(ethylene). Therefore, in fact, a total of nine combinations was tested. Unfortunately, in the paper no direct comparison of bond strength ratios to predicted levels based on the surface energies was shown. Clear effects of the coating with polyethylene were observed and discussed.

Wasiuddin [55] compared results of a modified pull-off test (PATTI) to SFE free energy calculations for several aggregate binder combinations and binders modified with sasobit and liquid anti strip. The results did not correlate well. In this study, a new moisture conditioning procedure was proposed, able to produce adhesive failure which can be quantified by the modified pull-off test. For this purpose, plain glass plates, coated with bitumen, were placed overnight in an oven at 64 °C in submerged condition. This conditioning resulted in most cases in an adhesive failure, but for some of the modified samples the failure was still either a mixed-mode (between cohesive and adhesive failure) or a fully cohesive failure. The authors also noted that in the submerged condition, water can develop randomly distributed, micro and macro scale cylindrical holes in the bitumen film. These holes penetrate up to the asphalt–substrate interface and start replacing asphalt film from the substrate surface, causing adhesive failure or stripping. Maybe this observation is related to the high temperature of 64 °C, at which sample conditioning took place.

Liu et al. [41] tested three limestone types and three granite types of aggregates with various binders. Differences in moisture sensitivity of the mixtures were observed based on the test method used: the boiling water test (BWT) and the rolling bottle test (RBT) were the most discriminative tests while the static immersion and the ultrasonic test were the least sensitive. The results for the total water immersion test had mixed success. The BWT and RBT ranked limestone combinations better compared to the granite combinations. The correspondence between SFE calculated and measured water-sensitivity was partially correct, the mixtures that performed worst or best in RBT and BWT were identified. Furthermore, the authors concluded that the magnitude of the work of debonding in the presence of water was found to be aggregate type dependent which suggests that the physicochemical properties of aggregates may play a more significant role in the generation of moisture damage, than bitumen properties.

Grenfell et al. [26] evaluated two aggregates, limestone and granite, using three binders. In this study the rolling bottle tests were well predicted by SFE calculations. All four aggregate–bitumen bond energy parameters (ER1, ER2, ER1*SSA, and ER2*SSA) could be used to predict moisture sensitivity of asphalt mixtures. Threshold values of (0.75 for ER1, 0.50 for ER2, 0.50 for ER1*SSA, and 0.35 for ER2*SSA) were defined to separate 'good' from 'poor' moisture damage performing aggregate–bitumen combinations. The authors concluded that surface energy properties of the materials combined with the parameters obtained by conventional moisture sensitivity assessment techniques can contribute towards the development of a material screening protocol for determining the best combinations of bitumen and aggregates for the local road material providing better bitumen–aggregate adhesion and less susceptibility to moisture damage/stripping.

Apeagyei et al. [71] followed the indirect stiffness versus conditioning time; generally, stiffness decreased with conditioning time (and consequently water content) except at low exposure times where low water content was associated with a slight but repeatable increase in stiffness. In this paper, the reversibility of moisture damage was investigated, this is the stiffness degradation in wet specimens being fully recovered upon specimen drying. The effect of moisture exposure on durability was found to be reversible as moisture conditioned asphalt mixtures that had lost up to 80% of the initial stiffness upon drying fully recovered. This was associated with a plasticization process, softening the bulk mastic, and moving the critical stress concentration location from the interfacial region of the aggregate-mastic bond into the bulk mastic. Micro-CT scans of the asphalt mixtures internal structure suggest moisture diffusion was mainly restricted to the bulk mastic and not the aggregate–mastic interface. The results suggest that cohesive rather than adhesive failure dominate the durability of asphalt mixtures under the long-term moisture exposure used in this study. In another paper by Apeagyei et al. [15] the worse moisture resistance of the granite mastics compared with the limestone mastic bonds could be explained, in part, by the dominant mineral phases in the granite. The three dominant minerals in granite, namely albite, feldspar, and quartz have been associated with poor adhesion and interfacial failure. A good correlation was found between the thermodynamic work of adhesion and debonding, and the practical work of adhesion of the aggregate–mastic bonds. This suggests that physical adsorption controls the moisture damage in aggregate–mastic bonds. In this paper the authors also noted an increase in cohesive strength with moisture exposure, which was related to a plasticization effect.

Zhang et al. [22] noted that differences in the water sensitivity tests could be explained by the water absorption and mineral compositions of the aggregates. They concluded that based on the peel test, the moisture absorption and mineralogical compositions of aggregate were considered as two important factors to moisture sensitivity. This phenomenon suggests that in a susceptible asphalt mixture, the effect of aggregate may be more influential than the effect of bitumen. The SATS test and the peel test showed similar moisture sensitivity results demonstrating the good correlation between these two mechanical tests. However, the surface energy tests, and the mechanical tests cannot correlate in terms of moisture sensitivity evaluation. For the aggregates considered in this research, the surface-energy-based method did not correlate well with the peel test for moisture sensitivity evaluation.

Hamedi and Moghadas Nejad [31] tested combinations of eight aggregates and four binders. Four SFE calculated parameters were used, of which two showed a good correlation with the laboratory tests: EP1 defined as the difference between the surface energy of adhesion of asphalt–aggregate and water–aggregate divided by the surface energy of adhesion of asphalt–aggregate. Additionally, EP2, which is the ratio between the surface energy of adhesion of asphalt–aggregate (in dry conditions) to the difference between surface energy of adhesion in dry and wet conditions. Both parameters showed a good correlation with the tensile strength ratio. The R squares for these parameters were 0.82 and 0.81, respectively. The authors observed that these relations worsened slightly when including SSA, which they related to difficulties in the measurement of SSA. The authors also noted that the biggest shortcoming of thermodynamics is that it does not dictate rate; it shows in which direction things will go, when in equilibrium conditions. In the papers [45,60], 24 combinations were investigated. In this study, apart from the SSA of the aggregates, the authors also tested the AFT (asphalt film thickness) and the permeability of the asphalt mixtures. A multiple regression model was applied to predict the moisture susceptibility of asphalt mixtures based on thermodynamic parameters and the three other parameters (permeability, SSA, and AFT). A good model was obtained, showing that the combination of these parameters can predict the occurrence of moisture damage in asphalt mixtures, measured by the modified Lottman indirect tension test. The results of the coefficients also showed that the highest positive impact on asphalt mixture strength against moisture is caused by SSA, indicating that this parameter is certainly an important one [45]. In the paper by Azarion et al. [60], the authors noted that a comparison between variables based on the coefficients is not possible since the variables all had different sizes. The model showed that all parameters were significant.

Porot el al. [46] found that according to SFE calculations all combinations should have been well resistant to moisture damage, which did not correspond to the experimental test results. In this project, four aggregates and three binders were tested using various test methods.

Table 4. Overview of papers investigating relations between laboratory tests and predicted performance based on surface energy calculations (the findings are summarized in the text).

Materials	Lab Tests	Ref. nr.
Four binder–aggregate combinations; two binders (AAD-1 and AAM-1); two aggregates (a Texas limestone and a river gravel), the gravel aggregate was tested with and without hydrated lime. Mixtures with 4% air voids.	- Repeated compressive loading, in controlled stress mode, at 40 °C, 1 Hz haversine loading; - Water conditioning prior to the testing.	[71]
Twelve binder–aggregate combinations: Four binders (AAB, AAD, ABD, AAE), and three aggregates (granite, basalt, gravel). Mixtures with 4% air voids.	- Dynamic modulus in compression and in tension; dynamic creep; fatigue: Number of cycles to reach 1% permanent μ strain; - Water conditioning prior to testing; submerging in deionized water for 24 h at 50 °C, followed by air drying for 24 h.	[20,28,30]
Six binder aggregate combinations: Three binders (AAB, AAD, and ABD).Two aggregates: Limestone (good field performance in terms of resistance to moisture damage) + andesite (poor observed field performance).	- Uniaxial binder pull-off; samples were all prepared with a film thickness of 30 μm and tested at 23 °C, at a loading rate of 0.01 mm/s (aggregates surface slightly polished); - Binder–aggregate samples were submerged in distilled water for time periods of 0, 12, 24, and 48 h.	[24]
Twenty-five binder–aggregate combinations: Five stones, five binders.	Rolling bottle test (EN 12697-11 clause 5 [72]).	[53]
Three binder–aggregate combinations: One binder (60/70 pen grade) + three aggregates (limestone, granite, and quartzite), each aggregate was tested in three forms; unmodified and coated with two types of polyethylene (HDPE, LDPE), 0.43% by dry weight of aggregate. Mixtures with 4% air voids.	- Repeated unconfined, compressive loading, in controlled stress mode at 25 °C and at 1 Hz under haversine loading; - Preconditioning: AASHTO T283 [73].	[21]
Four binder–aggregate combinations, an unmodified PG 64-22, and a PmB PG 76-22. Binders were also tested after adding Sasobit and an amine-based liquid antistrip additive at 1% and 2%. Two aggregates: Limestone + novaculite.	- A modified PATTI test, pull-off tensile strength; - Plain glass plates, coated with bitumen, placed overnight in an oven at 64 °C in submerged condition.	[55]
Two unmodified binders (40/60 +160/220); 40/60 binder was modified with liquid antistripping agents (0.5 wt%)and four amine-based and one non-amine antistripping agent;three limestone + three granite aggregates.	- Static immersion test (ASTM D1664 [74]); - Rolling bottle test (EN 12697-11 [72]); - Boiling water test (ASTM D3625-96 [75]); - A total water immersion test; - An ultrasonic method.	[41]
Four aggregates: Two limestone + two granite; three binders: 15, 50 + 100 pen grades; aggregates similar to [41].	- Rolling bottle test (EN 12697-11 [72]); - Saturated ageing tensile stiffness (SATS) (combined ageing and water damage).	[26]
One limestone aggregate, two mineral fillers (granite and limestone), a 40/60 pen binder compaction to three nominal air void levels (4%, 6%, and 8%).	- Indirect tensile stiffness (20 °C); - Indirect tensile strength (ITS) tests (20 °C); - Water immersion at 60 °C, up to 70 days.	[71]
- Limestone and granite fine aggregate: Passing the 1 mm sieve and retained on 0.125 mm sieve. - Limestone and granite mineral fillers satisfying EN 1097-7-2008 [76]. - A 40/60 pen grade.	- Mineralogy; - Aggregate and mastic moisture absorption; - Mastic adhesion strength and mastic cohesion; - Water conditioning, depended on the test method.	[15]

Table 4. *Cont.*

Materials	Lab Tests	Ref. nr.
Two binders (40/60, 70/100); five aggregates (two limestones + three granite).	- Tensile force during a binder peel-off test; - SATS tests [26]; - Whole specimens were submersed in water at 20 °C for 7 and 14 days.	[22]
Four binders (AC 60-70); eight aggregates; (three limestones, two granite, two sandstones and quartzite) one liquid antistrip additive.	Modified Lottman indirect tension test; procedure (AASHTO T283 [73]).	[31]
Twenty-four combinations: Three aggregates (limestone, granite and quartzite, each with and without nano CaCO3) and two binders (AC 60-70 and 85-100; with and without nano ZnO and Wetfix BE).Mixtures with air void 7%.	- Modified Lottman indirect tension test procedure (AASHTO T283 [73]) [45]; - Number of cycles to failure wet/dry [60].	[45,60]
Four aggregates, various empirical degrees of stripping (limestone, basal, greywacke, and granite) and three binders, two unmodified, and one polymer modified.	- Rolling bottle test (EN 12697-11 clause 5 [72]); - Boiling water test (EN 12697-11 clause 7 [72]); - Bitumen bond strength (ASTM D 4541 [77]).	[46]

In conclusion, Table 4 demonstrates that a large number of laboratory test procedures have been used to validate SFE component calculations, including binder pull-off and peeling tests, mastic cohesion strength and adhesion tests, asphalt mixture tensile tests, stiffness, dynamic modulus and fatigue tests, rolling bottle and boiling water tests. In some cases, very good relations between predicted water sensitivity levels, based on the SFE concept and measured performances were obtained, [15,20,26,28–30,40,51], while in other cases relations were less good or even nonexistent as reported in [22,41,46,53,55]. It is clear that SFE parameters are important, but the overview of findings suggests that very often other properties need to be included when estimating the moisture susceptibility in laboratory tests or in practice. An important property of the aggregate, indicated by many papers [9,19,20,29–31,44,45,78], is the SSA, which is related to the aggregate roughness or texture. Of course, this property will be of less importance in tests where the aggregate surface is polished. Other parameters that have been considered include the binder film thickness and asphalt mixture porosity [24,45]. Again, in tests where the binder film thickness is constant, this parameter will not show up. Similar observations can be made for the water conditioning step, if the water needs to penetrate a large aggregate substrate before reaching the aggregate binder bond, probably the water permeability and water uptake of the aggregate will start to become important. In addition, the temperature at which the water conditioning is conducted will also have an influence on the results. Temperatures as high as 64 °C have been used, and due to the high temperature susceptibility of bitumen, one can expect a different behavior at high temperatures as compared to a conditioning conducted at 20 °C. For example, Wasiuddin [55] noted that where 64 °C was used during the water conditioning, binder flow, and cavitation were observed, which is probably related to this high temperature conditioning. Correlations between SFE and rolling bottle tests have been found to be weak [46,55]. The obvious reason for this, is the presence of a mechanical action in the rolling bottle test, which is not covered by SFE calculations. In addition, most likely in the rolling bottle test the mechanical action dominates over the SFE driven equilibrium conditions. Hirsch and Friemel-Göttlich [53] explained the lack of correlation by the formation of a good adhering but brittle film breaking under the rolling action and the formation of a strong cohesive, but less good adhering film. Furthermore, a number of papers also showed the importance of the filler type, which may need to be included. Finally, a reversibility of moisture damage has also been demonstrated.

The overview indicates that the set of properties, needed to obtain good relations between predicted and tested performances, depends very much on the experimental test setup, including the water conditioning regime.

As explained in several papers [15,46,51,55,79], water may cause a change in the damage location; in the dry state, the damage typically occurs in the binder or mastic phase, being the weakest bond, referred to as cohesive failure, while under wet conditions, the damage occurs in the aggregate–binder interface, referred to as adhesive failure. Depending on the time given, and the degree of water penetration, mixed behavior is also possible. Apeagyei et al. [15] observed a weakening of the mastic due to the presence of water. In their paper, it was postulated that the damage in the presence of water would still be cohesive and take place in the weakened mastic phase. If the failure location changes between wet and dry states, the use of a wet versus dry ratio becomes questionable. This ratio can be high because of a high moisture resistance, or because the water did not reach the aggregate–binder interface, or because already in the dry state the cohesive properties had a low value. In fact, the percentual area of the binder aggregate bond exposed to water may be a better indicator, as explained by Kim et al. [25].

To validate the SFE concept, in a first step it is legitimate to remove as much as possible other parameters, so that the significance of SFE parameters can be demonstrated. In further steps other parameters such as SSA, binder film thickness, and porosity could be included, to further evaluate their importance. Studies evaluating all or at least a large number of parameters are needed to further clarify the relationships between each of these properties, and to possibly rank them according to their importance. In the study of Hamedi and Moghadas Nejad [45], where indirect tensile tests on asphalt cores were conducted, it seems that the SSA of the aggregate is certainly one of the decisive parameters.

3.3.2. Correlations between Calculated Bond Strengths and Field Experience

When evaluating field behavior, external parameters come into play; which examine the conditions outside of the asphalt mixture system, such as traffic and weather fluctuations.

In the papers [19,20,28,36], bond strength calculations and the various energy ratios were compared for eight mixtures with known field performance. The comparison of field performance with the bond energy parameters, ER1 and ER2, and the energy ratios multiplied with the SSA shows that all ratios can distinguish good from poor performing field mixtures. For each ratio, threshold values of the bond energy parameter could be derived. The authors concluded that bond energy parameters can be used to segregate mixtures based on their moisture sensitivity, but that these data cannot be used for qualitative comparisons between various parameters. Since, unlike laboratory tests, it is difficult to control and quantify the moisture sensitivity of field mixtures on a uniform scale due to the differences in environmental and field conditions that influence these mixtures. Masad et al. [36] combined the use of surface energy components with DMA data on mastics are described to predict fatigue and healing in further detail. This approach is not discussed in this report. Bhasin et al. [19] used the same data but, in this case, they were ranked according to the ratio of the total free energy ratios and the free energy ratios calculated separately for the polar components. In this paper [19], for both energy ratios threshold values to identify moisture susceptible binder aggregate combinations were proposed. The authors concluded that the portion of the bond energy that results from the interaction of the acid component of binder and the base component of aggregate contributes the most to the total adhesive bond strength of the mixture.

4. Summary of Experimental Studies Based on the Hamaker Approach

The Hamaker equation is applicable to nonpolar materials and materials which have polar but no directional bonds, as for example hydrogen bonds. Three papers have been published that evaluate the adhesion between bitumen and aggregate both in the dry and wet state.

Lyne et al. [32] calculated Hamaker constants of a typical bitumen with several aggregates, based on dielectric and refractive index data, which were obtained from the literature. These Hamaker constants were calculated, for the intervening medium air and also for the medium water. When changing the intervening medium from air to water, the part of the Hamaker constant related to polar interactions increased, while the part related to dispersive interactions decreased a lot. It is expected that molecular forces related to dipoles increase with polar liquids such as water, while dispersive interactions decrease. For air as an intervening medium, the polar part contributed 1% of the total Hamaker value. For water as an intervening medium, the polar part contributed 9% to 25 % of the total interaction. The calculation showed that if water comes in between the aggregate and the bitumen, the interaction forces between bitumen and aggregate decrease by 80%–90%. This reduction is the main reason for stripping. The authors also showed that the dispersive part, derived from refractive index data is dominant in the bitumen stone interaction, in air, and also in water. As the dispersive terms are dominating, the data also imply that adhesion increases (dry as well as in water) as the refractive index of the aggregate and the bitumen are larger. In this paper [32], the ranking of the field performance of several aggregates, derived from literature data [80], corresponded to the ranking based on the calculation of the Hamaker constant.

Lyne et al. [65,66] focused on the variation in the dispersive component of minerals via their refractive indices, which were found in the literature. In this study, it was assumed that the bitumen–air–aggregate and the bitumen–water–aggregate adhesive interaction can be represented by the dispersive component of the Hamaker's constant. The data indicated that aggregates and minerals that have a refractive index, higher than a cutoff value of around 1.6 are expected to be less susceptible to stripping. This is represented in Figure 3.

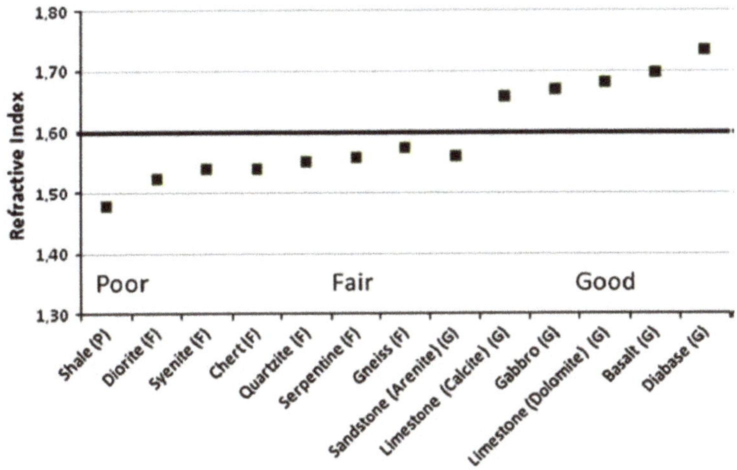

Figure 3. High refractive index value for aggregates and minerals classified according to their degree of resistance to stripping (P—poor, F—fair, and G—good) [65,80].

In this paper [66], the authors also related stripping and refractive index values to the elemental composition of the minerals. Some relations were found: (1) alkali metals tend to lower the refractive index and (2) alkali earth metals and transition metals tend to maintain the refractive index above 1.6. The authors concluded that the elemental composition of minerals has a clear influence on the refractive index and therefore also on the dispersive interaction ability with bitumen. However, it was also indicated that other mechanisms that are not dependent on an interaction may still occur, such as alkali metals, which form water-soluble salts, and which will contribute to stripping when present in higher concentrations.

Lyne et al. [65,66] evaluated several bitumen aggregate combinations. In this study, the refractive indexes were measured using ellipsometry, which allowed calculating the dispersive component of the Hamaker constant. Three aggregates and seven binders were included in the study. Regarding the binders, the refractive indices varied only in the second decimal place, between n = 1.550 and n = 1.598, while the spread in refractive index among the three aggregate samples was much larger. This observation corresponds to the observation that the stone surface determines the adhesion behavior more compared to the bituminous binders. From the refractive index data on bitumen, the cohesive energy of the binders was estimated as the dispersive component of the Hamaker equation in the medium air and water. In the paper, adhesive and cohesive energies were compared, based on the refractive index of the aggregate. If the aggregate had a refractive index that is higher than the binder, the link bitumen–aggregate would be stronger compared to bitumen–bitumen, and cohesive failure of the bitumen will be more likely. If the aggregate has a refractive index of the same magnitude as the bituminous binder, then an adhesive failure will be as likely as a cohesive failure. The critical refractive index value, discriminating good from fair aggregates was reported previously to be 1.6. This value is very close to the refractive index measured on bitumen and is, therefore, a threshold for cohesive versus adhesive failure. This observation indicates that stripping is related to a bitumen–aggregate adhesion and happens when the bitumen–aggregate adhesion is weaker than the bitumen cohesion.

5. Conclusions

When summarizing the test results of the SFE components of bitumen, typically the SFE is composed of a large dispersive part, sometimes accompanied by a small acid and/or base component. There is some variation between different binders, but it is small, and it is not clear how this relates to other properties. Softer binders tend to have lower polar components and slight differences between naphthenic and paraffinic bitumen have been observed. The effect of bitumen modifications is very small. Regarding test precision, it seems to be very difficult to measure the small parts, acid and base accurately. When comparing the different methods to determine the surface energy components, rather large differences between different test methods have been noted. It was also reported that the sample state has an effect a microtome-cut surface of bitumen is different compared to an air-cooled bitumen surface [53], and [54] showed differences between water-treated and air-cooled samples. These observations indicate that for a multicomponent material such as bitumen, care is needed when properties are derived from measurements made on the samples' surface. As bitumen consists of a very large numbers of different molecules, it could very well be that an air-cooled surface, and this is the one that is investigated in most studies, is different from the surface composition that adheres to the aggregate. Indications of chemical and microstructural differences between air-cooled and fractured bitumen surfaces have been noted [81,82] and recently microscopy studies have demonstrated differences in bitumen surfaces depending on the type of contact liquid [83]. In addition, also for polymers variations in surface energies depending upon the contact medium have been observed by Awaja et al. [84]. This seems to point out that SFE measurements on air-cooled surfaces are different from the state of the binder as it is when against a stone surface.

With respect to aggregates, it was observed that there is a much larger variation between the SFE components of aggregates compared to the variation seen for bitumen. Aggregates typically have large base components, although when looking at minerals there are specific minerals that have only small acid and base components. There are basically two test methods commonly used for aggregates, the universal sorption test and the sessile drop method. In both cases, the sample preparation of the aggregates is critical. For example, there is still a question if possible absorbed water layers or other impurities on the aggregate surface have been removed or need to be removed, which has not yet been fully answered. In turn, the time after cutting or heating the aggregates may also have an effect. For the sessile drop test, the flatness or the absence of sufficient flatness has been reported to influence the test result. In this test, only a very small area of the aggregate is tested. As aggregates are known to be nonuniform, and can also differ between the individual stone particles, a lot of repeats are needed to obtain a full view of the variation. In the universal sorption device, a larger amount of aggregates is tested, averaging internal variations. Regarding the test precision, it seems, as for bitumen, difficult to measure the smallest SFE parts accurately.

Regarding the calculation of bond strengths, small differences in acid and base components of the bitumen can have a large effect on the adhesive bond strengths in dry and in wet conditions, especially for aggregates with a large mostly base component. Consequently, small errors in the acid base determination of the bitumen can have a large effect.

Regarding the validation, a number of studies have shown good correlations between laboratory tests and the predicted behavior, while there are also papers where correlations were less good. It is clear that, in addition to SFE based calculations, other parameters need to be taken into account in order to make good predictions. Other parameters such as the SSA, the binder film thickness, the asphalt permeability have been proposed. In this respect, it is not clear how important and which role SFE based properties can play, within the mixture of all possible influences.

In another approach, the change in bitumen drops after they have obtained good adhesion were followed while being submerged in water. This approach seems promising, but also in this case questions about the effect of binder viscosity, temperature, and the influence of aggregate flatness need to be addressed. Finally, the Hamaker approach, which is mainly relating bitumen–aggregate adhesion to dispersive interactions [32,65,66], seems to be able to establish a ranking between different aggregate types. Nevertheless, for this approach, a laboratory validation including various bitumens is not yet available. Moreover, also in this case, possible sample and/or more suitable surface preparation effects have not been studied in detail. Additionally, again, also the Hamaker approach relates only to equilibrium conditions.

In general, fundamental questions with regard to moisture damage mechanisms as they take place in asphalt mixtures remain. It is still not fully clear which mechanism, or combination of mechanisms induce moisture damage, and how these depend on conditions such as temperature, the particular mix design, binder aging, the traffic, the water exposure time, and possibly the presence of other pavement failures. The use of SFE based properties can be regarded as a promising element, in this complex interplay of properties.

Author Contributions: Writing—original draft preparation and editing, H.S. and S.V.; writing—review and editing, P.K.D.M. All authors have read and agreed to the published version of the manuscript.

Funding: The research was partly funded by the CEDR Transnational Road Research Programme Call 2013: Energy Efficiency—Materials and Technology through the project FunDBitS (Functional Durability-related Bitumen Specification). Funding countries included Germany, Norway, UK, Austria, The Netherlands, and Slovenia.

Conflicts of Interest: The authors declare no conflict of interest.

References

1. Lu, Q.; Harvey, J.T.; Monismith, C.L. Investigation of Conditions for Moisture Damage in Asphalt Concrete and Appropriate Laboratory Test Methods: Summary Report, University of California, Pavement Research Center UC Davis and Berkeley. Available online: http://www.ucprc.ucdavis.edu/PDF/UCPRC-SR-2005-01.pdf (accessed on 26 January 2020).
2. Skaf, M.; Pasquini, E.; Revilla-Cuesta, V.; Ortega-Lopez, V. Performance and durability of porous asphalt mixtures manufactured exclusively with electric steel slag. *Materials* **2019**, *12*, 3306. [CrossRef] [PubMed]
3. Ingrassia, L.P.; Lu, X.; Ferrotti, G.; Canestrari, F. Renewable materials in bituminous binders and mixtures: Speculative pretext or reliable opportunity? *Resour. Conserv. Rec.* **2019**, *144*, 209–222. [CrossRef]
4. Ma, J.; Sun, D.; Pang, Q.; Sun, G.; Hu, M.; Lu, T. Potential of recycled concrete aggregate pretreated with waste cooking oil residue for hot mix asphalt. *J. Clean. Prod.* **2019**, *221*, 469–479. [CrossRef]
5. Terrel, R.; Al-Swailmi, S. Water Sensitivity of Asphalt–Aggregate Mixes: Test Selection. In *SHRP Report A-403*; National Research Council: Washington, DC, USA, 1994; p. 183. Available online: http://onlinepubs.trb.org/onlinepubs/shrp/SHRP-A-403.pdf (accessed on 10 October 2019).
6. Caro, S.; Masad, E.; Bhasin, A.; Little, D.N. Moisture susceptibility of asphalt mixtures, Part 1: mechanisms. *Int. J. Pavement Eng.* **2008**, *9*, 81–98. [CrossRef]

7. Abuawad, I.M.A. Mechanical and Surface Free Energy Characterization of Asphalt Concrete for Moisture Damage Detection. Ph.D. Thesis, University of Illinois at Urbana-Champaign, Champaign County, IL, USA, 2016; p. 179. Available online: https://pdfs.semanticscholar.org/7260/8d5a5564df3fbeec1c50857d5b759fa86c72.pdf (accessed on 13 February 2020).
8. Cheng, D.; Little, D.N.; Lytton, R.; Holste, J.C. Surface energy measurement of asphalt and its application to predicting fatigue and healing in asphalt mixtures. *Transp. Res. Rec.* **2002**, *1810*, 44–53. [CrossRef]
9. Little, D.L.; Allen, D.H.; Bhasin, A. *Modeling and Design of Flexible Pavements and Materials*; Springer International Publishing AG: Cham, Switzerland, 2018; p. 693. [CrossRef]
10. Caro, S.; Masad, E.; Bhasin, A.; Little, D.N. Moisture susceptibility of asphalt mixtures, Part 2: Characterisation and modelling. *Int. J. Pavement Eng.* **2008**, *9*, 99–114. [CrossRef]
11. Bagampadde, U.; Isacsson, U.; Kiggundu, B.M. Classical and Contemporary Aspects of Stripping in Bituminous Mixes. *RMPD* **2004**, *5*, 7–43. [CrossRef]
12. Ensley, E.K. Multilayer adsorption with molecular orientation of asphalt on mineral aggregate and other substrates. *J. Appl. Chem. Biotechnol.* **1975**, *25*, 671–682. [CrossRef]
13. Yao, H.; Dai, Q.; You, Z. Chemo-physical analysis and molecular dynamics (MD) simulation of moisture susceptibility of nano hydrated lime modified asphalt mixtures. *Constr. Build. Mater.* **2015**, *101*, 536–547. [CrossRef]
14. Terrel, R.L.; Shute, J.W. *Summary Report on Water Sensitivity. Report SHRP-A/IR-89-003*; Strategic Highway Research Program, National Research Council: Washington, DC, USA, 1989; p. 102. Available online: http://onlinepubs.trb.org/Onlinepubs/shrp/shrp-A-IR-89-003/SHRP-AIR-003.pdf (accessed on 10 October 2019).
15. Apeagyei, A.K.; Grenfell, J.R.A.; Airey, G.D. Moisture-induced strength degradation of aggregate–asphalt mastic bonds. *RMPD* **2014**, *15*, 239–262. [CrossRef]
16. Van Oss, C.J.; Chaudhury, M.K.; Good, R.J. Interfacial Lifshitz-van der Waals and polar interactions in macroscopic systems. *Chem. Rev.* **1988**, *88*, 927–941. [CrossRef]
17. Vansteenkiste, S.; Soenen, H. Correlation between Bitumen and Asphalt Properties. In *Project report FunDBitS. CEDR Call 2013: Energy Efficient–Materials and Technology*; 2016; p. 51. Available online: https://www.cedr.eu/download/other_public_files/research_programme/call_2013/energy_efficiency/fundbits/D.2e_bitumen-aggregate-interaction.pdf (accessed on 6 September 2019).
18. Hefer, A.W. Adhesion in Bitumen-Aggregate Systems and Quantification of the Effects of Water on the Adhesive Bond. Ph.D. Thesis, Texas A&M University, College Station, TX, USA, 2004; p. 222. Available online: http://hdl.handle.net/1969.1/1457 (accessed on 9 September 2019).
19. Bhasin, A.; Masad, E.; Little, D.; Lytton, R. Limits on Adhesive Bond Energy for Improved Resistance of Hot-Mix Asphalt to Moisture Damage. *Transp. Res. Rec.* **2006**, *1970*, 3–13. [CrossRef]
20. Little, D.N.; Bhasin, A. Using surface energy measurements to select materials for asphalt pavement. In *Final Report for National Cooperative Highway Research Program RRD 316*; National Cooperative Highway Research Program: Washington, DC, USA, 2007. [CrossRef]
21. Arabani, M.; Hamedi, G.H. Using the surface free energy method to evaluate the effects of polymeric aggregate treatment on moisture damage in hot-mix asphalt. *J. Mater. Civ. Eng.* **2011**, *23*, 802–811. [CrossRef]
22. Zhang, J.; Airey, G.D.; Grenfell, J.; Apeagyei, A.K. Moisture sensitivity examination of asphalt mixtures using thermodynamic, direct adhesion peel and compacted mixture mechanical tests. *Road Mater. Pavement Des.* **2016**, *19*, 120–138. [CrossRef]
23. Alvarez, A.E.; Ovalles, E.; Caro, S. Assessment of the effect of mineral filler on asphalt–aggregate interfaces based on thermodynamic properties. *Constr. Build. Mater.* **2012**, *28*, 599–606. [CrossRef]
24. Howson, J.; Masad, E.; Little, D.; Kassem, E. Relationship between bond energy and total work of fracture for asphalt binder-aggregate systems. *Road Mater. Pavement Des.* **2012**, *13*, 281–303. [CrossRef]
25. Kim, S.-H.; Jeong, J.-H.; Kim, N. Use of Surface Free Energy Properties to Predict Moisture Damage Potential of Asphalt Concrete Mixture in Cyclic Loading Condition. *KSCE J. Civ. Eng.* **2003**, *7*, 381–387. [CrossRef]
26. Grenfell, J.; Ahmad, N.; Liu, Y.; Apeagyei, A.; Large, D.; Airey, G. Assessing asphalt mixture moisture susceptibility through intrinsic adhesion bitumen stripping and mechanical damage. *RMPD* **2014**, *15*, 131–152. [CrossRef]
27. Kloubek, J. Development of methods for surface free energy determination using contact angles of liquids on solids. *Adv. Colloid. Interfac.* **1992**, *38*, 99–142. [CrossRef]

28. Bhasin, A. Development of Methods to Quantify Bitumen-Aggregate Adhesion and Loss of Adhesion Due to Water. Ph.D. Thesis, Texas A&M University, College Station, TX, USA, 2006; p. 158. Available online: http://hdl.handle.net/1969.1/5934 (accessed on 9 September 2019).
29. Bhasin, A.; Howson, J.; Masad, E.; Little, D.N.; Lytton, R.L. Effect on modification processes on bond energy of asphalt binders. *Transp. Res. Rec.* **2007**, *1998*, 29–37. [CrossRef]
30. Bhasin, A.; Little, D.N.; Vasconcelos, K.L.; Masad, E. Surface free energy to identify moisture sensitivity of materials for asphalt mixes. *Transp. Res. Rec.* **2007**, *2001*, 37–45. [CrossRef]
31. Hamedi, G.H.; Moghadas Nejad, F. Using energy parameters based on the surface free energy concept to evaluate the moisture susceptibility of hot mix asphalt. *RMPD* **2015**, *16*, 239–255. [CrossRef]
32. Lyne, A.L.; Birgisson, B.; Redelius, P. Interaction forces between mineral aggregates and bitumen calculated using the hamaker constant. *RMPD Spec. Issue Asph. Technol.* **2010**, *11*, 305–323. [CrossRef]
33. Israelachvili, J.N. Van der Waals forces in biological systems. *Q. Rev. Biophys.* **1973**, *6*, 341–387. [CrossRef]
34. Hefer, A.; Little, D. Towards Quantification of Adhesion and Water Stripping in Bituminous Materials Using Modern Surface Energy Theory. In Proceedings of the 24th South African Transport Conference (SATC2005), Pretoria, South Africa, 12 July 2005; pp. 115–124. Available online: https://repository.up.ac.za/bitstream/handle/2263/6314/015.pdf;sequence=1 (accessed on 9 September 2019).
35. Hefer, A.; Little, D.N. Adhesion in Bitumen-Aggregate Systems and Quantification of the Effects of Water on the Adhesive Bond. In *Technical Report ICAR-505-1*; Texas A&M University: College Station, TX, USA, 2005; p. 218. Available online: http://hdl.handle.net/2152/35371 (accessed on 9 September 2019).
36. Masad, E.; Zollinger, C.; Bulut, R.; Little, D.N.; Lytton, R.L. Characterization of HMA moisture damage using surface energy and fracture properties. *Assoc. Asph. Paving Technol. Proc. Tech. Sess.* **2006**, *75*, 713–754.
37. Wasiuddin, N.M.; Zaman, M.M.; O'Rear, E.A. Effect of Sasobit and Aspha-min on wettability and adhesion between asphalt binders and aggregates. *Transp. Res. Rec.* **2008**, *2051*, 80–89. [CrossRef]
38. Howson, J.E.; Bhasin, A.; Masad, E.; Lytton, R.; Little, D. Development of a Database for Surface Energy of Aggregates and Asphalt Binder. In *Report FHWA/TX-09/5-4524-01-1*; Federal Highway Administration/Texas Deptartment of Transportation: Austin, TX, USA, 2009; p. 76. Available online: http://tti.tamu.edu/documents/5-4524-01-1.pdf (accessed on 9 September 2019).
39. Bahramian, A. Evaluating surface energy components of asphalt binders using wilhelmy plate and sessile drop techniques. In *Degree Project in Highway and Railway Engineering*; KTH: Stockholm, Sweden, 2012; p. 40. Available online: https://www.diva-portal.org/smash/get/diva2:601925/FULLTEXT01.pdf (accessed on 9 September 2019).
40. Moghadas Nejad, F.; Hamedi, G.H.; Azarhoosh, A.R. Use of surface free energy method to evaluate effect of hydrate lime on moisture damage in hot-mix asphalt. *J. Mater. Civ. Eng.* **2013**, *25*, 1119–1126. [CrossRef]
41. Liu, Y.; Apeagyei, A.; Ahmad, N.; Grenfell, J.; Airey, G. Examination of moisture sensitivity of aggregate–bitumen bonding strength using loose asphalt mixture and physico-chemical surface energy property tests. *Int. J. Pavement. Eng.* **2014**, *15*, 657–670. [CrossRef]
42. Luo, R.; Zhang, D.; Zeng, Z.; Lytton, R. Effect of surface tension on the measurement of surface energy components of asphalt binders using the Wilhelmy Plate Method. *Constr. Build. Mater.* **2015**, *98*, 900–909. [CrossRef]
43. Shafabakhsh, G.H.; Faramarzi, M.; Sadeghnejad, M. Use of surface free energy method to evaluate the moisture susceptibility of sulfur extended asphalts modified with antistripping agents. *Constr. Build. Mater.* **2015**, *98*, 456–464. [CrossRef]
44. Hamedi, G.H. Evaluating the effect of asphalt binder modification using nanomaterials on the moisture damage of hot mix asphalt. *RMPD* **2016**, *18*, 1375–1394. [CrossRef]
45. Hamedi, G.H.; Moghadas Nejad, F. Evaluating the effect of mix design and thermodynamic parameters on moisture sensitivity of hot mix asphalt. *J. Mater. Civ. Eng.* **2017**, *29*, 04016207. [CrossRef]
46. Porot, L.; Soenen, H.; Besamusca, J.; Apeagyei, A.; Grenfell, J.; Vansteenkiste, S.; Chailleux, E.; Gaudefroy, V.; Chaturabong, P.; Tozzo, C.; et al. Bituminous Binder. In *Testing and Characterization of Sustainable Innovative Bituminous Materials and Systems*; RILEM State-of-the-Art Reports 24, PartI; Porot, L., Di Benedetto, H., Canestrari, F., Marsac, P., Tebaldi, G., Eds.; Springer International Publishing: Cham, Switzerland, 2018; pp. 15–74. [CrossRef]

47. Kakar, M.R.; Hamzah, M.O.; Akhtar, M.N.; Saleh, J.M. Evaluating the Surface Free Energy and Moisture Sensitivity of Warm Mix Asphalt Binders Using Dynamic Contact Angle. *Adv. Civ. Eng.* **2019**, *19*, 15. [CrossRef]
48. Koc, M.; Bulut, R. Characterization of warm mix asphalt additives using direct contact angle measurements and surface free energies. In Proceedings of the Transportation Research Board 93rd Annual Meeting, Washington, DC, USA, 12–16 January 2014.
49. Aguiar-Moya, J.P.; Salazar-Delgado, J.; Baldi-Sevilla, A.; Leiva-Villacorta, F.; Loria-Salazar, L. Effect of aging on adhesion properties of asphalt mixtures with the use of bitumen bond strength and surface energy measurement tests. *Transp. Res. Rec.* **2015**, *2505*, 57–65. [CrossRef]
50. Ji, J.; Liu, L.; Suo, Z.; Zhai, P.; Yang, X.; You, Z. Adhesion Evaluation of Asphalt-Aggregate Interface Using Surface Free Energy Method. *Appl. Sci. Basel* **2017**, *7*, 156. [CrossRef]
51. Moraes, R.; Velasquez, R.; Bahia, H. Using bond strength and surface energy to estimate moisture resistance of asphalt-aggregate systems. *Constr. Build. Mater.* **2017**, *130*, 156–170. [CrossRef]
52. Korn, S. Beurteilung der Benetzbarkeit und des Adhäsionsvermögens von Bitumen mittels Kontaktwinkelmessungen. Ph.D. Thesis, Hamburg University of Applied Sciences, Hamburg, Germany, 2004.
53. Hirsch, V.; Friemel-Göttlich, B. Bestimmung des adhäsiven Potentials von Bitumen und Gesteinsoberflächen mit Hilfe der Kontaktwinkel-messmethode. In *Berichte der Bundesanstalt für Straßenwesen Straßenbau (BAsT)*; Heft S 59; Bergisch Gladbach, Germany, 2009; Available online: https://bast.opus.hbz-nrw.de/opus45-bast/frontdoor/deliver/index/docId/71/file/S59.pdf (accessed on 9 September 2019).
54. Khan, A.; Redelius, P.; Kringos, N. Evaluation of adhesive properties of mineral-bitumen interfaces in cold asphalt mixtures. *Constr. Build. Mater.* **2016**, *125*, 1005–1021. [CrossRef]
55. Wasiuddin, N.M.; Saltibus, N.E.; Mohammad, L.N. Novel moisture-conditioning method for adhesive failure of hot- and warm-mix asphalt binders. *Transp. Res. Rec.* **2011**, *2208*, 108–117. [CrossRef]
56. Wistuba, M.P.; Grothe, H.; Grönniger, J.; Handle, F. Adhesion of bitumen: Screening and evaluating laboratory testing techniques. Proceedings of 5th Eurasphalt & Eurobitume Congress, Istanbul, Turkey, 13–15 June 2012; p. 12.
57. Radenberg, M. Einfluss der chemischen, rheologischen und physikalischen Grundeigenschaften von Straßenbaubitumen auf das Adhäsionsverhalten unterschiedlicher Gesteinskörnungen. In *Schlussbericht*; 16639 N/1, Bochum, Germany, 2014; p. 104. Available online: http://www.lvw.ruhr-uni-bochum.de/mam/content/forschung/projekte/bericht_aif_igf_16639_n-1.pdf (accessed on 9 September 2019).
58. Miller, C.M. Adhesion and the Surface Energy Components of Natural Minerals and Aggregates. Master's Thesis, Texas A&M University, College Station, TX, USA, 2010; p. 218. Available online: http://hdl.handle.net/1969.1/ETD-TAMU-2010-08-8237 (accessed on 9 September 2019).
59. Miller, C.; Little, D.N.; Bhasin, A. Surface energy characteristics of natural minerals and their impact of aggregate-bitumen bond strengths and asphalt mixture durability. *Transp. Res. Rec.* **2012**, *2267*, 45–55. [CrossRef]
60. Azarion, Y.; Shirmohammadi, H.; Hamedi, G.H.; Saedi, D. Model for Predicting Moisture Susceptibility of Asphalt Mixtures Based on Material Properties. *J. Mater. Civ. Eng.* **2019**, 31. [CrossRef]
61. Porot, L.; Besamusca, J.; Soenen, H.; Apeagyei, A.; Grenfell, J.; Sybilski, D. Bitumen/aggregate affinity – RILEM Round Robin test. In *8th RILEM International Symposium on Testing and Characterization of Sustainable and Innovative Bituminous Materials*; Canestrari, F., Partl, M.N., Eds.; Springer: Dordrecht, The Netherlands, 2016; RILEM 11; pp. 153–164. [CrossRef]
62. Rudawska, A.; Jacniacka, E. Analysis for determining surface free energy uncertainty by the Owen-Wendt method. *Int. J. Adhes. Adhes.* **2009**, *29*, 451–457. [CrossRef]
63. Cui, S.; Blackman, B.R.K.; Kinlock, A.J.; Taylor, A.C. Durability of asphalt mixtures: Effect of aggregate type and adhesion promoters. *Int. J. Adhes. Adhes.* **2014**, *54*, 100–111. [CrossRef]
64. Horgnies, M.; Darque-Ceretti, E.; Fezai, H.; Felder, E. Influence of the interfacial composition on the adhesion between aggregates and bitumen: Investigations by EDX, XPS and peel tests. *Int. J. Adhes. Adhes.* **2011**, *31*, 238–247. [CrossRef]
65. Lyne, A.L.; Redelius, P.; Collin, M.; Birgisson, B. Characterization of stripping properties of stone material in asphalt. *Mater. Struct.* **2013**, *46*, 47–61. [CrossRef]

66. Lyne, A.L.; Krivosheeva, O.; Birgisson, B. Adhesion between bitumen and aggregate: Implementation of spectroscopic ellipsometry characterisation and estimation of Hamaker's constant. *Mater. Struct.* **2013**, *46*, 1737–1745. [CrossRef]
67. Schellenberger, W. Haftung zwischen Bitumen und Mineralstoff [Adhesion between bitumen and mineral aggregates]. In *Asphalt*; Germany, 2004; p. 8.
68. Nehrings, A. Affinitätsproblem zwischen Gestein und Bitumen Lösungsansatz: Einsatz von Haftvermittlern. In *Zeitschrift für Asphalt*; Germany, 2009; p. 8. Available online: https://docplayer.org/33746588-Affinitaetsproblem-zwischen-gestein-und-bitumen-loesungsansatz-einsatz-von-haftvermittlern.html (accessed on 9 September 2019).
69. Aguiar-Moya, J.P.; Loria-Salazar, L.; Salazar, J.; Villegas, E.; Corrales-Azofeifa, J.P.; Hajj, E.Y. Evaluation of Adhesion Properties of Costa Rican Asphalt Mixtures Using Bitumen Bond Strength and Contact-Angle Measurement Tests. Available online: https://www.lanamme.ucr.ac.cr/repositorio/bitstream/handle/50625112500/545/TRB%202013%20Evaluation%20of%20Adhesion%20Properties%20of%20Costa%20Rican%20Asphalt%20Mixtures.pdf?sequence=1&isAllowed=y (accessed on 9 September 2019).
70. Aranowski, R.; Wojewodka, P.; Blazejowski, K. Determination of binder-aggregate adhesion by contact angle measurement. In *Asphalt Pavements*; Kim, Y.R., Ed.; Taylor&Francis Group: London, UK, 2014; pp. 617–624. [CrossRef]
71. Apeagyei, A.K.; Grenfell, J.R.A.; Airey, G.D. Durability of asphalt mixtures exposed to long-term moisture conditioning. In Proceedings of the TRB 93rd Annual Meeting, Washington, DC, USA, 12–16 January 2014; pp. 1–16. Available online: https://www.researchgate.net/publication/260002328_Durability_of_asphalt_mixtures_exposed_to_long-term_moisture_conditioning (accessed on 27 January 2020).
72. EN 12697-11. Bituminous Mixtures–Test Methods for Hot Asphalt–Part 11: Determination of the Affinity between Aggregates and Bitumen. European Committee for Standardization: Brussels, Belgium, 2012.
73. AASHTO T283. *Standard Method of Test for Resistance of Compacted Asphalt Mixtures to Moisture-Induced Damage*; AASHTO: Washington, DC, USA, 2014.
74. ASTM D 1664-80. *Test Method for Coating and Stripping of Bitumen Aggregates Mixtures*; ASTM International: West Conshohocken, PA, USA, 1985.
75. ASTM D 3625-96. *Standard Practice for Effect of Water on Bituminous-Coated Aggregate Using Boiling Water*; ASTM International: West Conshohocken, PA, USA, 2005.
76. EN 1097-7:2008. Tests for mechanical and physical properties of aggregates. In *Determination of the Particle Density of Filler. Pyknometer Method*; European Committee for Standardization: Brussels, Belgium, 2009.
77. ASTM D 4541-17. *Standard Test Method for Pull-Off Strength of Coatings Using Portable Adhesion Testers*; ASTM International: West Conshohocken, PA, USA, 2017.
78. Moghadas Nejad, F.; Asadi, M.; Hamedi, G.H.; Esmaeeli, M.R. Using hydrophobic coating on aggregate surfaces to reduce moisture damage in asphalt mixture. *J. Mater. Civ. Eng.* **2018**, *30*, 04018238. [CrossRef]
79. Zhang, J.; Airey, G.D.; Grenfell, J.; Apeagyei, A.K.; Barrett, M. Development of a composite substrate peel test to assess moisture sensitivity of aggregate–bitumen bonds. *Int. J. Adhes. Adhes.* **2016**, *68*, 133–141. [CrossRef]
80. Cordon, W.A. *Properties, Evaluation and Control of Engineering Materials*; McGraw-Hill College: New York, NY, USA, 1979; p. 550. ISBN 9780070131231.
81. Lu, X.; Sjövall, P.; Soenen, H. Structural and chemical analysis of bitumen using time-of-flight secondary ion mass spectrometry (TOF-SIMS). *Fuel* **2017**, *199*, 206–218. [CrossRef]
82. Blom, J.; Soenen, H.; Katsiki, A.; Van den Brande, N.; Rahier, H.; Van den bergh, W. Investigation of the bulk and surface microstructure of bitumen by atomic force microscopy. *Constr. Build. Mater.* **2018**, *177*, 158–169. [CrossRef]
83. Ramm, A.; Downer, M.C.; Sakib, N.; Bhasin, A. Morphology and kinetics of as-phalt binder microstructure at gas, liquid, and solid interfaces. *J. Microsc.* **2019**, 109–117. [CrossRef]
84. Awaja, F.; Gilbert, M.; Kellya, G.; Foxa, B.; Pigram, P.J. Adhesion of polymers. *Prog. Polym. Sci.* **2009**, *34*, 948–968. [CrossRef]

infrastructures

Article

Recycled Waste Powders for Alkali-Activated Paving Blocks for Urban Pavements: A Full Laboratory Characterization

Piergiorgio Tataranni

Department of Civil, Chemical, Environmental and Materials Engineering, University of Bologna, 40136 Bologna, Italy; piergiorg.tataranni2@unibo.it

Received: 31 October 2019; Accepted: 21 November 2019; Published: 22 November 2019

Abstract: Paving blocks are today a popular paving solution for urban surfaces. Considering the wide variety of products currently on the market, it is possible to build pavements that differ in terms of functionality, bearing capacity, skid resistance, visual impact, and aesthetic integration with the surrounding landscape. Interlocking concrete paving block is the most common construction technology considering its low cost and its easy installation. Different wastes and second-hand materials have recently been tested in order to completely or partially replace the raw materials used for the production of paving blocks. In this paper, a waste basalt powder is used for the production of alternative paving blocks through the alkali-activation process. Two different synthetic blocks were produced, with and without aggregates. Taking into account the EN 1338 standard for concrete paving blocks, a complete laboratory characterization is proposed for the two experimental blocks. Tests highlighted positive results and downsides that need to be optimized in order to convert the laboratory production to an industrial scale.

Keywords: paving blocks; alkali-activated materials; urban pavements; waste powders; recycled materials

---

1. Introduction

Recent studies have highlighted the dramatic development of the urban land cover phenomenon, given by the actual era of unprecedented global urbanization [1,2]. The growth in the size of cities has completely changed the original concept of urbanization, making the modern urban area a complex system of paved surfaces [3,4]. Everyday people spend countless hours of their lives in the road network and considering the multitude of activities carried out on urban pavements, these can no longer be treated as simple infrastructures [5,6]. The intricacy of the modern urbanization has led to a differentiation in the urban pavement network, which is currently composed by lanes for powered vehicles, special lanes, bike, and pedestrian lanes, parking areas, sidewalks and squares [7]. Studies revealed that around the 95% of road users wish to have a clear and instant visual identification of the different paths, which compose the urban roads network [8].

The need to differentiate the pavements according to the final intended use has created different paving solutions, in terms of construction technology and materials [9]. From the traditional bituminous pavements, the new design solutions encompass the application of special asphalt concretes (porous or colored asphalt mixtures), paving blocks, cobblestone pavements or special ultrathin surface layers [10].

Paving blocks represent a suitable alternative to cobblestone or bituminous sidewalks, bike or pedestrian lanes and to historic pavements, especially in old cities centers [11,12]. These are commonly employed as paving solution due to the relatively low production and laying costs [13]. Furthermore, considering the possible use of a wide range of materials and craft different shapes, paving blocks have a large applicability in civil constructions [14]. The most common paving blocks are produced in cement concrete, where the mix design is a function of the final performance required for the block. Lightweight concrete is often used for pedestrian and outdoor pavements. Porous concrete is generally required for permeable pavements (i.e., parking areas) and high-performance cement concrete is suitable for heavy load traffic pavements or heavy load storage areas [15–18].

According to the latest estimates, the constructions sector is responsible for 36% of global energy use and around 40% of CO2 emissions [19]. Taking into account the growing interest for environmental issues such as the limitation of non-renewable resources and the emission of greenhouse gasses related to human activities, the construction's sector has been strongly affected by eco-friendly policies. In the last years, an increasing demand for alternative and sustainable materials has been registered to promote and to develop the so-called novel "green constructions" [20–23]. The recycling of waste materials seems to be a viable solution for the production of new construction materials. The re-uses of wastes, industrial byproducts and second-hand materials can couple the advantages given by the conservation of resources to the inclusion of materials destined for landfills in the production cycle of a new product [24]. This approach is perfectly in line with the circular economy concept, where the objective is the reduction of the environmental footprint, also related to the construction's sector. Furthermore, when scientifically proven, the re-use of waste materials does not compromise the construction standards [25,26]. Thus, researchers from all over the world are focusing on experimental applications of wastes as construction materials, being the recycling the new frontier of the civil engineering [27].

The paving blocks market is not further from this phenomenon. The cement concrete is the most common constitutive material for modular elements, and the Portland cement production is today under investigation from an environmental point of view [28]. Andrew calculated the CO2 emission from cement production in 2017 as 1.48 Gt, corresponding to about 8% of the carbon dioxide globally produced [29]. These emissions derived from the combined action of the chemical reaction involved in the Portland cement production (formation of clinker) and the power needed to heat the raw materials over 1000 °C. Over the years, attempts have been made to partially or completely substitute the Portland cement with sustainable materials in order to reduce the environmental footprint of the concrete production [30,31]. The literature shows several applications of alternative materials, as paving block constituents. Most of the studies concentrated on the substitution of natural aggregates with recycled materials [32]. Different researches evaluated the possible addition of Construction and Demolition Wastes (CDW) within concrete paving blocks [33] and positive outcomes were verified for the replacement of fine aggregates with recycled materials (i.e., dragged sediments, waste marble, ceramic tiles, etc.) [34–36]. However, a relatively low number of studies focused on the use of byproducts or waste cementitious materials as binding agents, in order to reduce the cement content of the final product [37–39]. The advantage given by the replacement or the reduction of Portland cement with alternative materials would be remarkable, considering the impact of the cement production and the waste disposal on the environment.

Thus, in the case presented here, alternative paving blocks were produced through the alkali-activation process of a waste basalt powder, without the addition of Portland cement. Starting from the laboratory characterization of the alkali-activated paste, two different versions of modular elements were cast: with and without aggregates. The evaluation of the physical, mechanical, and functional properties of the paving blocks was based on laboratory tests suggested by the EN 1338 standard, which specifies the requirements and test methods for concrete modular elements.

2. Materials and Methods

Two different experimental paving blocks were tested: one (labelled PBP) entirely produced with alkali-activated waste basalt powder and a second one (labelled PBA) with the same synthetic paste but with the addition of aggregates according to a specific grading distribution.

The alkali-activation process is a synthesis between two groups of materials: precursors and activators. The result of this process is a cementitious-like material with final properties and performance related to the chemical composition of its constituents [40]. Thus, the properties of activators and precursors are fundamental for the quality of the final alkali-activated material (AAM). Well-established literature verified suitable mechanical performance for AAMs produced with precursors rich in silica and alumina, in strong alkaline conditions generated by specific activators [41,42]. AAMs are today considered a sustainable alternative to Portland cement, considering the relatively low environmental footprint of the production process [43]. Furthermore, if properly designed, the chemical and mechanical property of the material, as well as its durability, are considerably higher if compared to traditional cement concrete.

2.1. Precursors

In this experimental application, a waste basalt powder (B) and metakaolin (M) were used as precursors according to a specific mix design.

B is a material completely passing the 0.005 mm sieve and it is a waste from the basalt extraction process in quarries. Today, this material is landfilled and its re-use can represent an eco-friendly solution to its disposal. Furthermore, the use of basalt in the alkali-activation process has been scientifically proven by several studies [44,45].

M is obtained by the thermal treatment of kaolin and its adoption for the synthetic process dates back to the first AA applications. Considering the chemical composition of M, it is widely used in order to improve the mechanical and durability properties of the final product [46].

The chemical properties of both precursors are summarized in Table 1.

Table 1. Chemical properties of basalt and metakaolin.

Compound	Unit	Basalt	Metakaolin
SiO_2	% p/p	45.3	55.2
CaO	% p/p	8.8	0.2
Na_2O	% p/p	1.7	0.6
Al_2O_3	% p/p	21.6	40.3
Fe_2O_3	% p/p	8.5	1.4
SO_3	% p/p	<0.1	0.2
MgO	% p/p	2.0	0.1
P_2O_5	% p/p	0.7	<0.1
TiO_2	% p/p	0.2	1.5
ZnO	% p/p	<0.1	<0.1
K_2O	% p/p	9.7	0.2

2.2. Activators

The activators are needed in order to create the strong alkaline environment suitable for the chemical reaction. Taking as a reference, the well-established literature review and previous experimental studies, the liquid mix used as an activator was a blend of Sodium Silicate (SS) and Sodium Hydroxide (SH). SS is a commercial product with SiO_2/Na_2O ratio equal to 1.99, while SH was prepared with a molarity fixed at 10.

Being the chemical properties crucial for the performance of the final material, different activator blends were produced in terms of ratio between SS and SH.

2.3. Research Plan

The research plan can be divided into two steps: the first is related to the characterization of the alkali-activated paste, while the second phase is about the laboratory characterization of the experimental paving blocks.

The evaluation of the quality of the alkali-activated paste was based on mechanical tests. It is worth noting, that there are no specific test methods or standardized procedures for the characterization of AAMs. Thus, the mechanical analysis was carried out in terms of compressive strength on cubic samples, in compliance with the EN 1015-1 standard, which is traditionally taken as a reference for hardened mortar.

Once the correct mix design for the AAM was defined, two different mixes for paving blocks were prepared, with and without aggregates. The material was casted in plastic rectangular, specific for the production of interlocking modular elements. The following physical, mechanical and functional characterization was based on tests specified in the EN 1338 standard. This European Standard identifies the material requirements and the test protocols and methods for concrete paving blocks. Considering the wide range of applications of modular elements, their performance requirements are defined by the standard in terms of classes and associated marking designations.

Therefore, according to the reference classes, a concrete paving block is considered suitable for its specific application (i.e., road pavement, pedestrian use, parking areas, etc.).

The following tests were carried out on the experimental samples:

- Shape and dimensions;
- Weathering resistance in terms of water absorption;
- Tensile splitting strength;
- Abrasion resistance;
- Slip/skid resistance.

Based on data and on the resulting classification, the experimental paving blocks could be suitable for specific real applications.

3. Alkali-Activated Material Characterization

As previously stated, the characterization of AAMs is generally based on mechanical tests performed in compliance with reference standards for common construction materials, due to the lack of specific tests methods.

In the case under study, the evaluation of the quality of the AA pastes was based on the compressive strength of 40 × 40 mm cubic samples. The reference standard is the EN 1015-1, which is commonly used for hardened mortars. According to the aforementioned standard, the compressive strength is calculated applying an increasing load (from 50 N/s to 500 N/s) on cubic samples, so that the failure occurs in a range from 30 to 90 s. The maximum load is registered and used for the calculation of the compressive strength.

Different mixtures were prepared, according to the following variables:

- Dosage of waste basalt powder and metakaolin;
- Dosage of SS and SH;
- Precursors/Activators ratio;
- Curing time and procedure.

The right mix design was chosen in terms of workability, mechanical properties and low environmental impact curing conditions.

After several trials, the mix design considered the 70% of waste basalt powder and 30% of metakaolin as precursors mix. The SS/SH ratio for the activator was fixed equal to 4, while the precursors/activators ratio was 0.75. Once mixed, the AA paste was casted in 40 × 40 mm cubic molds and one day in the oven at 70 °C was chosen as optimized curing conditions. In order to have a complete mechanical characterization of the AA mixture, the compressive strength tests were carried out after 3, 7, 14, 21 and 28 days of curing.

The mechanical results are presented in Table 2.

Table 2. Compressive strength results for the alkali-activated (AA) cubic samples.

Compressive Strength (MPa)	Day 3	Day 7	Day 14	Day 21	Day 28
Sample 1	45	61	60	59	63
Sample 2	49	61	62	61	64
Sample 3	48	63	57	59	64
Sample 4	46	59	57	57	65
AVG.	**47**	**61**	**59**	**58**	**64**

The AA mixture highlighted a remarkable compressive strength if compared to common Portland cement concrete after just 3 days of curing. However, a light deflection of the mechanical properties is registered for samples cured from day 7 to day 28. This reduction does not affect the performance of the material and, based on results, the complete curing of the experimental mixture can be considered as concluded after 7 days. Furthermore, a slight variation in mechanical results is a common phenomenon for AAMs, due to the high influence of mixing and casting operations on the final performances of the material.

4. Paving Blocks Characterization

Once defined the final mix design and the curing conditions, the AA paste was used for the production of the experimental paving blocks. Two versions of the modular element were produced. The PBP was entirely made in AA mixture, while PBA was mixed with aggregates, according to a specific grading distribution, and the experimental AA paste as binder.

Figure 1 shows the particles size distribution, which fits with the gradation band commonly used for concrete paving blocks. Common limestone aggregates suitable for construction materials were used. They were mixed together with the AA paste according to a specific ratio in order to have a suitable workability of the final material.

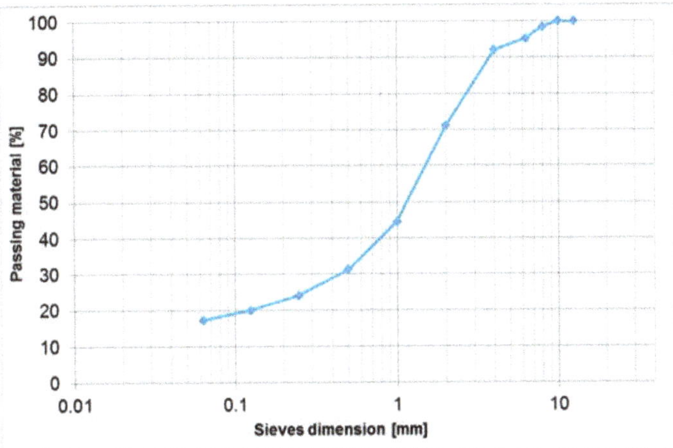

Figure 1. Grading distribution for Paving Block Aggregates (PBA).

The aggregates were mixed with the AA paste in order to improve the mechanical properties of the final mixture and to evaluate whether or not the presence of aggregates could affect the performance of the experimental paving blocks.

Once mixed or not with aggregates, the resulting material was casted in plastic molds, specific for the production of concrete interlocking modular elements. As defined during the mix design phase, the material was cured into the oven at 70 °C for 24 h. The obtained paving blocks were than de-molded with the injection of compressed air. The samples were than stored for seven days before being tested in order to achieve the maximum of their mechanical properties, as verified during the preliminary AAMs characterization.

The plastic mold and the final samples are shown in Figure 2.

(**a**) (**b**)

Figure 2. (**a**) Plastic mold; (**b**) Final paving blocks after de-molding.

In the following paragraphs, based on EN 1338 standard, the laboratory characterization of the experimental paving blocks is presented.

4.1. Geometrical Properties, Visual Aspect And Physical Properties

The geometrical measurement of the paving blocks is considered a compulsory test in order to verify the production consistency of the paving blocks. According to the Annex C of the EN 1338 standard, the thickness of a block is measured to the nearest millimeter. The maximum difference between the readings is calculated and recorded. Five experimental paving blocks were tested for each product.

The standards specify permissible deviations based on the product dimensions. In the case under the study, the paving blocks dimensions, the differences between measurements and the permissible deviations are reported in Table 3.

Table 3. Paving blocks measurements and deviations.

Measurement	Length (mm)	Width (mm)	Thickness (mm)
PBP	200	100	60
PBP deviations	±1	±1	±2
PBA	200	100	60
PBA deviations	±2	±2	±2
Permissible deviations	±2	±2	±3

According to the results, both experimental products are in line with the requirements suggested by the standard. This is a further confirmation of the workability of the AA mixture, which allows the complete filling and the perfect adhesion of the material to the mold profile. Furthermore, the addition of aggregates does not substantially affect these properties.

However, the visual inspection of the paving blocks highlighted the presence of a small amount of surface bubbles. This is mainly due to the casting operations that should be improved in order to dissipate the air trapped into the mixture during the production.

In terms of physical characterization, as imposed by the standard, the paving blocks are classified according to their weight per square meter. The PBP has a theoretical weight of 105 kg/m^2, while the presence of aggregates makes the PBA equal to 115 kg/m^2. The most common concrete paving blocks range between 120–180 kg/m^2.

4.2. Water Absorpition

According to the EN 1338 standard (Annex E), the weathering resistance is determined in terms of freeze-thaw resistance or water absorption. In the case under study, the water absorption was evaluated.

This test is very important for footpath paving materials considering their exposure to weather conditions. Still, this also represents an indirect evaluation of the air voids content of the material, being the porosity directly responsible for the level of saturation of the paving blocks when submerged in water. To evaluate the water absorption, in compliance with the standard, the specimens are immersed in potable water at a temperature of (20 ± 5) °C until a constant mass is reached. Once saturated, the paving blocks are than oven dried to constant mass. The water absorption is than calculated as the ratio between the block weight before and after saturation.

Average results are summarized in Table 4.

Table 4. Water absorption tests results and standard limits.

Paving Blocks	Water Absorption (%)
PBP	14 ± 1
PBA	10 ± 0
EN 1338 limit	<6% (Class 1—Mark B)

It is worth noting, that the EN 1338 standard transfers to a national level, the durability requirements in terms of classes of weathering resistance. However, the maximum suggested water absorption limit is fixed to 6%. Both experimental paving blocks exceed the proposed limit. This is related to the porosity of the material, which was also highlighted from the visual analysis of the paving blocks' surfaces. However, the presence of aggregates within the AA mixture seems to reduce the water absorption. The improvement of the mixing and casting operations might reduce the porosity of the paving blocks and consequently their water absorption.

4.3. Tensile Splitting Strength

The tensile splitting strength represents the only test required by the EN 1338 (Annex F) standard for the mechanical characterization of the concrete paving blocks. According to the standard, the load is applied through two steel blades of a specified size on a sample, and it is progressively increased at a rate equal to 0.05 MPa/s. Consequently, the failure load is registered and the area of the failure planes is calculated. The tensile splitting strength is than calculated according to the following equation:

$$T = 0.637 \cdot k \cdot \frac{P}{S} \quad (1)$$

where k is a correction factor dependent on the block thickness, P is the failure load, while S is the area of the failure.

The failure load per unit length (F) is also required and calculated as the ratio between the failure load and failure length measured at the top and at the bottom of the paving block. According to the test procedure, the samples are kept in a water bath at 20 °C for 24 h before testing.

Three samples were tested for each product and results are presented in Table 5.

As for the reference standard limits, the average characteristic tensile splitting strength required should not be less than 3.6 MPa, but none of the samples can register a mechanical value lower than 2.9 MPa and a failure load per unit length lower than 250 N/mm.

In the case under study, only the PBA exceeds the standard requirements. This is due to the presence of aggregates within the AA mixture, which improves the structural properties of the paving block, as well as its cohesion, if compared to PBP. In this case, the absence of a lithic skeleton makes the structure weak for tensile splitting strength even if the compressive strength verified in the previous lab characterization on cubic samples was considerably high.

Table 5. Tensile splitting test results and limits.

Paving Blocks	T (MPa)	F (N/mm)
PBP 1	2.4	256
PBP 2	1.7	182
PBP 3	2.2	237
Avg. PBP	2.1	225
PBA 1	4.2	456
PBA 2	3.6	389
PBA 3	3.2	344
Avg. PBA	3.7	396
EN 1338 limit	>3.6 MPa	>250 N/mm

4.4. *Abrasion Resistance*

The abrasion resistance is generally considered the ability of a surface to withstand the friction action. This is an important property for paving materials considering that it is directly related to their durability, as well as to the functional properties of the surface.

This property of the paving blocks is evaluated through the Wide Wheel Abrasion test. According to the standard, the abrasive force is generated by an abrasive material, which flows on a rotating wheel that acts on the paving blocks surface (EN 1338, Annex G) (Figure 3).

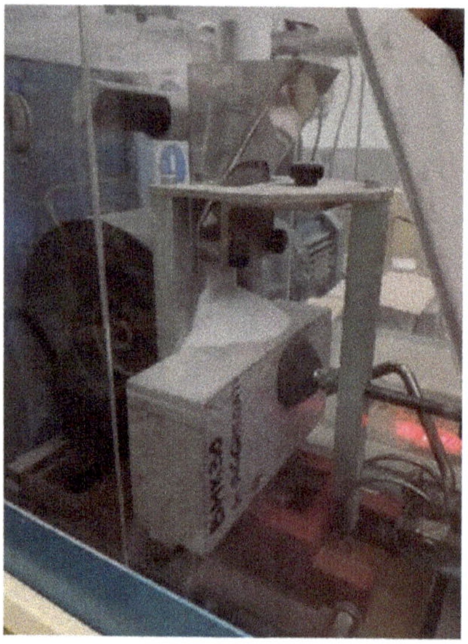

Figure 3. Wide Wheel Abrasion test on Paving Block Paste (PBP) sample.

The abrasive material is a corundum powder with a specific particles size and its flow onto the abrasion wheel with a minimum rate of 2.5 L/min. The wheel is made of specific Brinnel hardness steel and it rotates on the paving block surface according to 75 revolutions in 60 s. Furthermore, the sample is clamped in order to stay always in contact with the abrasive wheel during the test. After 75 revolutions, the surface of the paving block is cleaned and the dimension of the groove is registered. The EN 1338 standard classifies and marks the concrete paving blocks according to the groove dimensions.

Three samples were tested for every material and the results are presented in Table 6.

Table 6. Abrasion tests results and classification according to the EN 1338 standard.

Paving Blocks	Groove Dimensions (mm)	EN 1338 Classification
PBP 1	22.4	-
PBP 2	23.6	-
PBP 3	25.7	-
Avg. PBP	24	Class 1—Mark F
PBA 1	22.7	-
PBA 2	22.7	-
PBA 3	23.6	-
Avg. PBA	23	Class 3—Mark H

According to the results, the two experimental products have different class and mark, considering the higher abrasion resistance properties for the PBA. For traditional concrete paving blocks, the abrasion resistance is strictly related to the curing conditions, to the surface finishing or to the mix design (i.e., aggregates hardness, binder quality, and aggregates-paste ratio). In the case presented here, the only variable between the materials was the presence of aggregates for PBA, which seems to improve the abrasion resistance characteristics of the paving blocks.

4.5. Slip/skid Resistance

The skid resistance can be considered as one of the most important functional properties of a paved surface being directly responsible for the safety of the users. In compliance with the EN 1338 standard (Annex I), the slip/skid resistance is evaluated in terms of Unpolished Slip Resistance Value (USRV). This is a measure of the quality of the paving blocks and it determines whether the particular surface finish is appropriate for the proposed application.

The pendulum friction tester is proposed for the evaluation of the skid resistance. The friction force offered by a wetted surface to a rubber slider sliding on it, is measured in terms of reduction in length of the slider swing using a calibrated scale on the equipment (Figure 4).

Figure 4. Pendulum friction tester during the skid resistance tests on concrete paving blocks.

Considering that the friction force can be affected by the materials temperature, the test equipment, complete with rubber slider, is kept at 20 °C for at least 30 min immediately before performing the test. At the same time, the samples are kept in a water bath at 20 °C for 30 min. Three paving blocks were tested for each material and the results are shown in Table 7.

Table 7. URSV tests results.

Paving Blocks	USRV
PBP 1	60
PBP 2	62
PBP 3	63
Avg. PBP	62
PBA 1	56
PBA 2	51
PBA 3	53
Avg. PBA	54

The results highlighted higher skid resistance for the PBPs. In this case, the presence of aggregates does not improve the performance of the material, which instead are reduced. All in all, the URSV values are remarkably high for both experimental materials. It is worth noting, that the EN 1338 standard does not fix limitations or threshold values in terms of slip/skid resistance. The most common paving blocks have a minimum URSV value equal to 35 and some UK technical guidelines identify the range between 40 and 79 USRV as "low potential for slip" [47]. Thus, the experimental paving blocks show considerable skid resistance properties.

In the case under examination, the surface texture was giving by the mold, that had specific texture on the walkable surface of the modular element. The presence of aggregates has probably reduced the viscosity of the AA mixture, as a consequence, its perfect distribution on the mold profile needed to imprint the texture on the cured surface.

5. Conclusions

An alternative interlocking modular element produced with alkali-activated waste basalt powder is presented in this paper. Two different products were designed, produced and tested, with and without aggregates. Taking into account the EN 1338 standard for concrete paving blocks, a complete laboratory characterization of the experimental blocks is proposed.

Based on the data presented in this paper, several conclusions and comments can be drawn:

The chemical predisposition of the waste basalt powder for the alkali-activation process has been verified. This is further confirmation of previous research presented by the author. Furthermore, the adopted mix design seems to be suitable for the production of a mixture with adequate workability and mechanical properties to be used as a paste for the casting of modular elements.

The alternative paving blocks without aggregates (PBP) do not comply with some of the requirements of the EN 1338 standard. Despite the AA paste ensuring a perfect casting, the tensile splitting strength, as well as the water absorption of the samples, are limited. The mechanical properties are not enough, despite the high compressive strength results obtained during the AA paste characterization.

The presence of aggregates within the paving block is fundamental in order to achieve the tensile splitting strength required by the EN 1338 standard. As a downside, the addition of particles within the AA paste slightly reduces the workability and casting operations and limits the full adhesion of the mortar to the mold profile. This phenomenon was confirmed through the verification of the shape, superficial texture and dimensions of the paving blocks. However, the dimension's deviations were in line with the acceptance limits imposed by the reference standard.

In the light of the presented laboratory results, the production of paving blocks through the alkali-activation of waste basalt powder seems to be a viable alternative for interlocking modular elements. However, the presence of aggregates according to a specific grading distribution is needed to achieve the required mechanical properties. Furthermore, the optimization of the particles distribution and the paste/aggregates ratio would improve the workability properties of the mixture and the casting operations. Still, the use of vibrating tables during the molding phase could improve the quality of the final product, reducing the presence of trapped air in the mixture. It may convert the laboratory production to an industrial scale.

In an eco-friendly perspective, further studies will have to focus on the production and characterization of AA paving blocks from waste powders with recycled aggregates in order to achieve fully recycled and sustainable modular elements.

Funding: This research received no external funding.

Acknowledgments: Basalti Orvieto srl supplied the waste basalt powder for the research.

Conflicts of Interest: The authors declare no conflict of interest.

References

1. Artmann, M.; Inostroza, L.; Fan, P. Urban sprawl, compact urban development and green cities. How much do we know, how much do we agree? *Ecol. Indic.* **2019**, *96*, 3–9. [CrossRef]
2. Seto, K.C.; Güneralp, B.; Hutyra, L.R. Global forecasts of urban expansion to 2030 and direct impacts on biodiversity and carbon pools. *Proc. Natl. Acad. Sci. USA* **2012**, *109*, 16083–16088. [CrossRef] [PubMed]
3. Anastasiadou, K.; Vougias, S. "Smart" or "sustainably smart" urban road networks? The most important commercial street in Thessaloniki as a case study. *Transp. Policy* **2019**, *82*, 18–25. [CrossRef]
4. Ng, C.P.; Law, T.H.; Jakarni, F.M.; Kulanthayanc, S. Relative improvements in road mobility as compared to improvements in road accessibility and urban growth: A panel data analysis. *Transp. Res. Part A Policy Pract.* **2018**, *117*, 292–301. [CrossRef]
5. Gullotta, T.M.; Mistretta, M.; Praticò, F. A life cycle scenario analysis of different pavement technologies for urban roads. *Sci. Total Environ.* **2019**, *673*, 585–593. [CrossRef] [PubMed]
6. Tsiotas, D.; Polyzos, S. The topology of urban road networks and its role to urban mobility. *Transp. Res. Procedia* **2017**, *24*, 482–490. [CrossRef]
7. Lyu, X.; Han, Q.; de Vries, B. Procedural modeling of urban layout: Population, land use, and road network. *Transp. Res. Procedia* **2017**, *25*, 3333–3342. [CrossRef]
8. Harsha, G.; Thoradeniya, B.; Halwatura, R. Public perceptions on urban outdoor constructions and their materials. In *Proceedings of the 8th International Conference of Faculty of Architecture Research Unit (FARU)*; University of Moratuwa: Colombo, Sri Lanka, 2015; pp. 1–12.
9. Celauro, C.; Corriere, F.; Guerrieri, M.; Lo Casto, B. Environmentally appraising different pavement and construction scenarios: A comparative analysis for a typical local road. *Transp. Res. Part D Transp. Environ.* **2015**, *34*, 41–51. [CrossRef]
10. Di Mascio, P.; Moretti, L.; Capannolo, A. Concrete block pavements in urban and local roads: Analysis of stress-strain condition and proposal for a catalogue. *J. Traffic Transp. Eng.* **2019**. [CrossRef]
11. Mudiyono, R.; Nor, H.M.; Hainin, M.R.; Ling, T.-C. Performance of concrete paving blocks pavement on sloped road section. *Int. J. Pavement* **2007**, *6*, 136–145.
12. Mendoza, J.-M.F.; Oliver-Solà, J.; Gabarrell, X.; Rieradevall, J.; Josa, A. Planning strategies for promoting environmentally suitable pedestrian pavements in cities. *Trans. Res. D Trans. Environ.* **2012**, *17*, 442–450. [CrossRef]
13. Sun Poon, C.; Chan, D. Paving blocks made with recycled concrete aggregate and crushed clay brick. *Constr. Build. Mater.* **2006**, *20*, 569–577. [CrossRef]
14. Hengl, H.L.; Füssl, J. The influence of superelevated profiles of paving block structures on their load-bearing behavior. *Eng. Struct.* **2016**, *117*, 195–203. [CrossRef]
15. Bhuvaneshwari, P.; Priyadharshini, U.; Gurucharan, S.; Mithunram, B. Strength characteristics of lightweight concrete blocks using mineral admixtures. *IOP Conf. Ser. Earth Environ. Sci.* **2017**, *80*, 012003. [CrossRef]
16. Elizondo-Martinez, E.-J.; Andrés-Valeri, V.C.; Jato-Espino, D.; Rodriguez-Hernandez, J. Review of porous concrete as multifunctional and sustainable pavement. *J. Build. Eng.* **2020**, *27*, 100967. [CrossRef]
17. Giustozzi, F. Polymer-modified pervious concrete for durable and sustainable transportation infrastructures. *Constr. Build. Mater.* **2016**, *111*, 502–512. [CrossRef]
18. Shan, J.S.; Li, H.; Jiang, H.W. Mechanical characteristics and design method of interlocking concrete block pavement. *J. Traffic Transp. Eng.* **2015**, *15*, 9–17.

19. International Energy Agency and the United Nations Environment Programme. *2018 Global Status Report: Towards a Zero-Emission, Efficient and Resilient Buildings and Construction Sector*; International Energy Agency and the United Nations Environment Programme: Katowice, Poland, 2018.
20. Kutnik, M.; Suttie, E.; Brischke, C. Durability, efficacy and performance of bio-based construction materials: Standardisation background and systems of evaluation and authorisation for the European market. In *Performance of Bio-based Building Materials*; Elsevier: Amsterdam, The Netherlands, 2017; pp. 593–610. [CrossRef]
21. Tataranni, P.; Sangiorgi, C.; Simone, A.; Vignali, V.; Lantieri, C.; Dondi, G. A laboratory and field study on 100% Recycled Cement Bound Mixture for base layers. *Int. J. Pavement Res. Technol.* **2018**, *11*, 427–434. [CrossRef]
22. Franzoni, E.; Volpi, L.; Bonoli, A.; Spinelli, R.; Gabrielli, R. The environmental impact of cleaning materials and technologies in heritage buildings conservation. *Energy Build.* **2018**, *165*, 92–105. [CrossRef]
23. Van den Heede, P.; de Belie, N. Environmental impact and life cycle assessment (LCA) of traditional and "green" concretes: Literature review and theoretical calculations. *Cem. Concr. Compos.* **2012**, *34*, 431–442. [CrossRef]
24. Sormunen, P.; Kärki, T. Recycled construction and demolition waste as a possible source of materials for composite manufacturing. *J. Build. Eng.* **2019**, *24*, 100742. [CrossRef]
25. Chen, W.; Jin, R.; Xu, Y.; Wanatowski, D.; Li, B.; Yan, L.; Pan, Z.; Yang, Y. Adopting recycled aggregates as sustainable construction materials: A review of the scientific literature. *Constr. Build. Mater.* **2019**, *218*, 483–496. [CrossRef]
26. Yap, S.P.; Goh, Y.; Mo, K.H.; Ibrahim, H.A. Recycling of Construction and Demolition Wastes into Renewable Construction Materials. In *Reference Module in Materials Science and Materials Engineering*; Elsevier: Amsterdam, The Netherlands, 2019.
27. Robayo-Salazar, R.A.; Fernando Rivera, J.; Mejía de Gutiérrez, R. Alkali-activated building materials made with recycled construction and demolition wastes. *Constr. Build. Mater.* **2017**, *149*, 130–138. [CrossRef]
28. Xiao, J.; Li, W.; Fan, Y.; Huang, X. An overview of study on recycled aggregate concrete in China (1996–2011). *Constr. Build. Mater.* **2012**, *31*, 364–383. [CrossRef]
29. Andrew, R.M. Global CO_2 emissions from cement production, 1928–2017. *Earth Syst. Sci. Data* **2018**, *10*, 2213–2239. [CrossRef]
30. Tayeh, B.A.; Hasaniyah, M.W.; Zeyad, A.M.; Yusuf, M.O. Properties of concrete containing recycled seashells as cement partial replacement: A review. *J. Clean. Prod.* **2019**, *237*, 117723. [CrossRef]
31. Rashad, A.M. An overview on rheology, mechanical properties and durability of high-volume slag used as a cement replacement in paste, mortar and concrete. *Constr. Build. Mater.* **2018**, *187*, 89–117. [CrossRef]
32. Silva, R.V.; de Brito, J.; Dhir, R.K. Use of recycled aggregates arising from construction and demolition waste in new construction applications. *J. Clean. Prod.* **2019**, *236*, 117629. [CrossRef]
33. Soutsos, M.N.; Tang, K.; Millard, G.S. Use of recycled demolition aggregate in precast products, phase II: Concrete paving blocks. *Constr. Build. Mater.* **2011**, *25*, 3131–3143. [CrossRef]
34. Wang, L.; Chen, L.; Tsang, D.C.W.; Li, J.S.; Baek, K.; Hou, D.; Ding, S.; Sun Poon, C. Recycling dredged sediment into fill materials, partition blocks, and paving blocks: Technical and economic assessment. *J. Clean. Prod.* **2018**, *199*, 69–76. [CrossRef]
35. Gancel, O.; Ozel, C.; Koksal, F.; Erdogmus, E.; Martinez-Barrera, G.; Brostow, W. Properties of concrete paving blocks made with waste marble. *J. Clean. Prod.* **2012**, *21*, 62–70. [CrossRef]
36. Wattanasiriwech, D.; Saiton, A.; Wattanasiriwech, S. Paving blocks from ceramic tile production waste. *J. Clean. Prod.* **2009**, *17*, 1663–1668. [CrossRef]
37. Ganjian, E.; Jalull, G.; Sedeghi-Pouva, H. Using waste materials and by-products to produce concrete paving blocks. *Constr. Build. Mater.* **2015**, *77*, 270–275. [CrossRef]
38. Kumar, A.; Kumar, S. Development of paving blocks from synergistic use of red mud and fly ash using geopolymerisation. *Constr. Build. Mater.* **2013**, *38*, 865–871. [CrossRef]
39. Agyeman, S.; Obeng-Ahenkra, N.K.; Assiamah, S.; Twumasia, G. Exploiting recycled plastic waste as an alternative binder for paving blocks production. *Case Stud. Constr. Mater.* **2019**, *11*, e00246. [CrossRef]
40. Provis, J.L.; Van Deventer, J.S.J. Introduction to geopolymers. In *Geopolymers*; Elsevier: Amsterdam, The Netherlands, 2009; pp. 1–11.

41. Pacheco-Torgal, F.; Castro-Gomes, J.; Jalali, S. Alkali-activated binders: A review: Part 1. Historical background, terminology, reaction mechanisms and hydration products. *Constr. Build. Mater.* **2008**, *22*, 1305–1314. [CrossRef]
42. Provis, J.L. Alkali-activated materials. *Cem. Concr. Res.* **2018**, *114*, 40–48. [CrossRef]
43. Rakhimova, R.N.; Rakhimov, R.Z. Toward clean cement technologies: A review on alkali-activated fly-ash cements incorporated with supplementary materials. *J. Non-Cryst. Solids* **2019**, *509*, 31–41. [CrossRef]
44. Tataranni, P.; Besemer, G.M.; Bortolotti, V.; Sangiorgi, C. Preliminary Research on the Physical and Mechanical Properties of Alternative Lightweight Aggregates Produced by Alkali-Activation of Waste Powders. *Materials* **2018**, *11*, 1255. [CrossRef]
45. Tataranni, P.; Sangiorgi, C. Synthetic Aggregates for the Production of Innovative Low Impact Porous Layers for Urban Pavements. *Infrastructures* **2019**, *4*, 48. [CrossRef]
46. Palomo, A.; Blanco-Varela, M.T.; Granizo, M.L.; Puertas, F.; Vazquez, T.; Gruzteck, M.W. Chemical stability of cementitious materials based on metakaolin. *Cem. Concr. Res.* **1999**, *29*, 997–1004. [CrossRef]
47. UK Slip Resistance Group. *The Assessment of Floor Slip Resistance*; Issue 2, RAPRA; UK Slip Resistance Group: Coventry, UK, 2000.

infrastructures

MDPI

Article

Characterization of Asphalt Mixtures Produced with Coarse and Fine Recycled Asphalt Particles

Saeed S. Saliani [1], Alan Carter [1], Hassan Baaj [2] and Pejoohan Tavassoti [2,*]

1 Department of Construction Engineering, École de Technologie Supérieure, Montreal, QC H3C 1K3, Canada; saliani.saeed@gmail.com (S.S.S.); Alan.Carter@etsmtl.ca (A.C.)
2 Department of Civil and Environmental Engineering, University of Waterloo, Waterloo, ON N2L 3G1, Canada; hassan.baaj@uwaterloo.ca
* Correspondence: ptavassotikheiry@uwaterloo.ca

Received: 10 July 2019; Accepted: 21 October 2019; Published: 26 October 2019

Abstract: Utilizing recycled asphalt pavements (RAP) in pavement construction is known as a sustainable approach with significant economic and environmental benefits. While studying the effect of high RAP contents on the performance of hot mix asphalt (HMA) mixes has been the focus of several research projects, limited work has been done on studying the effect of RAP fraction and particle size on the overall performance of high RAP mixes produced solely with either coarse or fine RAP particles. To this end, three mixes including a conventional control mix with no RAP, a fine RAP mix (FRM) made with 35% percent fine RAP, and a coarse RAP mix (CRM) prepared with 54% of coarse RAP were designed and investigated in this study. These mixes were evaluated with respect to their rutting resistance, fatigue cracking resistance, and low temperature cracking performance. The results indicate that although the CRM had a higher RAP content, it exhibited better or at least the same performance than the FRM. The thermal stress restrained specimen testing (TSRST) results showed that the control mix performed slightly better than the CRM, while the FRM performance was adversely affected with respect to the transition temperature midpoint and the maximum tensile stress temperature. Both of the RAP incorporated mixes exhibited better rutting resistance than the control mix. With regard to fatigue cracking, the CRM performed better than the FRM. It can be concluded that the RAP particle size has a considerable effect on its contribution to the total binder content, the aggregate skeleton of the mix, and ultimately the performance of the mix. In spite of the higher RAP content in the CRM versus FRM, the satisfactory performance observed for the CRM mix indicates a great potential in producing high RAP content mixes through optimizing the RAP particle size and content. The results also suggest that the black curve gradation assumption is not representative of the actual RAP particles contribution in a high RAP mix.

Keywords: hot mix asphalt; recycled asphalt; RAP gradation; complex modulus; fatigue cracking; permanent deformation; thermal cracking resistance

---

1. Introduction

Utilizing reclaimed asphalt pavement (RAP) in hot mix asphalt (HMA) is proven to be a green alternative to produce environment-friendly asphalt mixes. Adding RAP in asphalt mixes is suggested not only to conserve the aggregates and bitumen, but also to have at least the same performance [1]. Recycling of the existing mineral aggregates and asphalt binder in RAP particles would be of great benefit to the environment by saving the nonrenewable materials. Milled pavements are considered to be valuable materials after reaching the end of pavement service life. At a minimum level, RAP can play the role of virgin mineral aggregates in order to conserve the energy and save the environment. However, the ideal goal is to maximize reusing the waste materials in new pavement construction projects in a way that the same or even better performance as compared to the conventional materials can be achieved.

1.1. High RAP Mixes

Although there is no recognized unanimity about the limit of the maximum amount of RAP in HMA, RAP percentage in HMA has been limited by many agencies, mainly, due to the unproven performance of high RAP mixes and also lack of a unified mix design [2]. The review of the previous research on RAP indicates use of up to 100% RAP in HMA mixes. However, most of the plant-produced 100% RAP hot mix asphalt projects date back to 1991 and earlier [3–5]. In 1997, the Federal Highway Administration's RAP expert task group developed guidelines for the design of Superpave HMA containing RAP [6]. In the same year, another study by Kandhal and Foo [7] recommended a three-tier process to deal with RAP in asphalt concrete, where a RAP content of 25% and more was defined as high RAP mix, requiring detailed evaluations [7]. RAP limitation was also supported by the findings of the NCHRP research report 9–12 [8]. In spite of several research projects conducted on RAP incorporated mixes, still there is not a clear vision about the interaction of RAP and virgin materials in details. Different scenarios can be considered about the interaction of virgin and aged binder: (1) there is no interaction between old and virgin materials, so RAP could be assumed as a black rock. In other words, the aged binder in RAP does not significantly contribute the total binder content. As the rheology of RAP may be affected by facing preheated aggregate and hot virgin binder, this assumption would most probably be different from what happens in reality. (2) All of the aged binder in RAP blends into the mix and with virgin materials effectively. Again, it is not clear whether this assumption is close to reality or not. Therefore, further research is needed to figure out the rate of interaction between the used and new materials and the significant parameters affecting this phenomenon. Previous study showed that depending on the RAP size and aggregate gradation, the available binder content in RAP would vary [9]. They have also concluded that there is a significant difference between large and small particles with respect to transition of the asphalt binder from RAP to virgin aggregates.

There was no guidance until early 1990's for implementation of RAP in HM, but based on experimental studies, FHWA Asphalt Mixture Expert Task Group defined the interim recommendations [6]. Based on the performance of Marshal Mixes with RAP, and mixes designed according to the Superpave system, AASHTO Standards MP2 (now M323) describes how to design HMA with RAP [10].

Recycled asphalt mixes consist of complex bituminous material. Further, sometimes unknown milling processes make it difficult to study RAP with predictable properties. Therefore, many issues arise due to RAP variability when high percentages of RAP are used in a mix. One solution for this issue might be using RAP in different layers of pavement structure. An example of such application is discussed by Pratico et al. who describe the feasibility of building a two-layer porous asphalt (TLPA) by recycling from permeable European mixes (PEM) RAP, when highly variable RAP stockpiles are involved [11].

1.2. RAP Binder Characteristics

RAP mainly consists of aggregates and aged binder. When incorporating RAP in a new HMA, the aged binder can affect the mix behavior in long-term because of the diffusion of RAP and virgin binder. Generally, adding high RAP contents into new mixes can increase the stiffness significantly [12]. Asphalt as a petroleum-based product is an organic material which can be subjected to short term chemical oxidation due to the combined effect of heat and atmospheric oxygen during the mixing and hauling process. Characterization of oxidation is of utmost importance because this phenomenon makes the material brittle [13]. This becomes even more crucial when the RAP content in a mix is more than 25%, which according to the conventional definitions mentioned earlier is known as high RAP mix. Therefore, depending on the RAP content, presence of aged binder can change the mixture performance. During the past few decades, several studies have focused on characterizing RAP and on finding the proper way of using it in producing asphalt mixtures. For example, Cosentino et al. [14]

concluded that the controlling factors in the performance of RAP are dependent on where the RAP is obtained and its gradation. However, there are still several aspects of using RAP in HMA that require further investigation such as the impact of RAP source, content, gradation, conditioning, etc.

According to McDaniel et al. [8], less than 15% RAP has no impact on the blended binder performance in mix. Between 15% and 25% RAP, the virgin binder grade is commonly decreased by one grade (6 °C) on both ends (e.g., a PG 64-22 is changed to a PG 58-28). For more than 25% of RAP, binder needs to be graded using the performance-graded binder tests. Therefore, for the mixes with high RAP content, full characterization of binder is needed.

The aging level of RAP binder film thickness might be changed according to depth where the mix is placed at and presence of oxygen. Bressi et al. [15] proposed a methodology to detect the existence of a cluster phenomenon (Figure 1) and they also proposed a first approach to show a different aging level in the RAP binder film thickness (i.e., partial differential aging). Stephens et al. [16] also investigated the asphalt films properties for the coarse aggregates. They concluded that it would be more prone to blending with virgin aggregates than asphalt film around fine aggregates. The aged binder recovered from the coarse and the fine particles was compared by conducting a series of Dynamic Shear Rheometer (DSR) tests. They concluded that there is no correlation between variation in the binder stiffness and the asphalt coating of coarse or fine aggregates. The main issue in this domain was referred to its exposure to heat and air during production, which is a random process and does not correlate with either the aggregate size or related film thickness.

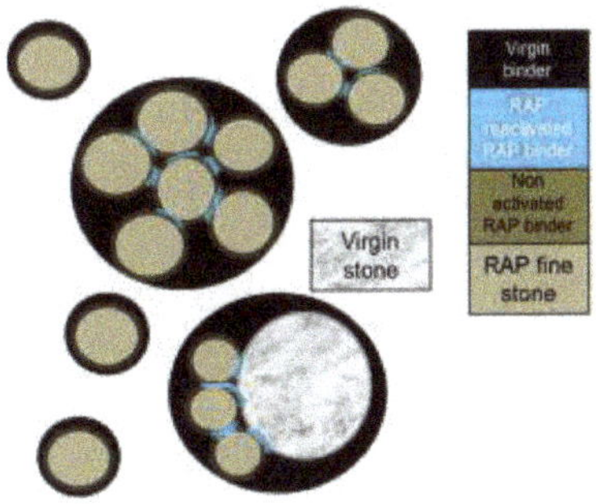

Figure 1. Schematic diagram of cluster phenomenon [15].

1.3. *Effect of RAP Particle Size on Binder-Blending Phenomenon*

In addition to the binder rheology in RAP particles several other parameters can affect the final performance of a RAP produced mix. Saliani et al. [9] showed active binder in coarse RAP is significantly higher than the fine RAP. In their work, Saliani et al. mainly focused on virgin aggregate surface area and correlation with cutting or melting the aged binder from RAP particles. Saliani et al. [9] concluded that virgin aggregate surface area is another factor that can have an impact on the interaction of recycled and virgin material. In addition to aggregate surface area, film thickness in RAP particles are not the same suggesting that more investigations are needed to characterize the film thickness properties properly.

Gardiner [17] concluded that the complex modulus is not solely controlled by the stiffness of the binder, whereas several other factors including the gradation and angularity of the aggregate have impact on it. Mixing method, heating temperature, and mixing duration of RAP need to be optimized to ensure the complete blending of the old and new binders so that the plant production process can be better simulated in the lab [18]. However, all of the aforementioned studies concluded that the stiffness of coarse and fine RAP would be the same, but their binder contribution to the new mix depends on some other factors, such as virgin aggregates surface area, mixing temperature, and RAP preheating temperature (if applicable).

Several researchers studied the blending of RAP and virgin binders. Chen [19] found that not only RAP does not act like a black rock but also a significant blending occurs between RAP and the virgin binder. However, Huang [20] studied the blending of RAP with virgin HMA mixtures for a given type of screened RAP. They concluded that aged binder in RAP formed a stiffer layer coating the RAP aggregate particles than the virgin binder (see Figure 2).

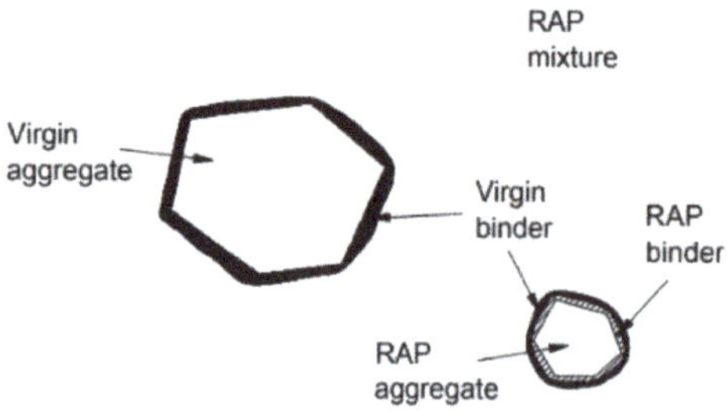

Figure 2. Composite-layered system in recycled asphalt pavement (RAP) Virgin [20].

Composite analyses indicated that the layered system in RAP (Figure 3) helps in reducing the stress concentration in HMA mixtures microstructure. The aged binder mastic layer was actually serving as a cushion layer in between the hard aggregate and the soft binder mastic [20].

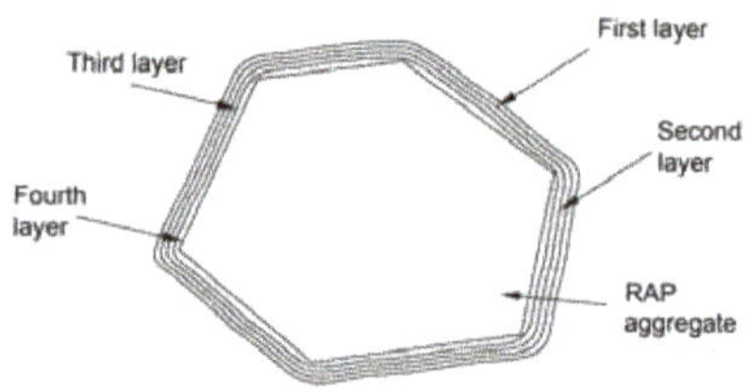

Figure 3. Layers of asphalt binder coating RAP aggregate [20].

As there is aged binder in RAP particles, bitumen additives might be applied to rehabilitate the aged binder which is not the case in this study. On the other hand, since there is high RAP content in HMA, rejuvenators are recommended [21–23]. Additionally, green additives are recommended as cost effective and environmentally friendly alternatives. The green additive is obtained by a simple method from two low-cost and eco-friendly pre-cursors to restore the mechanical properties of the oxidized bitumen, acting on the structure of the bitumen, having a restructuring effect on the altered colloidal network of the aged bitumen binder [24].

In addition to the binder characterization, mix performance needs to be evaluated when a higher RAP content is used. The indirect tension and semicircular bending test results which were conducted by Huang et al. showed that RAP increases the mixture stiffness [25]. All RAP materials in Huang et al. study was screened through the No. 4 sieve to acquire a consistent gradation which can compare to the fine aggregates group in this study. There is a possibility to increase the stiffness by adding fine RAP but it depends on job mix formula with inclusion of RAP. Huang et al. assumed that RAP binder totally contributes to the mix, an assumption which still needs to be verified.

Traditionally, black curves and white curves have been used for sieves analysis of RAP incorporated mixes. Black curves are the gradation of RAP particles from fractioned RAP and white curves are the gradation of recovered RAP aggregate after binder extraction. For a given mix, these two curves are significantly different. Al-Qadi et al. [26] compared these two curves and concluded that; black curve tends to indicate higher amounts of large particles and lower amounts of fine particles. Therefore, to avoid the detrimental effects caused by unexpected extra fine particles, black curves are not suggested for use in job mix formula calculations. Using the white curve is common practice, however it is not the only approach being used. It should be noted that neither black curve nor white curve represents the actual gradation of the RAP material, and the real gradation lies somewhere in between [27].

Previous research has shown that the bitumen recovered from coarse particles differ from fine particles [28]. They concluded that RAP bitumen participation in hot mix process significantly depends on RAP size. The aging rate in fine particles is also faster than coarse particles. Therefore, it can be concluded that the RAP particle size would affect the properties of the RAP incorporated mix and can affect the overall performance of the pavements.

2. Research Goal, Scopes, and Objectives

Review of the literature on RAP indicates that there is not a consensus on several aspects of the RAP binder contribution to the new mix. Therefore, it is difficult to come up with a synthesis of previous work that would be unanimously acceptable. In summary, it can be concluded that the coarse and the fine RAP fractions have relatively similar stiffness, but their contribution to the new mix is different. Therefore, the goal of this project is to understand the interaction of coarse RAP and fine RAP binder in HMA more precisely, through studying mixes prepared with either fine RAP fraction or coarse RAP fraction separately. To this end, various empirical and thermo-mechanical tests are adopted to validate the impact of RAP fractions on the HMA mixes.

In this study, the RAP particles are separated in two groups by following the LC 21-040 protocol: particles passing sieve number 5 (5 mm), called fine RAP, and those retained on sieve number 5 are considered as the coarse RAP. The main objective of this research is to characterize RAP particles according to the particle sizes used to produce them. Generally, fine RAP is expected to possess a higher binder content, while there has not been any clear evidence to support this claim yet. The presence of such extra amount of binder (as compared to the coarse RAP) may potentially enhance the pavement resistance to cracking. Previous studies showed that in high RAP content mixes, the amount of binder (or mastic) that would diffuse into virgin binder from the fine RAP particles is less than that from the coarse RAP particles. Therefore, it was deemed necessary to further investigate the mix design and mix performance to characterize the fine and coarse RAP mixes more comprehensively. This research aims to characterize the mixes produced using the fine RAP and Coarse RAP in terms of stiffness, fatigue cracking, permanent deformation, and thermal cracking resistance.

3. Materials and Experimental Methods

Generally, limits have been set on the maximum allowable amount of RAP in HMA mixes to avoid the loss of performance due to the impact of more aged binder content, which is present in RAP particles. However, many aspects of RAP incorporated mixes have been investigated by several past studies, the effect of increased RAP content using only coarse or fine RAP particles has not been fully explored. Therefore, in this study it was hypothesized that coarse RAP mix characteristics is different from fine RAP mix and such difference can affect the mix performance. The results might be used to identify the functional class for proper use of RAP mixes in pavement structural design in different layers. Three mixes were designed in this study, including a control mix, a fine RAP mix, and a coarse RAP mix. It was assumed that all of the RAP binder would contribute to the total binder content of the mix. Consequently, it was assumed that all three mixes have the same binder content. Furthermore, the black curve was used for the aggregate gradation in the mix design process.

Four performance tests were used to characterize the mixes in this study. These tests were selected to evaluate the mixes from the major aspects of the pavement materials characteristics, that is, high temperature performance, low temperature cracking, and fatigue cracking. They can be classified as two categories of empirical and thermo-mechanical tests. Resistance to rutting (MLPC wheel tracking rutting tester or French Wheel Tracking Test) is used in this study as the empirical test. The thermo-mechanical tests utilized in this research are as follows:

- Complex (dynamic) modulus test,
- Uniaxial tension–compression test for resistance to fatigue cracking, and
- Thermal stress restrained specimen testing (TSRST) to evaluate resistance to low temperature cracking.

3.1. Mix Design and Volumetrics

A control mix was designed and prepared with only virgin aggregates and virgin asphalt binder. The control mix was a 20 mm-dense graded HMA, commonly used as a base course in Quebec. The design binder content using a PG 64-28 was determined to be 4.5% by the weight of the total mix. A bituminous mixture with a nominal maximum aggregate size of 20 mm, called Grave Bitumen (aka GB20) is mostly used in binder course layer of pavements in Quebec. The selected virgin binder (PG 64-28) is a medium grade asphalt binder that can be used in warm climates. The aggregate size and gradation were selected based on the LC method specifications. This virgin asphalt concrete mix will be referred to as the control mix hereafter in this paper. The LC Test Methods Compendium, produced by the General Directorate of the Pavement Laboratory of the Ministère des Transports du Québec, presents all the test methods used to measure the characteristics of materials used in the construction and maintenance of infrastructures. Additionally, two more mixes (fine and course) were also designed for comparison and validation purposes. Fine RAP mix contained 35% fine RAP content (passing sieve 5 mm) and coarse RAP mix contained 54% coarse RAP content (retained on sieve 5 mm). Initially, it was assumed that all of the aged binder in the RAP could blend into the virgin binder, so that the total binder content could be assumed to be the same for all of the mixes. These mixes would be referred to as Fine RAP (FRM) and Coarse RAP (CRM) in this paper. It should be noted that regardless of the RAP fraction sizes, the target gradation of the mixes were kept the same.

More experimental mixes were initially designed to study the active RAP binder content and their participation in the mixt. The RAP content varied from 0% to 59%. The gradation curves of the experimental mixes were all in accordance with the Ministère des Transports, de la Mobilité durable et de l'Électrification des Transports specification. The mix gradations are shown in Figure 4. It can be seen that the mixes have the same black curve gradation. First, there was a question of whether to use the white or black curve gradation. With respect to having the same binder content and aggregate gradation as the fixed variable, the black curve was chosen in this research. Consequently, all mixes would have the same binder content, same aged binder content and same black curve, but different RAP content.

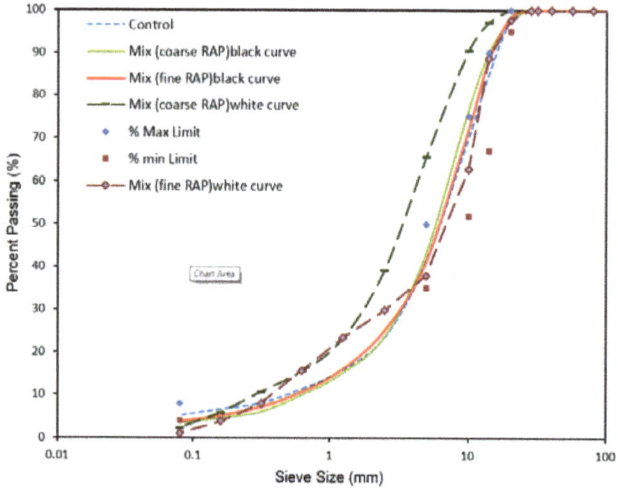

Figure 4. Mixes gradation.

The differences between the two RAP mixes are the RAP particle size and content. This experimental program was designed to investigate the impact of RAP size on Hot Mix Asphalt (HMA). It was envisioned that there is no advantage in looking solely into the RAP content by itself, rather considering the RAP binder content and RAP mix gradation. This plan was designed to achieve the same RAP binder content in both mixes, as well as keeping the final mix gradation the same. In this project, RAP content translates to RAP binder content and RAP size, where the former was kept constant in both mixes and studying the effect of the latter was set as the main objective of this research.

The Superpave^{TM} mix design method was followed in this study. The only design criterion was the binder content at this level of the study (see Table 1 Design Criterion). Using the design binder content of 4.5%, the control mix was prepared with 4.5% virgin binder content, while for the RAP mixes the RAP binder contribution was considered in determining the needed virgin binder content. The recovered RAP binder from chemical extraction showed that the Coarse RAP (CR) had 4.3% binder, whereas the Fine RAP had 6.7%. In order to do mix design, sufficient amounts of coarse and fine RAP were chosen to have the same recycled binder ratio in the final mixes. Table 1 presents the summary of the virgin and aged binder contributions in the fine and coarse RAP mixes.

Table 1. Design criterion.

	A = % RAP Content	B = % RAP Binder	% Recycled Binder Ratio = A*B	C = % Total Binder	% Virgin Binder = C- (A*B)
Coarse **RAP**	54%	4.3	2.3	4.5	2.2
Fine RAP	35%	6.7	2.3	4.5	2.2

The composition of control and experimental mixes, as well as the values of the volumetric obtained from the mix design, are shown in Table 2. It can be realized that most of the volumetric values are within the LC method specification.

The compaction process was done following to the LC 26-400 Fabrication d'éprouvettes au compacteur LCPC. The laboratory compacted cylindrical specimens were stored for a minimum of one month at room temperature in a sand bed prior to testing. Mechanical tests, including fatigue, and complex modulus were performed on cored specimens extracted from slabs as shown in the schematics in Figure 5. Samples were compacted by the French MLPC wheel compactor (Figure 6).

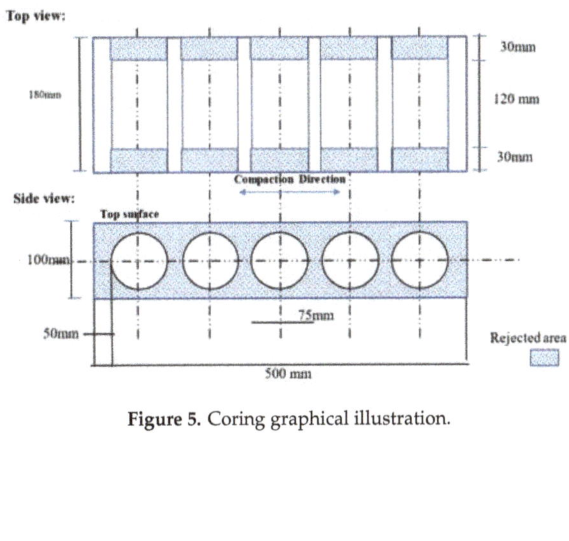

Figure 5. Coring graphical illustration.

Table 2. The composition of control and experimental mixes.

Mix Type	% RAP	% Air Void (@200 G)	% Void Mineral Aggregate	% Void Fill with Bitumen	Mix Specific Gravity
Control	0	2.3	14.6	66.5	2.520
Fine RAP	35	1.5	11.6	86.8	2.606
Coarse RAP	54	2.6	12.5	79.7	2.602

Figure 6. Photo of the Laboratoire Central des Ponts et Chaussées (LCPC) slab compactor.

3.2. Resistance to Rutting

All mixes were subjected to laboratory testing for resistance to rutting. The equipment used in this study was developed by France's Laboratoire Central des Ponts et Chaussées (LCPC) (see Figures 7 and 8). The test was standardized in Europe (EN 12697-22A1) and in the province of Quebec, Canada (LC 26-410). It is also commonly used for research purposes by the asphalt industry in other countries [29,30].

Figure 7. The French rutting test equipment.

Slab dimensions were 500 mm by 180 mm with a thickness of 100 mm. The level of compaction must correspond to what is obtained in the field. On roadways, the required minimum compaction level is usually 92%. For most hot mixes, laboratory-manufactured specimens at the 92% level may lead to rutting after compaction. Consequently, laboratory-prepared specimens are compacted to a greater value, approximately 95%. At this level, post-compaction is generally negligible [31]. Heating temperatures for mixing and compaction are indicated in the test method LC 26-003. This laboratory step was done according to AFNOR Standard P98-250-2 Préparation des mélanges hydrocarbonés; Partie 2: Compactage des plaques.

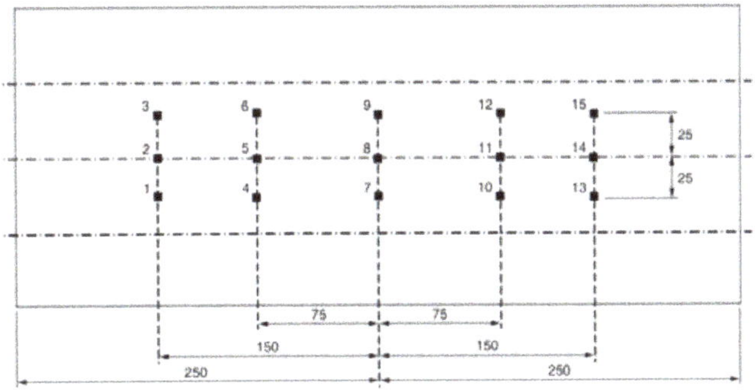

Figure 8. Measurement points location in mm (LC 21-410).

Prior to the rutting test, a preconditioning was done by rolling the pneumatic tire of the rutting tester across the specimen for 1000 cycles at the ambient laboratory temperature. The preconditioning helps with minimizing discrepancies due to the installation of the specimens in the mold. The slabs were then conditioned to reach the testing temperature of 60 °C. Once the temperature of 60 °C was reached, the rutting test was started, and rut depths were measured after 30, 100, 300, 1000, 3000, 10,000, and 30,000 cycles (as applicable). The rut is defined as the mean vertical displacement of the hot mix surface as compared to the mean height of the specimen before starting the test. As described in AFNOR P 98-253-1 Déformation permanente des mélanges hydrocarbonés; Partie 1: Essai d'orniérage, height measurements were taken at 15 locations over the slab area. The stress induced by the tires was maintained at 0.6 MPa during the tests.

Rutting generally progresses along a straight line when plotted on a logarithmic scale against the number of wheel passes. In order to have an acceptable rutting resistance, the rut depth, expressed as percentage of the specimen thickness, should be less than 10%. Yildirim et al. [32] characterized the rutting trend by post-compaction consolidation, creep slope, stripping slope, and stripping inflection point by a typical Hamburg Wheel Tracking Device Test (see Figure 9). Post-compaction consolidation is the deformation (mm) at 1000-wheel passes. Creep slope is the inverse of the rate of deformation in the linear region of plot between post compaction and stripping inflection point (if stripping occurs). Stripping inflection point is the number of wheels passes at the intersection of creep slope and stripping slope. Finally, stripping slope is defined as the inverse rate of deformation after the stripping inflection point.

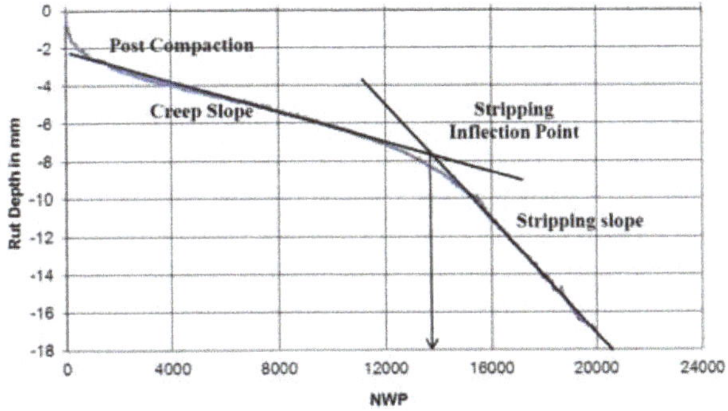

Figure 9. Typical Hamburg Wheel Tracking Device Test results [32].

Meunier [33] characterized the rutting trend from cyclic compression-tension test in three phases as shown in Figure 10. They concluded that the deformation increases rapidly in first phase. In phase two, the deformation increases by a constant rate per loading cycle. It should be noted that phase three marks the failure of the material and is usually considered less accurate for the purpose of prediction process than the previous two phases.

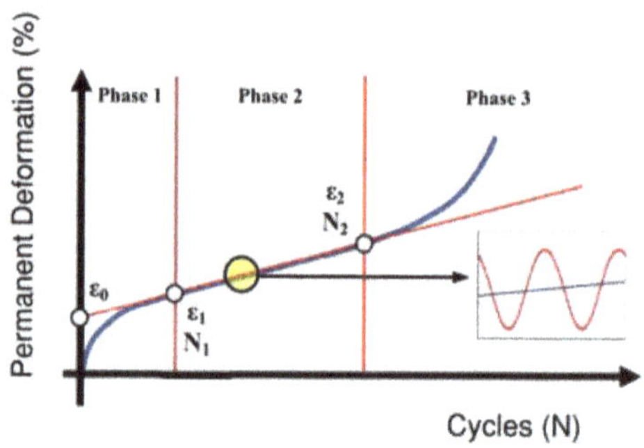

Figure 10. Evaluation of permanent deformation [33].

3.3. *Thermo-Mechanical Tests*

For the purpose of thermo-mechanical characterization of the mixes in this study, two main tests, namely uniaxial fatigue test and TSRST, were performed by means of a 25 kN servo-hydraulic system. Figure 11 shows a graphical illustration of the test set-up with the specimen and extensometers. Three extensometers were mounted on the specimens, 120° apart around the sample, to measure the axial strain during the tests.

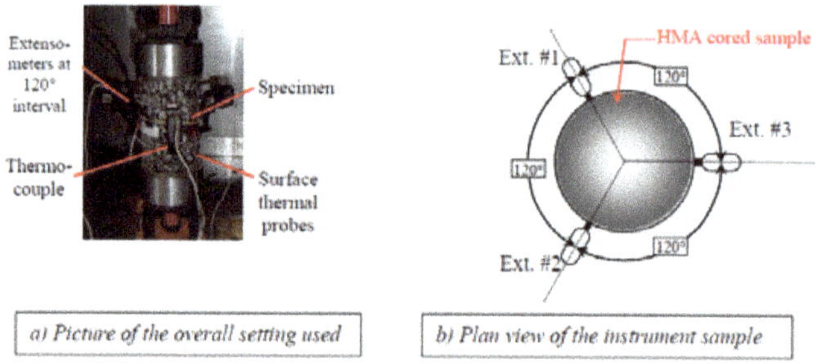

Figure 11. Schematics of the test setup used in this study.

The set-up was enclosed in an environmentally controlled chamber with three temperature probes, capable of cooling and heating within a range of −40 °C to 80 °C. The following sections provide more details about these two tests.

3.3.1. Fatigue Resistance

Fatigue characterization was performed by means of the uniaxial tension–compression (T–C) tests on cylindrical specimens in this study. The experimental test setup is almost the same as the complex modulus test. The fatigue test was performed at a single loading frequency of 10 Hz at 10 °C. The advantage of using this test over the other conventional fatigue tests is possibility of maintaining the homogeneous state of stress and strain in the sample during the testing process.

The cored samples from slabs were tested under uniaxial T–C condition and the axial strain values were measured using three extensometers. The average of recorded values was considered as the strain level in the sample. Data quality measures were used to ensure that the assumption of homogenous stress/strain condition has not been violated. To this end, reaching a difference of ±25% in the recorded values was considered as an indication of highly non-homogeneous conditions for the strain field within the sample. Therefore, in such cases the test should be considered no longer valid beyond that limit [34].

The graphical presentation of the fatigue test results is usually given by Wöhler curve or fatigue curve (see Figure 12). This curve shows the relation between the fatigue life (N_f) and the level of loading expressed by the initial strain (or stress) amplitude in a bi-logarithmic scale [35]. A particular value of strain called (ε_6) can be found to correspond to the value of the strain level that would lead to a fatigue life of 1,000,000 cycles. This value is commonly used to characterize the fatigue resistance of the bituminous mixes [36]. The fatigue resistance is determined through a series of laboratory tests in different magnitudes of solicitation under controlled conditions (temperature and frequency). As demonstrated by the log-log plot in Figure 12, Wöhler's Law is associated with a straight line, where fatigue behavior is characterized by two parameters: the slope (c_2) and the Y-intercept (c_1). Coefficients c_1 and c_2 depend on both, the material and the chosen failure criterion [37]. It should be noted that most of the fatigue-cracking models characterize fatigue failure in three stages: crack initiation, crack propagation, and fatigue induced fracture [38]. The classical fatigue failure criterion determines the fatigue life as the number of loading cycles that the specimen can take to the point that a 50 percent loss of the initial stiffness for homogeneous tests, or when a 50 percent loss of the initial sample rigidity for non-homogeneous tests is observed [39,40].

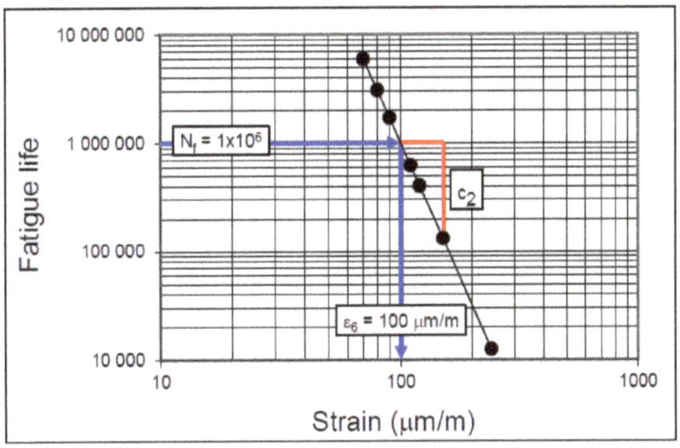

Figure 12. Typical fatigue test results from laboratory tests done on an asphalt mixture specimen [38].

Based on the Wöhler curve, the fatigue characteristics of asphalt mixtures can be expressed by Equation.

$$N_f = C_1(\varepsilon_0)^{(-c_2)} \tag{1}$$

where:

- N_f is fatigue life (number of cycles corresponding to the failure point for a given criterion);
- ε_0 is applied strain amplitude (mm/mm) at a given testing temperature (θi) under a specific testing frequency (f);
- C_1 is coefficient corresponding to the expected fatigue life for a strain amplitude of 1 mm/mm, at the given testing temperature and loading frequency;
- C_2 is slope of the Wöhler curve when it is associated with a straight line in the log N_f–log ε_0 domain.

3.3.2. Resistance to Low Temperature Cracking

TSRST simulates thermo-mechanical response of flexible pavements during the cooling period. The principle of the test is to restrain the tested specimen from any axial deformation by keeping the total height of the specimen constant throughout the testing period. As a result of decreasing the chamber temperature at a constant cooling rate of 10 °C/h, the magnitude of thermal stress in the specimen would increases until the failure of the specimen. It is also possible to calculate the axial stress as a function of the measured temperature.

Once at failure point, the stress would reach its peak value, referred to as the failure strength (r_f), whereas the corresponding temperature can be defined as the failure temperature (T_f). The slope of the stress-temperature curve increases progressively until a certain temperature where it remains quasi-constant (the stress temperature curve becomes linear). To estimate the value of the quasi-constant slope, the parameter drz/dT is calculated by linear fitting of the curve between the failure temperature and the transition temperature. Tapsoba et al. [41] assumed transition temperature (T_t) as the temperature where axial stress reaches 50% of the failure strength. It corresponds to the temperature where the material changes from ductile to brittle behavior and vice versa and will be used to evaluate the repeatability of TSRST.

4. Results and Discussion

Three mixes were investigated in this study, including a control mix, an HMA mix with inclusion of fine RAP (FRM) and an HMA mix with inclusion of coarse RAP (CRM). The FRM mix consisted of 35% RAP with 2.2% virgin binder and the CRM mix had 54% RAP and 2.2% virgin binder. The results of the experimental studies on these three mixes are as follow:

4.1. Rutting Resistance of FRM versus CRM

Permanent deformation of the mixes was evaluated at 60 °C using the French rut tester. All mixes (slabs of 100 × 180 × 500 mm) were subjected to repeated loading of a tire inflated to 0.6 MPa, mounted on a carriage that moves back and forth at 1 Hz with a load magnitude of 5 kN. Figures 13 and 14 show the results of rutting tests. Figure 13 indicates the percentage of permanent deformation by straight line for all mixes in the logarithmic scales. The results confirmed that all of the mixes exhibited deformation magnitudes less than 10% after 30,000 cycles. Therefore, it can be concluded that all of the mixes in this study were strong enough to resist the permanent deformation failure. It should be noted that the mixes had the same black curve gradation, but they showed different behavior under the cyclic wheel load. Therefore, as it was expected the black curve assumption was not found to be a reliable representation of the aggregate skeleton when RAP is incorporated.

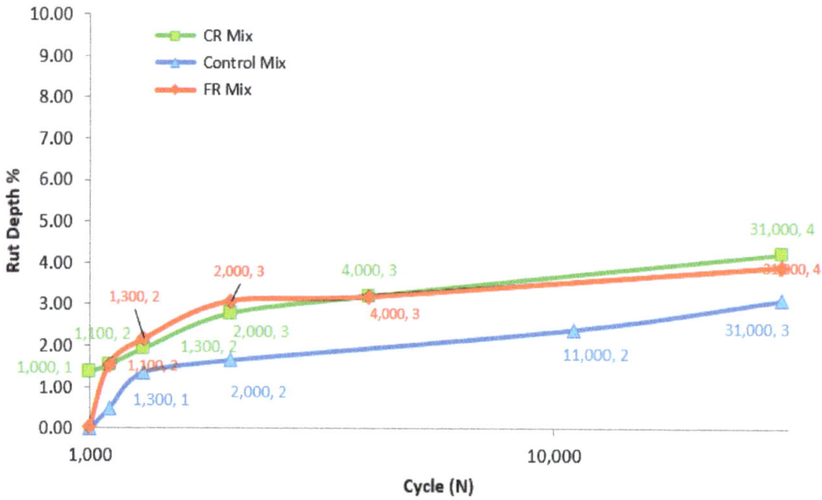

Figure 13. Permanent deformation result.

Figure 14. RAP mix slabs after rutting test.

A single sample of each mix type was adopted for permanent deformation validation. Basically, the first 1000 preconditioning cycles (aka cold runs) are assumed to capture the continued consolidation stage. There was a significant difference between FRM and CRM mixes. The rest of loading was performed at 60 °C. In addition, the binder exhibits a softer response at 60 °C than the cold cycles temperature. This difference in rutting might be caused by impact of aggregate gradation and air void content.

After 1000 hot cycles (post compactions), FRM was deformed almost as same as CRM. Both RAP mixes deformation were two times higher than control mix. This section could be characterized by S1 and S2. Parameter "S" represents the slope of the permanent deformation in Figure 13. For the FRM and the control mix, rut depth dramatically increased at first 300 cycles (S1) and continued at a constant slop, whereas for the CRM, these slopes increased at the same rate in both steps (i.e., S1 and S2).

The last stage represents the reaction of material to wheel passes loading which can be translated as rutting values. Table 3 shows the slop per section of rutting test. It can be seen that the CRM and control mix responded the same way to the load in rutting section, which was two times higher than FRM (see S4 in Table 3). The aggregate gradation plays the main role in rutting resistance. In this study, black curve was kept the same in all mixes.

Table 3. Slope for each section.

Slope	FR Mix	CR Mix	Control
S1 (0 to100 cycle)	1.490%	0.165%	0.484%
S2 (100 to 300 cycle)	0.310%	0.195%	0.439%
S3 (300 to1000 cycle)	0.131%	0.122%	0.042%
S4 (1000 to 30,000 cycle)	0.003%	0.005%	0.005%

CRM white curve showed finer than control mix but the black curve was almost same as control mix. CRM deformed as same as control mix at last stage but it could be compacted more than control mix at the beginning. Large aggregate gradation (D > 5 mm) in FRM black curve was same as the control mix but fine part of FRM white curve showed more fine content in gradation which was expected. However. It cannot be concluded whether it is more appropriate to use white or black curve up to this point.

In conclusion, rutting results can be divided in three phases: deformation at the end of 1000 cold cycles, 1000 hot cycles and 30,000 cycles. First phase which was called continued consolidation earlier, suggests that the CRM gradation and air void were different form FRM mix, which was found to be true; because CRM specimen had 8% air void but FRM specimen had 6%. Second phase which was called post compaction (S1, S2, S3), suggests that the specimen binder is soft enough to indicate the difference in aggregate gradation. Flatter slope can be translated to well packing phenomenon. It was recognized that the CRM could be packed better than FRM. Last section which was called rutting, showed the rutting resistance of mixes. The results indicate that the FRM was more rut-resistant than CRM and also than the control mix. The FRM and the control mix differed only in the fine part in aggregate gradation, especially magnified by the white curve. Thus, FRM mix had stiffer fine skeleton than control mix. According to black curve, CRM and control mix had the same gradation, but considering the white curve, CRM was much finer than control mix. Basically, coarser mixes have better rutting resistance as compared to the CRM. The CRM was expected to be weaker than the control mix but exhibited the same response as the control mix. In both FRM and CRM mixes, better or at least the same resistance as the control mix was recorded in spite of the fact that there is 54% (or 35%) recycle materials in the mix. The results indicate that the FRM prepared with 35% RAP exhibited almost similar performance as the CRM prepared with 54% RAP. It can be concluded that both of the RAP incorporated mixes exhibit satisfactory rutting resistance.

4.2. Fatigue Resistance Results

In this study, the classical method was used among the four types of failure criteria mentioned earlier. Table 4 provides the specimen details, the actual and target initial strain values, and the number of cycles to failure (aka fatigue life) for each specimen. The fatigue results are sensitive to the air void level, and hence it was attempted to maintain the same level of air void for all the specimens.

Table 4. Uniaxial T–C fatigue test conditions (10Hz, 10 °C).

Mix Type	Sample Name	Target Def, (µdef)	% Va	Real Def, (µdef)	Nf	Log Real Def, (def)	Log Nf II/III
Control Mix	S2C2	80	6.8	71	2,405,986	−4.15	6.38
	S1C5	100	6.8	99	61,464	−4.01	4.79
	S1C2	70	8.0	66	4,715,411	−4.18	6.67
	S2C7	90	6.5	84	2,500,000	−4.08	6.40
Coarse RAP Mix	CR1	60	7.2	53	1,324,415	−4.28	6,12
	CR3	40	7.6	30	2,145,327	−4.52	6.33
	CR2	70	8.4	61	26,211	−4.21	4.42
Fine RAP Mix	FR1	90	4.4	82	513	−4.09	2.71
	FR2	110	4.1	113	277	−3.95	2.44
	FR3	100	7.1	87	5103	−4.06	3.71
	FR4	60	5.3	59	79,608	−4.23	4.90
	FR5	50	6.3	48	22,163	−4.32	4.35
	FR6	70	5.5	58	46,000	−4.24	4.65

In Table 5, it should be noted that the classic failure criterion of 50% reduction in the initial stiffness was not found reliable, due to the fact that a significant loss of modulus has occurred during the first phase of the T–C test. In spite of the fact that some researchers use the 50, 100, 200, or even 1000 cycles to calculate the initial modulus, the results were not representative of the fatigue-induced damage. Therefore, the more scientific Wöhler approach was used to study the fatigue performance of the mixes in this study.

Table 5. The complex modulus properties of the mixes.

MIX	E* at −30 °C, 3 Hz (MPa)	E*/Sin(ϕ) at 40 °C and 0.03 Hz (MPa)	E*at 20 °C, 3 Hz (MPa)	E*|at 15 °C and 3 Hz (MPa)	E_0 (MPa)	E_∞ (MPa)	k
Control	37,000	99	4000	5100	30	40,000	0.18
CR	32,762	113	5700	7554	6	33,000	0.19
FR	31,214	27	3000	4915	5	32,000	0.16

Regression based fatigue equations were developed based on the test results to quantitatively characterize the mixes (Figure 15). In order to develop this chart, various fatigue samples were subjected to sinusoidal load at three different strain levels in order to be able to run a linear regression. The value of ε_6 corresponds to the strain level at which the asphalt mix would reach a fatigue related failure after 1 million cycles. For the sake of comparison, it can be noted that a standard asphalt base course material, made with straight run asphalt cement, usually exhibits ε_6 values in the range of 70 to 90 μm/m. ε_6 in this project is 81 μm/m for control mix. The value of R^2 shows the quality of linear assumption. ε_6 for CR in higher than that of FR. CR failed at 43.57 μm/m and FR failed at 28.94 μm/m. The slope of the trend line shows the degree of sensitivity of mix to deformation. Sharp slope is highly sensitive to deformation, it means that under a small change of deformation there would a huge difference in number of repetition that the mix can take until failure. The CR had less sensitivity to the changes in deformation, and it even surpasses the control mix with this regard.

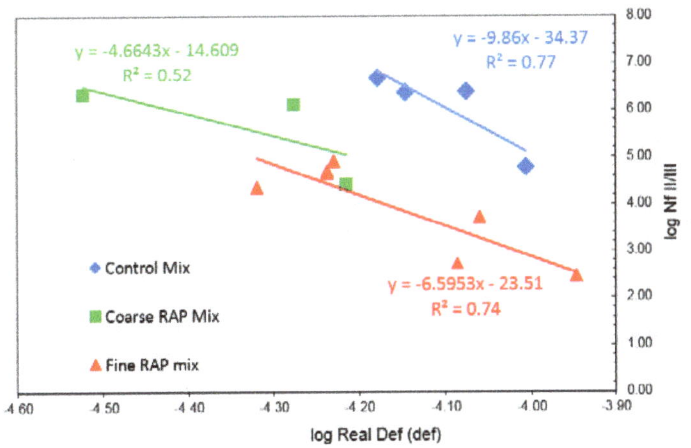

Figure 15. Wöhler Curve.

Basueny et al. [42] concluded that when the percentage of RAP in the mix is considerably high, the aged RAP binder creates a significant change in the mixture properties. Therefore, it can be concluded that the influence of RAP on the final HMA property also varies with the amount of RAP. The mixes in this study were supposed to have similar recycled binder replacement ratio and black curve gradation, however the CR mix resists to fatigue much better than the FR mix. It can be concluded that black curve assumption is not the best representation of RAP gradation. Virgin binder in CRM is mostly covering the fine natural aggregates and adhesion to CR. More unaged binders in mastic increase the resistance under the tension and compression repeated loads. FRM has more aged binder in fine part of the skeleton that caused weakness of fatigue resistance. Overall, the CRM exhibited a better fatigue performance than the FRM.

4.3. Resistance to Low Temprature Cracking through TSRST

In addition to fatigue cracking, another major concern for HMA mixes with RAP particles is their resistance to low temperature cracking. In general, RAP mixes are stiffer than conventional mixes, due to the highly oxidized nature of the aged binder in RAP particles. The values of the fracture temperature and the corresponding stress at failure, obtained from the TSRST tests for all the tested mixes, are presented in Figure 16.

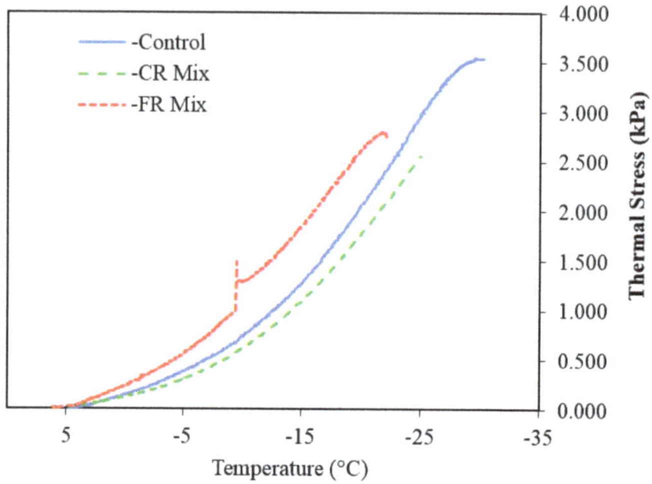

Figure 16. TRST results.

The description of the test progress and the associated data collected is as follows [43]:

- Slope no.1: This parameter represents the performance during the relaxation period.
- Slope no.2: This slope represents a value analogous to the modulus of elasticity in a diagram of stress versus deformation of an elastic material.
- Glass transition temperature (Tg): This temperature represents the end of the relaxation and the beginning of a linear change with temperature.
- Transition temperature midpoint: This temperature corresponds to the intersection of the tangent lines in section no.1 and the tangent line fitted to the section no.2 of the curve, and indicates the transition temperature between two stages of the simplified bilinear response of a material.
- Maximum tensile strength: This value represents the maximum stress applied to the test specimen just before it fails.
- Maximum stress temperature: This temperature is simply the value obtained when the maximum stress is reached.

Figures 17–19 show the TSRST values derived from the results for the purpose of comparison. The maximum tensile strength values were found to be 3548, 2558 and 2799 kPa for the control, CR and FR mixes, respectively. The maximum stress temperatures were measured as −30, −25, and −22 °C for the control, CR, and FR mixes, respectively. The Transition temperature midpoint of the control and CR mix is almost the same (i.e., −11 °C) but Transition temperature of the FR was very low (i.e., −5 °C). The results indicate that the CR mix performed better than the FR mix with respect to low temperature properties. The CR had lower Tg midpoint and lower failure temperature, however FR failure stress was slightly higher than the CR. In addition, the value of T_t, calculated according to Tapsoba et al. [41] study, was found to be the same for both mixes.

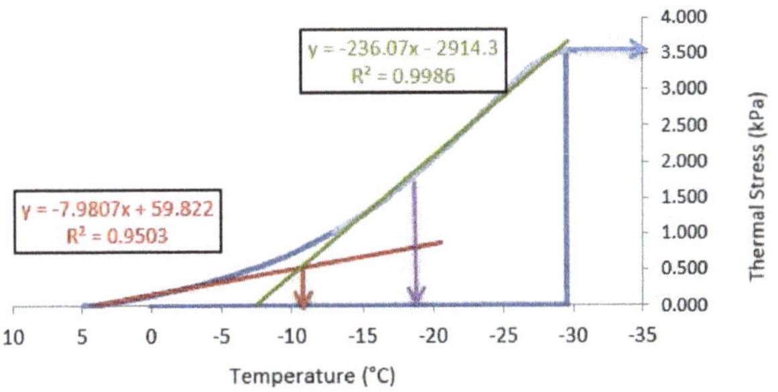

Figure 17. TSRST for control mix.

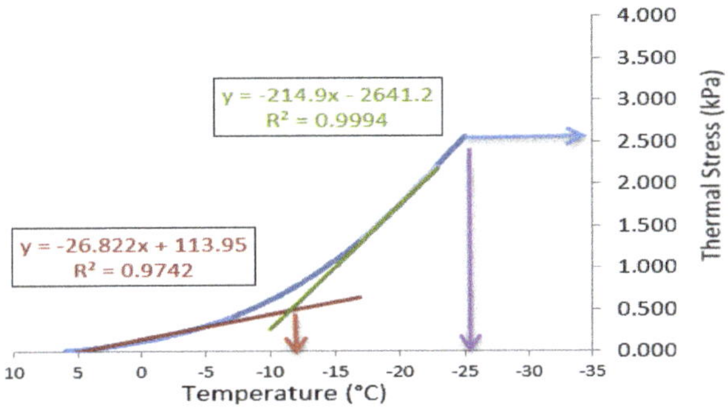

Figure 18. TSRST for coarse RAP (CR) mix.

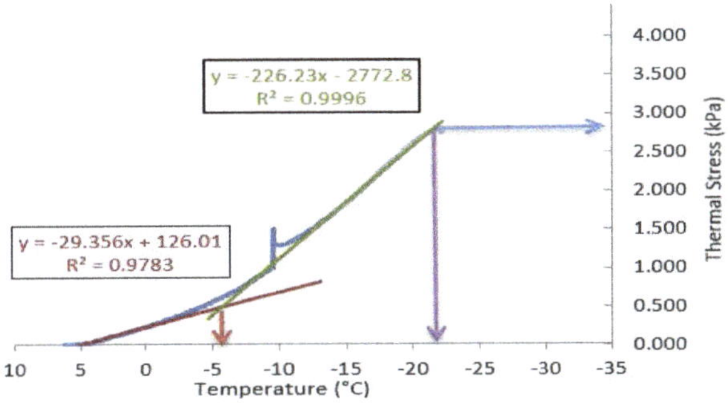

Figure 19. TSRST for fine RAP (FR) mix.

Void fills with bitumen (VFB) represents the effective bitumen content. The decrease of VFB indicates a decrease of effective bitumen film thickness between aggregates, which will result in higher low-temperature cracking and lower durability of bitumen mixture since bitumen perform the filling and healing effects to improve the flexibility of mixture.

4.4. Complex Modulus

Various criteria are available in order to compare the stiffness of different bituminous materials. Baaj et al. (2013) [44] suggested to looking into the stiffness of the materials in the following ways:

- The stiffness |E*| at −30 °C and 3 Hz: this value gives the material stiffness for a low temperature and a high-frequency condition.

- The values of |E*|/sin (ϕ) at 40 °C and 0.03 Hz: this ratio is used as an indicator of rutting resistance.
- The stiffness |E*| at 20 °C and 3 Hz frequency: The stiffness of the mix (resilient modulus) at this temperature is used in the AASHTO'93 empirical pavement design method.

In addition, Perraton et al. [45] also used the stiffness at 15 °C and 3 Hz. typically, standard bituminous base course materials have dynamic modulus values in the range of 5000 to 7000 MPa when tested under the same conditions at 15 °C.

Figure 20 indicates the Cole-Cole plot for all mixes from −35 °C to +35 °C. Two replicate specimens were used for each of the mixes. The measured data was modeled with the 2S2P1D model. There is a notable difference between the control mix and RAP mixes with respect to loss (or so-called imaginary) modulus. Several factors can affect the loss modulus of a bituminous material such as air void level, bitumen content, and bitumen type. The results indicate that, generally, the two RAP mixes are the same according to the Cole-Cole diagram presented in Figure 20. However, this plot cannot explicitly distinguish the differences in the bitumen characteristics.

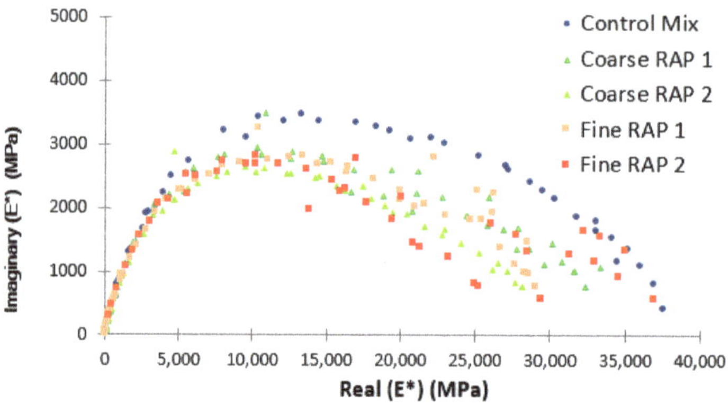

Figure 20. Complex modulus master curve in Cole-Cole plot.

The time-temperature superposition principle (TTSP) was applied to analyze the complex modulus test data. This principle was verified by several studies dealing with the unidirectional linear viscoelastic behavior of bituminous materials [46]. As shown in Figure 21, at high frequency, RAP mixes have lower stiffness than the control mix, and this difference becomes greater at lower frequencies. According to TTSP principle, high frequency could be translated to low temperature and low frequency could be translated to high temperature. Therefore, in a full range of temperature, RAP mixes were slightly softer than the control mix. However, the RAP content is not the same (i.e., 0%, 35%, and 54%). Therefore, the FRM with inclusion of 35% RAP content has almost the same behavior as the CRM with 54% RAP content. Consequently, it can be inferred that the binder contribution from 54% CR would be almost have the same effect as that of the 35% FR mix used in this study. In addition, FRM stiffness was found to be strongly sensitive to the testing conditions. Figure 3 shows that the same FRM specimens are variable at low frequency but all CRM mix specimens show consistent response, suggesting that they are less sensitive to the testing conditions.

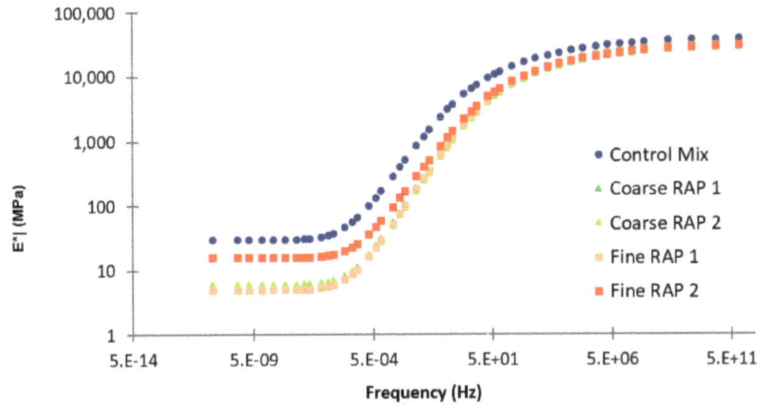

Figure 21. Master curve of the norm of complex modulus.

The rheological properties of the mixes can also be expressed in terms of phase angle. A phase angle (δ) value of 0 degrees means a purely elastic material and 90° means a purely viscous material. Figure 22 shows the master curve of the phase angle for the mixes investigated in this study. The RAP mixes exhibited more viscous response than the control even though they have less virgin bitumen content (i.e., 2.2%). The FR results varied significantly, which might be associated with the higher RAP bitumen content and some unexpected phenomena in the fine RAP particles like clustering and variability in the film thickness of the particles.

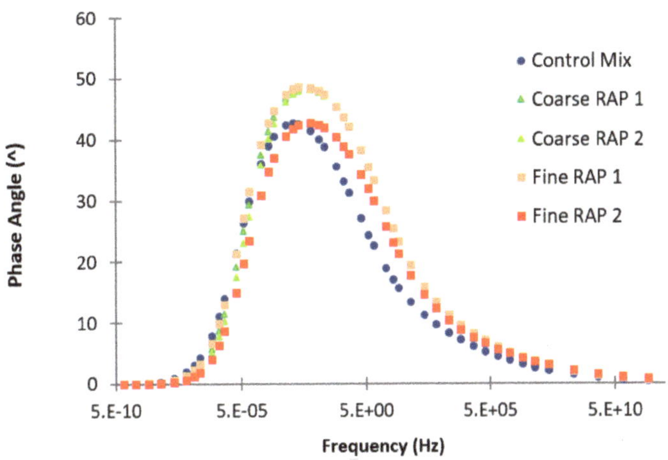

Figure 22. Master curve of the phase angle of complex modulus.

Figure 23 presents a summary of the 2S2P1D model parameters in the Cole-Cole model and the corresponding values of these parameters are listed in Table 5.

- h, k: exponents such as $0 < k < h < 1$, related to the ratio $E_{Imaginary}/E_{Real}$ when ϖ tends to 0 (resp. to infinity),
- E_0: the glassy modulus when $\varpi \approx \infty$
- E_{00}: the static modulus when $\varpi \approx 0$
- η: Newtonian viscosity

CRM is the same with control mix and FRM in black curve but it has more active aged bitumen. FR bitumen could not increase the FRM stiffness but CR bitumen was more active in mixture and increased the CRM stiffness.

The fact that the same or even better results could be achieved using the coarse RAP at a higher rate, as compared to the fine RAP, offers significant potential advantages in producing high RAP content mixes.

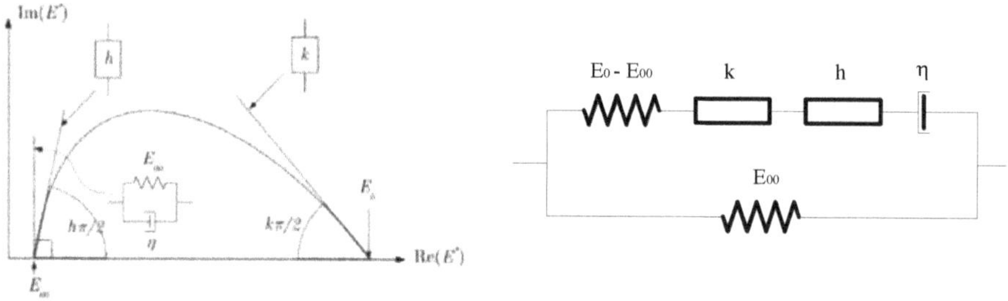

Figure 23. 2S2P1D model parameters in Cole-Cole model.

5. Conclusions

Review of the literature indicates that research on producing asphalt concrete mixes with only fine or coarse RAP particles has been scarce. The main goal of this study was to evaluate the effect of fine and coarse RAP fractions on performance of the high RAP content asphalt concrete mixes. Accordingly, three mixes were designed and investigated in this study, including a control HMA prepared with virgin materials, a Fine RAP mix (FRM) with 35% percent RAP, and a Coarse RAP mix (CRM) with 54% RAP. The total binder content considering the different contributions from the RAP particles was maintained the same for all the mixes. The following conclusions can be drawn based on the results of this study:

Black curve and white curve assumptions for the RAP particles were explored and it was concluded that the black rock assumption, which is commonly used, cannot be representative of the RAP contribution to the total binder content and the skeleton of the mix. The actual gradation lies somewhere in between these two extreme cases.

- The classifying the RAP particles in HMA is changing the role of virgin binder in the mix. Virgin binder role is covering the natural coarse aggregates and adhesion to aged bitumen of fine part in FRM. On the other hand in CRM, it is covering the natural fine aggregate.
- It is important to look into the binder contribution from RAP with respect to the RAP particle sizes, rather than solely considering the RAP content.
- The FRM and CRM specimens both passed the rutting resistance evaluation criterion of less than 10% deformation. The rutting performance of these high RAP mixes was found to be at least the same or even better than the control mix.
- While the control mix surpassed both the FRM and CRM with respect to fatigue life, it was found to be more sensitive to the changes in deformation.
- The transition temperature midpoint and the maximum tensile stress temperature for the CRM was much more desirable as compared to the FRM, and was slightly worse than the control mix.

Overall, the CRM mix, exhibited acceptable performance with respect to rutting, fatigue, and thermal cracking. It can be concluded that the RAP particle size can have a more significant effect on the mix performance than the RAP content. RAP content should be considered along with other important parameters such as RAP particle size and gradation, recycled binder ratio, and RAP binder content.

Author Contributions: S.S.S. conceived of the presented idea. A.C. and H.B. developed the theory and verified the analytical methods and supervised the project. S.S.S. wrote the original draft; and P.T. wrote, reviewed, and edited the manuscript and verified the analysis. All authors discussed the results and contributed to the final manuscript.

Funding: This research was funded by The Pavements and Bituminous Materials Laboratory (LCMB).

Acknowledgments: This work was supported by The Pavements and Bituminous Materials Laboratory (LCMB) and the Centre for Pavement and Transportation Technology (CPATT). The authors would like to thank the companies in Quebec that provided us with the materials for the project.

Conflicts of Interest: The authors declare no conflict of interest.

References

1. Kandhal, P.S.; Mallick, R.B. *Pavement Recycling Guidelines for State and Local Governments: Participant's Reference Book*; No. FHWA-SA-98-042; Washington, DC, USA, 1998.
2. Zaumanis, M.; Mallick, R.B.; Frank, R. 100% hot mix asphalt recycling: Challenges and benefits. *Transp. Res. Proced.* **2016**, *14*, 3493–3502. [CrossRef]
3. Federal Highway Administration. *Pavement Recycling Executive Summary and Report*; Federal Highway Administration: Washington, DC, USA, 1995.
4. Bloomquist, D.; Diamond, G.; Oden, M.; Ruth, B.; Tia, M. *Engineering and Environmental Aspects of Recycled Materials form Highway Construction*; Western Research Institute: Laramie, WY, USA, 1993.
5. Little, D.N.; Epps, J.A. Evaluation of Certain Structural Characteristics of Recycled Pavement Material. In Proceedings of the Association of Asphalt Paving Technologists, Louisville, KY, USA, 18–20 February 1980.
6. Bukowski, J.R. Guidelines for the Design of Superpave Mixtures Containing Reclaimed Asphalt Pavement (RAP). In Proceedings of the Memorandum, ETG Meeting, FHWA Superpave Mixtures Expert Task Group, San Antonio, TX, USA, 28 March 1997.
7. Kandhal, P.S.; Foo, K.Y. Designing Recycled Hot Mix Asphalt Mixtures Using Superpave Technology. In *Progress of Superpave (Superior Performing Asphalt Pavement): Evaluation and Implementation*; ASTM International: Washington, DC, USA, 1997.
8. McDaniel, R.S.; Soleymani, H.; Anderson, M.R.; Turner, P.; Peterson, R. *Recommended Use of Reclaimed Asphalt Pavement in the Superpave Mix Design Method*; NCHRP Web document 30 Transportation Research Board; National Research Council (US): Springfield, VA, USA, 2000.
9. Saliani, S.S.; Carter, A.; Baaj, H. Investigation of the Impact of Rap Gradation on the Effective Binder Content in Hot Mix Asphalt. In Proceedings of the Search Results CSCE 2016 Annual Conference, London, ON, Canada, 1–4 June 2016.
10. Basueny, A.; Perraton, D.; Carter, A. Laboratory study of the effect of RAP conditioning on thermo-mechanical properties of hot mix asphalt containing RAP. *Mater. Struct.* **2013**, *47*, 1425–1450. [CrossRef]
11. Praticò, F.G.; Vaiana, R.; Giunta, M.; Iuele, T.; Moro, A. Recycling PEMs Back to TLPAs: Is that Possible Notwithstanding RAP Variability. *Appl. Mech. Mater.* **2013**, *253*, 376–384. [CrossRef]
12. Tavassoti-Kheiry, P.; Solaimanian, M.; Qiu, T. Characterization of high RAP/RAS asphalt mixtures using resonant column tests. *J. Mater. Civ. Eng.* **2016**, *28*, 04016143. [CrossRef]
13. Petersen, J.C. A review of the fundamentals of asphalt oxidation: Chemical, physicochemical, physical property, and durability relationships. *Transp. Res. Circ.* **2009**, *E-C140*, P78.
14. Cosentino, P.J.; Kalajian, E.H.; Shieh, C.S.; Mathurin, W.J.K.; Gomez, F.A.; Cleary, E.D.; Treeratrakoon, A. *Developing Specifications for Using Recycled Asphalt Pavement as Base, Subbase or General Fill Materials, Phase II*; No. FL/DOT/RMC/06650-7754; Florida Institute of Technology: Melbourne, FL, USA, 2003.
15. Bressi, S.; Dumont, A.-G.; Pittet, M. Cluster phenomenon and partial differential aging in RAP mixtures. *Constr. Build. Mater.* **2015**, *99*, 288–297. [CrossRef]
16. Stephens, J.E.; Mahoney, J.; Dippold, C. *Determination of the PG Binder Grade to Use in a RAP Mix.Report No. JHR 00-278*; Connecticut Department of Transportation also investigated the asphalt films properties for the coarse aggregates; Connecticut Department of Transportation: Storrs, CT, USA, 2001.
17. Stroup-Gardiner, M.; Wagner, C. Use of reclaimed asphalt pavement in Superpave hot-mix asphalt applications. *Transp. Res. Rec.* **1999**, *1681*, 1–9. [CrossRef]
18. Hassan, R. Feasibility of Using High RAP Contents in Hot Mix Asphalt. In Proceedings of the 13th International Flexible Pavements Conference, Brisbane, Australia, September 2009.

19. Chen, J.S.; Huang, C.C.; Chu, P.Y.; Lin, K.Y. Engineering characterization of recycled asphalt concrete and aged bitumen mixed recycling agent. *J. Mater. Sci.* **2007**, *42*, 9867–9876. [CrossRef]
20. Huang, B.; Li, G.; Vukosavljevic, D.; Shu, X.; Egan, B.K. Laboratory investigation of mixing hot-mix asphalt with reclaimed asphalt pavement. *Transp. Res. Rec.* **2005**, *1929*, 37–45. [CrossRef]
21. Yu, X.; Zaumanis, M.; Dos Santos, S.; Poulikakos, L.D. Rheological, microscopic, and chemical characterization of the rejuvenating effect on asphalt binders. *Fuel* **2014**, *135*, 162–171. [CrossRef]
22. Moghaddam, T.B.; Baaj, H. The use of rejuvenating agents in production of recycled hot mix asphalt: A systematic review. *Constr. Build. Mater.* **2016**, *114*, 805–816. [CrossRef]
23. Król, J.B.; Kowalski, K.J.; Niczke, Ł.; Radziszewski, P. Effect of bitumen fluxing using a bio-origin additive. *Constr. Build. Mater.* **2016**, *114*, 194–203. [CrossRef]
24. Caputo, P.; Loise, V.; Ashimova, S.; Teltayev, B.; Vaiana, R.; Rossi, C.O. Inverse Laplace Transform (ILT) NMR: A powerful tool to differentiate a real rejuvenator and a softener of aged bitumen. *Colloids Surf. Physicochem. Eng. Asp.* **2019**, *574*, 154–161. [CrossRef]
25. Huang, B.; Zhang, Z.; Kingery, W.; Zuo, G. Fatigue Crack Characteristics of HMA Mixtures Containing RAP. In *Fifth International RILEM Conference on Reflective Cracking in Pavements*; RILEM: Limoges, France, 2004; pp. 631–638.
26. Al-Qadi, I.L.; Carpenter, S.H.; Roberts, G.; Ozer, H.; Aurangzeb, Q.; Elseifi, M.; Trepanier, J. *Determination of Usable Residual Asphalt Binder in RAP*; Illinois Center for Transportation (ICT): Rantoul, IL, USA, 2009.
27. Roque, R.; Yan, Y.; Cocconcelli, C.; Lopp, G. *Perform an Investigation of the Effects of Increased Reclaimed Asphalt Pavement (RAP) Levels in Dense Graded Friction Courses*; No. RFP-OT-1/12-9033-RC; Florida Department of Transportation; Research Center: Florida, USA, 2015.
28. Saliani, S.S.; Carter, A.; Baaj, H.; Mikhailenko, P. Characterization of Recovered Bitumen from Coarse and Fine Reclaimed Asphalt Pavement Particles. *Infrastructures* **2019**, *4*, 24. [CrossRef]
29. Gabet, T.; Di Benedetto, H.; Perraton, D.; De Visscher, J.; Gallet, T.; Bańkowski, W.; Olard, F.; Grenfell, J.; Bodin, D.; Sauzéat, C. French wheel tracking round robin test on a polymer modified bitumen mixture. *Mater. Struct.* **2011**, *44*, 1031–1046. [CrossRef]
30. Perraton, D.; Di Benedetto, H.; Sauzéat, C.; De La Roche, C.; Bankowski, W.; Partl, M.; Grenfell, J. Rutting of bituminous mixtures: Wheel tracking tests campaign analysis. *Mater. Struct.* **2011**, *44*, 969–986. [CrossRef]
31. Rahman, F.; Hossain, M. *Review and Analysis of Hamburg Wheel Tracking Device Test Data (No. KS-14-1)*; Department of Transportation. Bureau of Materials; Research: Manhattan, KS, USA, 2014.
32. Yildirim, Y. Polymer modified asphalt binders. *Constr. Build. Mater.* **2007**, *21*, 66–72. [CrossRef]
33. Meunier, M. Élaboration d'un outil de prédiction du comportement de la chaussée à l'orniérage. Thèse de doctorat en génie de la construction, Montréal, L'école de technologie supérieure.Moneron, Revue générale des routes et des aérodromes. Ph.D. Thesis, L'école de technologie supérieure, Montreal, QC, Canada, 2012.
34. Baaj, H.; Di Benedetto, H.; Chaverot, P. Effect of binder characteristics on fatigue of asphalt pavement using an intrinsic damage approach. *Road Mater. Pavement Des.* **2005**, *6*, 147–174. [CrossRef]
35. Tayebali, A.A.; Bor-wen, T.; Carl, L. *Monismith. Stiffness of asphalt-aggregate mixes*; No. SHRP-A-388; Strategic Highway Research Program, National Research Council: Washington, DC, USA, 1994.
36. Di Benedetto, H.; De La Roche, C. *State of the Art on Stiffness Modulus and Fatigue of Bituminous Mixtures*; Rilem Report; London, UK, 1998; pp. 137–180.
37. Perraton, D.; Di Benedetto, H.; Carter, A. Correspondances entre les coefficients des modèles de fatigue dans les méthodes mécanistiques-empiriques de dimensionnement de chaussées souples. *Can. J. Civ. Eng.* **2011**, *38*, 1287–1299. [CrossRef]
38. Olidis, C.; Hein, D. Guide for the Mechanistic-Empirical Design of New and Rehabilitated Pavement Structures Materials Characterizatio. In Proceedings of the 2004 Annual Conference of the Transportation Association of Canada, Quebec, QC, Canada, 19–22 September 2004.
39. Baaj, H. *Comportement à la Fatigue Des Matériaux Granulaires Traités Aux Liens Hydrocarbones*; Département Génie Civil et Bâtiment (DGCB); École Nationale des Travaux Publics de l'État, Lyon, INSA: Lyon, France, 2002.
40. Di Benedetto, H.; Ashayer Soltani, A.; Chaverot, P. Fatigue Damage for Bituminous Mixtures: A Pertinent Approach. *J. Assoc. Asph. Paving Technol.* **1996**, *65*, 142–158.
41. Tapsoba, N.; Baaj, H.; Sauzéat, C.; Di Benedetto, H.; Ech, M. 3D analysis and modelling of Thermal Stress Restrained Specimen Test (TSRST) on asphalt mixes with RAP and roofing shingles. *Constr. Build. Mater.* **2016**, *120*, 393–402. [CrossRef]

42. Basueny, A.; Carter, A.; Perraton, D.; Vaillancourt, M. Laboratory Evaluation of Complex Modulus and Fatigue Resistance of Asphalt Mixtures with RAP. In Proceedings of the 8th RILEM International Symposium on Testing and Characterization of Sustainable and Innovative Bituminous Materials, Nantes, France, 7–9 June 2016; pp. 521–532.
43. Carter, A.; Paradis, M. Laboratory Characterization of the Evolution of the Thermal Cracking Resistance with the Freeze-Thaw Cycles. Available online: http://data.abacus.hr/h-a-d/radovi_s_kongresa/nagoya_japan_2010/90061.pdf (accessed on 25 October 2019).
44. Baaj, H.; Ech, M.; Tapsoba, N.; Sauzeat, C.; Di Benedetto, H. Thermomechanical characterization of asphalt mixtures modified with high contents of asphalt shingle modifier (ASM®) and reclaimed asphalt pavement (RAP). *Mater. Struct.* **2013**, *46*, 1747–1763. [CrossRef]
45. Perraton, D.; Proteau, M.; Carter, A.; Meunier, M.; Dufresne, R. Development of high modulus asphalt mixes for cold climate. In Proceedings of the Fifty-Ninth Annual Conference of the Canadian Technical Asphalt Association (CTAA), Winnipeg, MB, Canada, 16–19 November 2014; pp. 249–268.
46. Delaporte, B.; Di Benedetto, H.; Chaverot, P.; Gauthier, G. Linear viscoelastic properties of bituminous materials: From binders to mastics. *Assoc. Asph. Paving Technol.* **2007**, *76*, 488–494.

infrastructures

MDPI

Article

Principal Component Neural Networks for Modeling, Prediction, and Optimization of Hot Mix Asphalt Dynamics Modulus

Parnian Ghasemi [1,*], Mohamad Aslani [2], Derrick K. Rollins [3] and R. Christopher Williams [1]

1 Department of Civil and Environmental Engineering, Iowa State University, Ames, IA 50011, USA
2 Department of Mathematics, Florida State University, Tallahassee, FL 32304, USA
3 Department of Chemical Engineering, Iowa State University, Ames, IA 50011, USA
* Correspondence: pghasemi@iastate.edu; Tel.: +1-515-708-7692

Received: 14 June 2019; Accepted: 10 August 2019; Published: 17 August 2019

Abstract: The dynamic modulus of hot mix asphalt (HMA) is a fundamental material property that defines the stress-strain relationship based on viscoelastic principles and is a function of HMA properties, loading rate, and temperature. Because of the large number of efficacious predictors (factors) and their nonlinear interrelationships, developing predictive models for dynamic modulus can be a challenging task. In this research, results obtained from a series of laboratory tests including mixture dynamic modulus, aggregate gradation, dynamic shear rheometer (on asphalt binder), and mixture volumetric are used to create a database. The created database is used to develop a model for estimating the dynamic modulus. First, the highly correlated predictor variables are detected, then Principal Component Analysis (PCA) is used to first reduce the problem dimensionality, then to produce a set of orthogonal pseudo-inputs from which two separate predictive models were developed using linear regression analysis and Artificial Neural Networks (ANN). These models are compared to existing predictive models using both statistical analysis and Receiver Operating Characteristic (ROC) Analysis. Empirically-based predictive models can behave differently outside of the convex hull of their input variables space, and it is very risky to use them outside of their input space, so this is not common practice of design engineers. To prevent extrapolation, an input hyper-space is added as a constraint to the model. To demonstrate an application of the proposed framework, it was used to solve design-based optimization problems, in two of which optimal and inverse design are presented and solved using a mean-variance mapping optimization algorithm. The design parameters satisfy the current design specifications of asphalt pavement and can be used as a first step in solving real-life design problems.

Keywords: hot mix asphalt dynamic modulus; principal component analysis; linear regression modeling; artificial neural network; receiver operating characteristic; optimization

1. Introduction

The stress-strain relationship for asphalt mixtures under sinusoidal loading can be described by the dynamic modulus, $|E^*|$, a function of material's components properties, loading rate, and temperature [1,2]. The dynamic modulus is one of the primary design inputs in Pavement Mechanistic-Empirical (M-E) Design to describe the fundamental linear viscoelastic material properties [3–5], and is one of the key parameters used to evaluate rutting and fatigue cracking distress predictions in Mechanistic-Empirical Pavement Design Guide (MEPDG) [5,6]. Although $|E^*|$ has a significant role in pavement design, the associated test procedure is time-consuming and requires expensive equipments, so extensive effort has been extended to predict $|E^*|$ from hot mix asphalt (HMA) material properties [7–9].

Predictive modeling is a process of estimating outcomes from several predictor variables using data mining tools and probability theory. An initial model can be formulated using either a simple linear equation or a more sophisticated structure obtained through a complex optimization algorithm [10].

There are several well-known predictive models for dynamic modulus, some of them are regression models, and some more recent ones have used techniques that include Artificial Neural Networks (ANN) and genetic programming [11]. Andrei et al. [12], used 205 mixtures with 2750 data points and revised the original Witczak model, and the developed model has subsequently been reformulated to use binder shear modulus rather than binder viscosity [13]. Christensen et al. [14], developed a new $|E^*|$ predictive model based on the law of mixtures. The data base used for training the model contained 206 $|E^*|$ measurements from 18 different HMA mixtures. Jamrah et al. [15], attempted to develop improved $|E^*|$ predictive models for HMA used in the State of Michigan. They observed a significant difference between measured and fitted $|E^*|$ values, especially at high temperatures and low frequencies. Alkhateeb et al. [16], developed a new predictive model from the law of mixtures to be used over broader ranges of temperature and loading frequencies, including higher temperatures/lower frequencies. The predictor variables used in that model were Voids in Mineral Aggregate (VMA) and binder shear modulus (G^*).

Sakhaeifar et al. [17], developed individual temperature-based models for predicting dynamic modulus over a wide range of temperatures. The predictor variables used in their model were aggregate gradation, VMA, Voids Filled with Asphalt (VFA), air void (V_a), effective binder content (V_{beff}), G^*, and binder phase angel (δ).

The existing dynamic modulus predictive models in the literature typically use two or more predictors from the following list: aggregate gradation, volumetric properties, and binder shear properties. These predictor variables are not necessarily an independent set of variables and thus it may not be appropriate for use in developing models. Since cross-correlated inputs in a dataset can unfavorably affect the accuracy of a predictive model by unduly affecting the estimation of their causative effects on the response variable, a pre-processing step of data evaluation would be useful for studying the quality of the input variables and their pair-wise correlations [18]. Principal Component Analysis (PCA) is a multivariate statistical approach that not only reduces the dimensionality of the problem but also converts a set of correlated inputs to a set of orthogonal (pseudo-)inputs using an orthogonal transformation [19]. During such a transformation, PCA maximizes the amount of information of the original dataset $\mathbf{X}$ by using a smaller set of pseudo-variables [20,21]. Another issue in all of the predictive models is extrapolation that can be risky because a model might behave differently outside of the convex hull that contains all of the data points used for its training. To avoid using points outside of this convex hull, a hyper-space containing all data points can be found and added as a constraint on the desired modeling problem.

Ghasemi et al. [22], developed a methodology for eliminating correlated inputs and extrapolation in modeling; they created a laboratory database of accumulated strain values of several asphalt mixtures and used the resulting framework to estimate the amount of permanent deformation (rutting) in asphalt pavement. Following their new PCA-based approach, this study focuses on developing a machine-learning based framework for predicting the dynamic modulus of HMA using orthogonal pseudo-inputs obtained from principal component analysis. Unlike most of the existing $|E^*|$ predictive models, the proposed framework uses different data sets for model training and performance testing. To avoid extrapolation, an n-dimensional hyperspace is developed and added as a constraint to the modeling problem. This study also claims to determine the optimal HMA design and design variables for a pre-specified $|E^*|$ by applying framework using an evolutionary-based optimization algorithm. It is worth pointing out that, unlike other predictive models, the proposed framework is not site-specific and also not limited to the materials used in the American Association of State Highway and Transportation Officials (AASHTO) road test, i.e., this framework can adjust itself based on the dataset presented to the framework. The need for a more robust and general framework for performance prediction in asphalt pavement also stems from the availability of the vast amount of experimental data in this field. In this work, the developed framework operates in such a spirit and improves the accuracy of available models via machine learning-based approaches.

The remainder of the document is organized as follows: Section 2 presents material and methodology, followed by Section 3 that covers results and discussion. Two examples of the proposed framework's applications are discussed in Section 4, followed by conclusions presented in Section 5.

2. Material and Methodology

Twenty-seven specimens from nine different asphalt mixtures (three replicates for each mixture group) were used in this study. Using AASHTO TP 79-13 the dynamic modulus test was performed at three temperatures (0.4, 17.1, and 33.8 °C) and nine loading frequencies (25, 20, 10, 5, 2, 1, 0.5, 0.2, 0.1 Hz). The maximum theoretical specific gravity (G_{mm}), the bulk specific gravity (G_{mb}), and the effective binder content (V_{beff}) were determined and used to calculate other volumetric properties of the asphalt mixtures.Asphalt binder shear properties were obtained from a dynamic shear rheometer (DSR) test. Using ASTM D7552-09(2014) the test was performed over a wide range of temperatures (−10 to 54 °C) and frequencies (0.1 Hz to 25 Hz), the same test temperatures and loading frequencies used in the mixture dynamic modulus test. It is important to note that this study uses a consistent definition of frequency, and that in order to predict the dynamic modulus value of an asphalt mixture for example at 4 °C and 25 Hz, for example, one should use as a model input the complex shear modulus of asphalt binder, $|G^*|$, at 4 °C and 25 Hz. A summary of the nine different mixture properties is given in Table 1. Using the laboratory test results on 27 specimens, a database of 243 data points was created for use in further modeling.

Table 1. General Mixture Properties of Nine Asphalt Mixtures Used in this Study.

	Mix 1	Mix 2	Mix 3	Mix 4	Mix 5	Mix 6	Mix 7	Mix 8	Mix 9
Binder performance grade	58–28	58–28	58–28	58–34	58–34	58–34	64–28	64–34	64–28
% V_{beff}	4.20	4.10	4.10	3.90	3.50	4.30	4.20	4.00	4.60
%VMA	13.50	13.50	13.60	13.10	12.50	13.90	13.70	13.40	14.40
% VFA	70.30	70.40	70.60	69.60	68.10	71.20	70.80	70.20	72.30
G_{mb}	2.32	2.31	2.31	2.32	2.31	2.32	2.31	2.32	2.31
G_{mm}	2.41	2.46	2.51	2.48	2.64	2.46	2.48	2.51	2.44
% V_a	4.01	3.99	3.99	3.98	3.98	4.03	4	3.99	3.98
% passing 3/4″	100.00	100.00	100.00	100.00	100.00	100.00	100.00	100.00	100.00
% passing 1/2″	93.90	96.40	87.20	93.50	95.10	96.40	94.10	94.40	94.20
% passing 3/8″	77.50	84.60	73.70	76.40	83.10	87.30	83.40	82.00	80.90
% passing #4	49.80	53.10	48.40	52.20	52.20	60.90	63.80	48.20	58.60
% passing #8	34.40	38.40	35.10	43.60	38.80	46.90	47.10	34.90	46.00
% passing #30	16.70	18.70	17.90	20.90	18.80	23.40	21.70	19.20	25.90
% passing #50	10.30	10.80	10.90	11.40	9.90	12.40	11.90	11.80	13.80
% passing #100	6.10	5.90	6.40	5.80	5.40	6.10	6.60	6.10	7.20
% passing #200	3.60	3.30	6.20	3.30	3.50	3.40	4.00	3.10	4.00

2.1. Preliminary Processing Step: Input Variable Selection

A parsimonious set of input variables is required to develop a model [20]. For a common model structure, one can represent the expectation function of the response as $y_i = f_i(\mathbf{x}_i, \theta)$, where y_i is the expected response variable at the ith measurement, $i = 1, \ldots, n$, $\mathbf{x}_i$ is the input vector at the ith measurement, and θ is the vector of unknown model parameters with $\theta = \left[\theta_1 \ldots \theta_q\right]^T$. It is assumed that the element in the ith row and jth column of the Jacobian matrix , $\mathbf{J}$, is $\frac{\partial \eta_i}{\partial \theta_j}$ i.e., $\mathbf{J} = \left\{\frac{\partial \eta_i}{\partial \theta_j}\right\}$. Note that the jth column represents θ_j and its column vector reflects the variation in the response space as θ_j varies over a specific set of experimental conditions. If j and k are two orthogonal columns, their correlation coefficient (r) must be zero, meaning that the information used to estimate θ_j is independent from the information used to estimate θ_k and vice versa. The benefit of using orthogonal input variables is that not only does it result in consolidation of causative effects of inputs on the output but it also maximizes parameter accuracy and therefore estimation accuracy of the predicted output.

According to the literature [11–14,16,17,23,24], the stiffness characteristic of an asphalt mixture presented by a dynamic modulus can be estimated by its component properties. In this study, the input variables vector (**x**) defines the asphalt mixture's component properties. A summary of the selected input variables and their ranges in the dataset is presented in Table 2 with the x_i's and y being the input and output variables, respectively.

Table 2. Selected Input Variables (x) And Output Variable (y) For Model Development.

Variable	Identity	Min.	Max.	Ave.	Std. Dev.
y	$Log\|E^*\|$	2.62	4.37	3.76	0.46
x_1	Cum. % retained on 3/4″	3.60	13.00	6.11	2.63
x_2	Cum. % retained on 3/8″	12.68	26.29	19.01	4.11
x_3	Cum. % retained on #4	36.20	51.76	45.86	5.319
x_4	Cum. % retained on #8	52.87	65.70	59.42	5.06
x_5	Cum. % retained on #30	74.06	83.30	79.63	2.76
x_6	Cum. % retained on #50	86.22	90.12	88.57	1.15
x_7	Cum. % retained on #100	92.81	94.59	93.83	0.48
x_8	% Passing from #200	3.07	6.18	3.81	0.89
x_9	$Log\|G^*\|$	−2.29	3.03	0.50	1.26
x_{10}	Phase angle (degree)	28.15	79.17	52.86	11.54
x_{11}	$\%V_{beff}$	3.50	4.60	4.10	0.29
x_{12}	%VMA	12.50	14.40	13.51	0.49
x_{13}	%VFA	68.10	72.30	70.40	1.08
x_{14}	$\%V_a$	3.98	4.01	3.99	0.01

Cross-correlation analysis is performed on the 14 selected predictor variables and the obtained pairwise correlation matrix is given in Table 3 along with the schematic heat map of the correlation matrix given by Figure 1. Correlation coefficients with absolute values of 0.5 or higher are displayed in bold red text. The corresponding cells in the correlation heat map are shown in dark blue and dark red as shown in Figure 1. According to Table 3, the absolute values of the 130 correlation coefficients are greater than 0.1, with 50 of them greater than 0.5, indicating that several of the input variables give an impression of being highly correlated. The correlation heat map also clearly indicates that a high level of correlation (dark blue and dark red cells) exists within the input variables. If the correlated input variables are detected, to enable accurate mapping of the inputs to the response variable, it would be useful to produce a smaller set of orthogonal pseudo-variables using the PCA method and use them in model development [20].

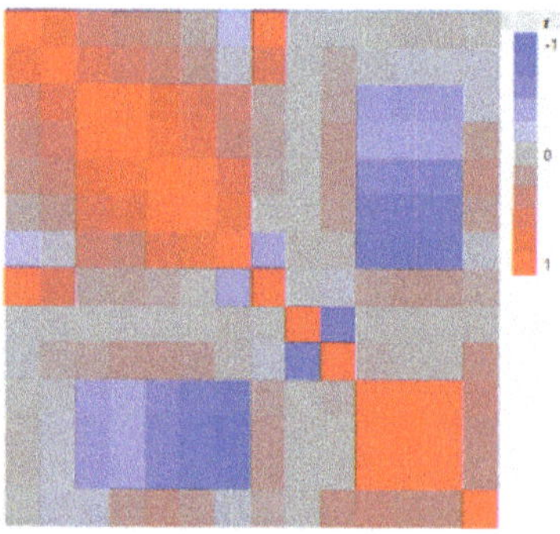

Figure 1. Heat map depict a visualization for the pairwise correlation matrix of the input variables. Each column i and row j shows correlation coefficient r_{ij}. Cells with colors otherthan gray indicate some level of correlation. As the colors get darker the level of correlation goes higher.

Table 3. Pairwise Correlation Matrix for the Selected Input Variables.

	x_1	x_2	x_3	x_4	x_5	x_6	x_7	x_8	x_9	x_{10}	x_{11}	x_{12}	x_{13}	x_{14}
x_1	**1**	**0.832**	0.412	0.366	0.294	0.119	−0.269	**0.905**	−0.044	−0.058	0.003	0.04	0.049	0.013
x_2	**0.832**	**1**	**0.597**	0.458	0.391	0.246	−0.109	**0.583**	−0.035	0.106	−0.061	−0.099	−0.089	−0.115
x_3	0.412	**0.597**	**1**	**0.918**	**0.756**	**0.596**	0.425	0.133	−0.019	0.154	−0.465	−0.485	−0.49	−0.111
x_4	0.366	0.458	**0.918**	**1**	**0.87**	**0.687**	0.375	0.169	−0.028	0.237	−0.388	−0.412	−0.424	0.212
x_5	0.294	0.391	**0.756**	**0.87**	**1**	**0.919**	**0.618**	0.112	−0.021	0.235	**−0.585**	**−0.631**	**−0.633**	0.3
x_6	0.119	0.246	**0.596**	**0.687**	**0.919**	**1**	**0.794**	−0.009	0.003	0.203	**−0.741**	**−0.796**	**−0.806**	0.209
x_7	−0.269	−0.109	0.425	0.375	**0.618**	**0.794**	**1**	−0.414	0.036	0.047	**−0.854**	**−0.886**	**−0.892**	−0.087
x_8	**0.905**	**0.583**	0.133	0.169	0.112	−0.009	−0.414	**1**	−0.032	−0.102	0.179	0.238	0.238	0.142
x_9	−0.044	−0.035	−0.019	−0.028	−0.021	−0.003	0.036	−0.032	**1**	**−0.808**	0.021	0.016	0.013	0.034
x_{10}	−0.058	0.106	0.154	0.237	0.235	0.203	0.047	−0.102	**−0.808**	**1**	0.09	0.024	0.014	0.3
x_{11}	0.003	−0.061	−0.465	−0.388	**−0.585**	**−0.741**	**−0.854**	0.179	0.021	0.09	**1**	**0.988**	**0.985**	0.372
x_{12}	0.04	−0.099	−0.485	−0.412	**−0.631**	**−0.796**	**−0.886**	0.238	0.016	0.024	**0.988**	**1**	**0.998**	0.321
x_{13}	0.049	−0.089	−0.49	−0.424	**−0.633**	**−0.806**	**−0.892**	0.238	0.013	0.014	**0.985**	**0.998**	**1**	0.301
x_{14}	0.013	−0.115	−0.111	0.212	0.3	0.209	−0.087	0.142	0.034	0.3	0.372	0.321	0.301	**1**

2.2. Orthogonal Transformation Using PCA

In multivariate statistics, PCA is an orthogonal transformation of a set of (possibly) correlated variables into a set of linearly uncorrelated ones, and the uncorrelated (pseudo-) variables, called principal components (PCs), are linear combinations of the original input variables. This orthogonal transformation is performed such that the first principal component has the greatest possible variance (variation within the dataset). This procedure is then followed for the second component, then the third component, etc. This means that each succeeding component in turn has the highest variance when it is orthogonal to the preceding components [25–28]. To help visualize the PCA transformation, a schematic dataset with three input variables is presented in Figure 2 (left). As shown in this figure, in conducting PCA the data points are transferred from the original 3D original input space (on the left) to a 2D principal component space (on the right).

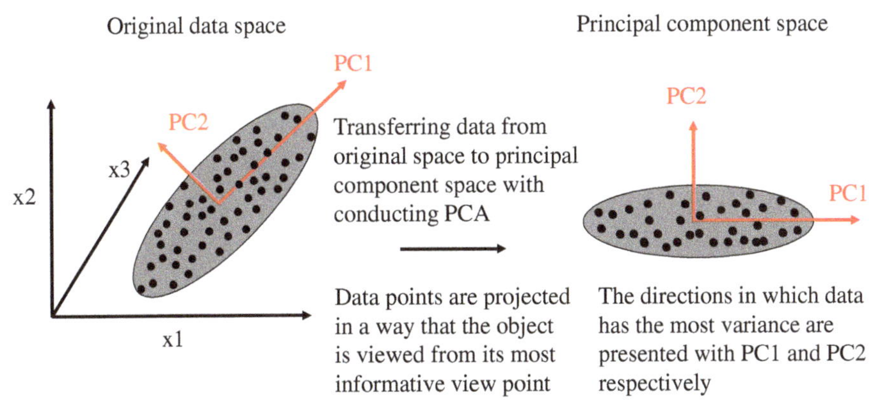

Figure 2. Schematic of the PCA transformation. Original data space presented on the left with 3 (input) variables transformed to a component space with lower dimension and pc_1 and pc_2 being the axes of the coordinate.

PCA can be performed either by eigenvalue decomposition of a data covariance (or correlation) matrix or by singular value decomposition. The process usually begins with mean centering the data matrix (and normalizing or using Z-scores) for each attribute as follows:

$$\mathbf{X} = \begin{bmatrix} x_{11} & x_{12} & \cdots & x_{1p} \\ x_{21} & x_{22} & \cdots & x_{2p} \\ \vdots & \vdots & \ddots & \vdots \\ x_{n1} & x_{n2} & \cdots & x_{np} \end{bmatrix} \quad \mathbf{Z} = \begin{bmatrix} \frac{x_{11}-\overline{x}_1}{s_1} & \frac{x_{12}-\overline{x}_2}{s_1} & \cdots & \frac{x_{1p}-\overline{x}_p}{s_p} \\ \frac{x_{21}-\overline{x}_1}{s_1} & \frac{x_{22}-\overline{x}_2}{s_2} & \cdots & \frac{x_{2p}-\overline{x}_p}{s_p} \\ \vdots & \vdots & \ddots & \vdots \\ \frac{x_{n1}-\overline{x}_1}{s_1} & \frac{x_{n2}-\overline{x}_2}{s_2} & \cdots & \frac{x_{np}-\overline{x}_p}{s_p} \end{bmatrix} \tag{1}$$

where for $k = 1$ to n and $j = 1$ to p, x_{kj} is the kth measurement for the jth variable, $\overline{x}_k$ is the sample mean for the kth variable, and s_k is sample standard deviation for the kth variable. As discussed in the previous section, highly correlated input variables lead to inflation of the standard error of estimate, negatively affecting the accuracy of the estimation. PCA will help us not only reduce the dimensionality of the modeling problem, but will also produce orthogonal pseudo-variables to be used in solving the problem. To perform PCA in this study we used eigenvalue decomposition of the correlation matrix of the data. The eigenvalues of the data correlation matrix are calculated, ranked, and sorted in descending order (representing their quota of the total variation within the dataset), as presented in Table 4. According to the eigenvalues, the first five PCs represent 95.8% of the existing variation within the dataset.

Recalling the fact that the PCs are linear combinations of the original input variables, the PCs can be defined as in Equation (2):

$$pc_i = \sum_{j=1}^{14} \alpha_{ij} x_j + \beta_i \tag{2}$$

where $i = 1$ to 14, the α_{ij} is the corresponding coefficients, the β_i are constants, and the x_j are the original input variables. Equation (2) can be stated in matrix notation as in Equation (3):

$$\mathbf{p} = \mathbf{Mz} + \mathbf{n} \tag{3}$$

where

$$\mathbf{p} = \begin{bmatrix} pc_1 \\ pc_2 \\ pc_3 \\ pc_4 \\ pc_5 \end{bmatrix}$$

$$\mathbf{M}^T = \begin{bmatrix} 0.03 & 0.19 & -0.08 & -0.06 & -0.09 \\ 0.03 & 0.11 & -0.04 & -0.05 & 0.04 \\ 0.06 & 0.04 & 0.00 & 0.00 & 0.10 \\ 0.06 & 0.05 & 0.02 & 0.04 & 0.07 \\ 0.13 & 0.06 & 0.05 & 0.09 & -0.01 \\ 0.33 & 0.01 & 0.09 & 0.16 & -0.20 \\ 0.71 & -0.54 & 0.05 & 0.13 & -0.20 \\ -0.03 & 0.52 & -0.20 & -0.07 & -0.53 \\ -0.01 & -0.05 & -0.4 & 0.43 & 0.12 \\ 0.00 & 0.01 & 0.06 & -0.02 & 0.00 \\ -1.26 & 0.64 & 0.52 & 0.61 & 0.67 \\ -0.75 & 0.37 & 0.21 & 0.28 & 0.31 \\ -0.34 & 0.17 & 0.09 & 0.12 & 0.14 \\ -4.41 & 18.38 & 47.08 & 76.24 & -40.48 \end{bmatrix} \tag{4}$$

$$\mathbf{n} = \begin{bmatrix} -55.95 \\ -58.54 \\ -218.20 \\ -352.79 \\ 174.78 \end{bmatrix}$$

Table 4. Eigenvalues of the Normalized Input Variables Matrix Sorted in Descending Order (Based on Their Contribution to Total Variation).

Number	Eigenvalue	Percent Variation	Cumulative Percent Variation
1	6.0225	43.018	43.018
2	3.2193	22.995	66.013
3	1.9746	14.104	80.118
4	1.4174	10.124	90.242
5	0.7850	5.607	95.848
6	0.3176	2.269	98.117
7	0.1091	0.779	98.896
8	0.0778	0.556	99.452
9	0.0549	0.392	99.844
10	0.0218	0.156	100

Further modeling efforts will be performed using the first five PCs.

2.3. Holdout Cross Validation

In prediction problems, cross validation will be used to estimate model accuracy. Cross validation is a model validation technique that can be used to prevent overfitting as well as to assess how the results of a statistical analysis can be generalized to an independent dataset. In this study, a holdout cross validation technique is used in which the given dataset is randomly assigned to two subsets, d_0 and d_1, the training set and the test set, respectively. Since the training set contains 80% of the data points and the test set contains 20% of the data points, 80% of the data points are used to train the model and the remainder are used to evaluate the trained model's performance.

2.4. Principal Component Regression (PCR)

Linear regression attempts to model the relationship between response variables and explanatory variables by fitting a linear equation to observed data. In regression analysis, the least-squares method is used to calculate the best fitting line for the observed data by minimizing the sum of the squares of the residuals (differences between the measured responses and the fitted values by a linear function of parameters).

All possible regression structures were considered for representing the relationship between the response variable, $log|E^*|$, and predictor variables (pc_1, pc_2, pc_3, pc_4, and pc_5). To estimate the values of the unknown coefficients in the model, the least-squares criterion of minimizing the sum of squared residuals (SSE) is used. Finally, after eliminating redundant terms, the reduced third order cubic and interaction terms were developed and selected as the best-fitted model. The developed model is called "Principal Component Regression (PCR)".

2.5. Principal Component Neural Network (PCNN)

A predictive model called "Principal Component Neural Network (PCNN)" is developed as briefly described in this section. ANNs are brain-inspired systems intended to replicate the way humans learn. Neural network structures consist of several layers, including input layers, output layers, and hidden layer(s), with nodes (neurons) in each layer [29–31]. A three-layer feed-forward neural network is developed for this study. It consists of an input layer of five neurons (five input variables), a hidden layer of 10 neurons, and an output layer of one neuron (one response variable). A trial-and-error procedure of optimizing the computational time and cost function is used to choose the number of hidden neurons. In this study supervised learning is used in which a training dataset, including inputs and outputs, is presented to the network. The network adjusted its weights in such a way that the adjusted set of weights produces an input/output mapping resulting in the smallest error. This iterative process is carried on until the sum of square residuals (SSE) increases. After the learning or training phase, the performance of the trained network must be measured against an independent (unseen) testing data [29,32]. Let the input of each processing node be pc_i, the adjustable connection weight be w_{ij}, and let the bias at output layer be b_0, so that the network transfer (activation) function is $f(.)$. The jth output of the first layer can be obtained using Equation (5)

$$v_j = f_1(pc_i, w_{ij}), \quad i = 1, ..., 5 \quad and \quad j = 1, ..., 10 \tag{5}$$

and the response will be

$$\hat{y} = f_2\left(f_1(pc_i, w_{ij})\right). \tag{6}$$

If we assume that

$$f_2(v_j, w_{Hj}) = b_0 + \sum_j v_j w_{Hj}, \tag{7}$$

and for each j,

$$f_1(pc_i, w_{ij}) = b_{Hj} + \sum_j pc_i w_{ij}, \tag{8}$$

then a feed-forward neural network can be formulated as follows:

$$\hat{y} = f_2\left\{b_0 + \sum_{j=1}^{n}\left[w_{Hj} \cdot f_1\left(b_{Hj} + \sum_{i=1}^{m} pc_i w_{ij}\right)\right]\right\}, \tag{9}$$

where pc_i is pseudo input parameter i, w_{ij} is the weight of connection between input variable i (for $i = 1$ to 5) and neuron j of the hidden layer, b_0 is a bias at the output layer, w_{Hj} is the weight of connection between neuron j of the hidden layer and output layer neuron, b_{Hj} is a bias at neuron j of the hidden layer (for $j = 1$ to 10), and $f_1(t)$ and $f_2(t)$ are transfer functions of the hidden layer and output layer, respectively.

It should be pointed out that iteration proceeds until the convergence criterion is met. Thus, similar to the linear regression model, the validation set is not used. The Bayesian Regularization algorithm is used to achieve network training efficiency.

2.6. *Effective Variable Space*

It is widely known that the use of an empirical predictive model outside the convex hull containing the data points is prohibited. In this context, *effective variable space* is referred to the space where the uncertainty of the developed models is bound to their already calculated thresholds. In other words, outside of this region, the extrapolated behavior of the models may not be predictable. To guard against extrapolation, Ghasemi et al. [22] concluded that the space containing input data could be interpreted as a symmetrical convex space, then demonstrated how this space can be used in the design procedure.

Following the approach in [22], a normal distribution is assumed for each input variable (x_i), resulting in their joint distribution being bi-variate normal, and such distributions are usually represented in form of a contour diagram. Since a contour curve on such a diagram contains the points on a surface with the same distance from the $x_i x_j$ plane, these points have a constant density function [33] (see Figure 3 for an example of such distribution). The cross section is obtained by slicing a bi-variate normal surface at a constant distance from the $x_i x_j$ plane. As indicated in Figure 3, the n-dimensional hyperspace is a hyper-ellipsoid with minimum volume (to avoid any gaps in the edges).

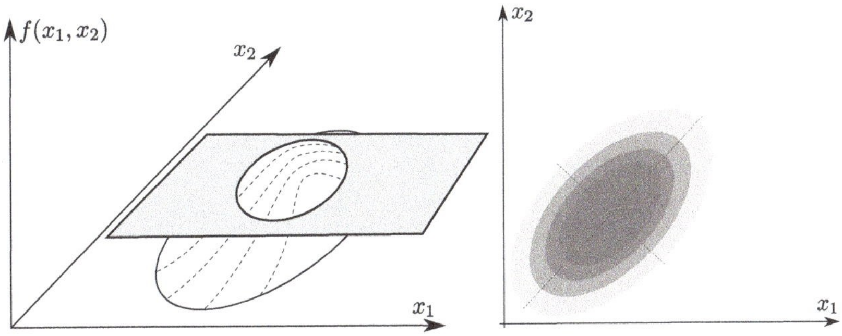

Figure 3. A schematic of a bivariate normal distribution (**left**) [33] and elliptic cross-section of this distribution (**right**). Projecting the cross-sectional area on the (x_1, x_2) plane results in an ellipsoid.

Khachiyan's work [34] formulates the problem of finding an approximate minimum volume enclosing ellipsoid ($\mathcal{E}$) given p data points in n-dimensions as an optimization problem. In Ghasemi et al. [22], the authors detailed the derivation of a procedure for solving this problem and obtaining its effective variable space. For brevity, the flowchart in Figure 4 summarizes this iterative method for finding the minimum volume enclosing ellipsoid. This algorithm was used to find two enclosing ellipsoids in the primary space (14-dimensional) and the pseudo space (5-dimensional) of the dataset. It should be pointed out that this space is independent of the predictive models and is used only to solve the optimal (and inverse) design problems.

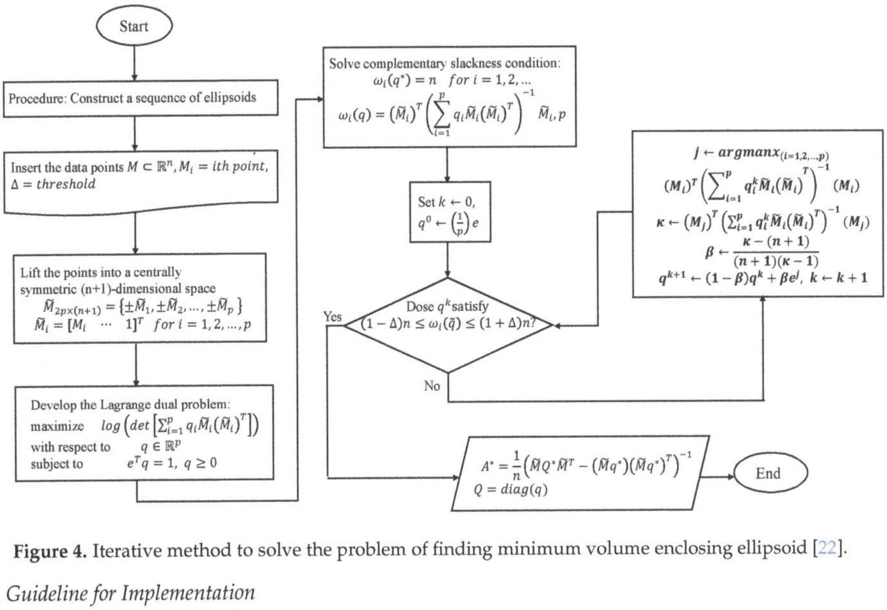

Figure 4. Iterative method to solve the problem of finding minimum volume enclosing ellipsoid [22].

2.7. Guideline for Implementation

A summary of the methodologies used to develop the framework is presented in Figure 5. The procedure begins with collecting experimental data from the laboratory, followed by the pre-processing step of input variable evaluation. The flowchart continues with the model development and the addition of a constraint on the n-dimensional input variable hyperspace to the modeling problem. The developed models can then be used to predict pavement performance, solve design-based optimization problems, etc. There are a number of aspects of the proposed framework that can be achieved using free and commercially available software like *MATLAB®, Python, and R packages*, and one may implement many parts of the framework in the language of their interest. For example, the algorithm to find the n-dimensional hyper-ellipsoid is very straightforward using the flowchart in Figure 4.

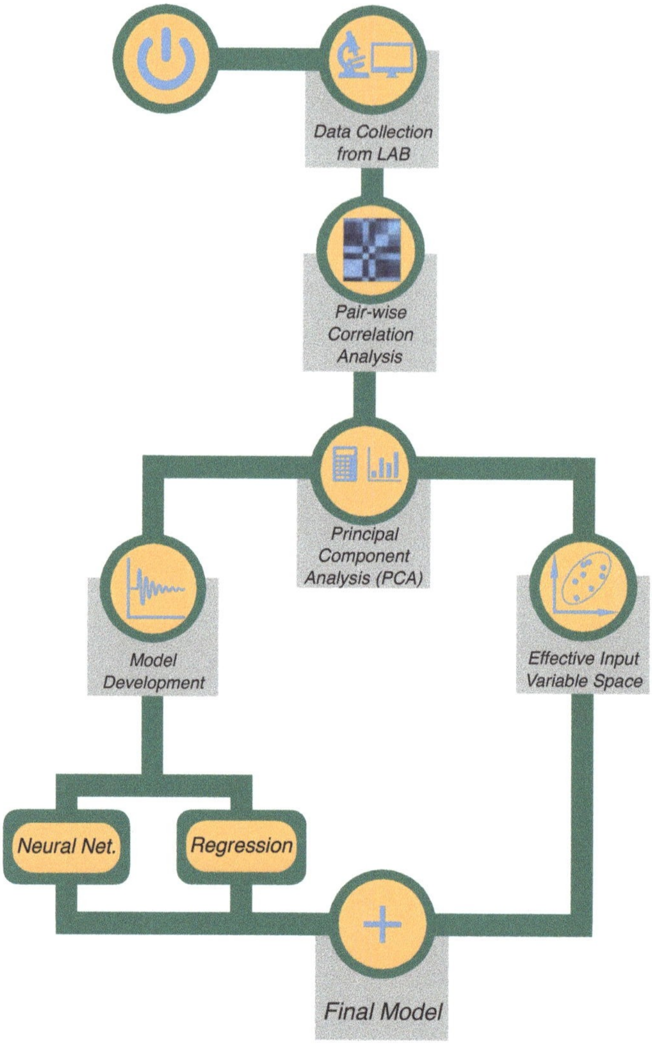

Figure 5. A summary of the methodologies used to develop the machine learning-based framework for predicting dynamic modulus (as an example of performance related property of asphalt mixture).

3. Developed Model Results, Performance, and Validation

The results produced by the developed models are presented in this section, and their capability to use empirical data to estimate the dynamic modulus of asphalt mixtures is evaluated.

3.1. Model Performance

The performance is first compared with the existing predictive models; modified Witczak, Hirsch, and Alkhateeb models selected from the literature are presented in Equations (10)–(13), respectively [13,14,16]

$$\begin{aligned}\log|E^*| = &-0.349 + 0.754(|G_b^*|^{-0.0052})(6.65 - 0.032\rho_{200} - 0.0027(\rho_{200})^2 + \\ &0.011\rho_4 - 0.0001(\rho_4)^2 + 0.006\rho_{3/8} - 0.00014(\rho_{3/8})^2 - 0.08V_a - 1.06(\frac{V_{beff}}{V_{beff} + V_a})) \\ &+ \frac{2.558 + 0.032V_a + 0.713(\frac{V_{beff}}{V_{beff}+V_a}) + 0.0124\rho_{3/8} - 0.0001(\rho_{3/8})^2 - 0.0098\rho_{3/4}}{1 + exp(-0.7814 - 0.5785\log|G_b^*| + 0.8834\log\delta_b)}\end{aligned} \quad (10)$$

where $|E^*|$ is dynamic modulus in psi; $|G^*|$ is the binder shear modulus in psi; δ_b is the binder phase angle in degrees; $\rho_{3/4}$ is the cumulative percent aggregate retained on the 3/4″ sieve (19 mm); $\rho_{3/8}$ is the cumulative percent aggregate retained on the 3/8″ sieve (9.5 mm); ρ_4 is the cumulative percent aggregate retained on the No. 4 sieve (4.75 mm); ρ_{200} is the percent aggregate passing the No. 200 sieve (0.075 mm); V_a is the percent air void in the mix; V_{beff} is the effective asphalt content; VMA is the percent of voids in the mineral aggregate, and VFA is the percent voids filled with asphalt,

$$|E_m^*| = P_c\left(4,200,000(1 - \frac{VMA}{100}) + 3|G^*|_b(\frac{VFA \times VMA}{10,000})\right) + \frac{(1 - P_c)}{\frac{1 - \frac{VMA}{100}}{4,200,000} + \frac{VMA}{3|G_b^*|(VFA)}} \quad (11)$$

where

$$P_c = \frac{(20 + 3|G_b^*|(VFA)/(VMA))^{0.58}}{650 + (3|G^*|_b(VFA)/(VMA))^{0.58}} \quad (12)$$

and $|E^*|_m$ is dynamic modulus of HMA in psi; P_c is the aggregate contact volume; VMA is the percentage of mineral aggregate voids in compacted mixture; and VFA is the percentage of voids filled with asphalt in the compacted mixture,

$$|E_m^*| = 3(\frac{100 - VMA}{100})(\frac{(90 + 1.45\frac{|G^*|_b}{VMA})^{0.66}}{1100 + (0.13\frac{|G_b^*|}{VMA})^{0.66}})|G_g^*| \quad (13)$$

where $|E_m^*|$, $|G_b^*|$, and $|G_g^*|$ (the complex shear modulus of binder in the glassy state, assumed to be 10^9 Pa.) are in Pa. Equation (14) shows the best reduced third-order (linear) regression model (PCR) fitting the measured response:

$$\begin{aligned}\hat{y} = &c_0 + c_1pc_1 + c_2pc_2 + c_3pc_3 + c_4pc_4 + c_5pc_5 \\ &+ c_6pc_1pc_2 + c_7pc_1pc_3 + c_8pc_1pc_4 + c_9pc_1pc_5 \\ &+ c_{10}pc_2pc_3 + c_{11}pc_2pc_4 + c_{12}pc_2pc_5 + c_{13}pc_3pc_4 \\ &+ c_{14}pc_3pc_5 + c_{15}pc_4pc_5 + c_{16}pc_1pc_2pc_3 + c_{17}pc_1pc_2pc_4 \\ &+ c_{18}pc_1pc_2pc_5 + c_{19}pc_1pc_3pc_4 + c_{20}pc_1pc_3pc_5 + c_{21}pc_2pc_3pc_4 \\ &+ c_{22}pc_1pc_4pc_5 + c_{23}pc_2pc_4pc_5 + c_{24}pc_3pc_4pc_5\end{aligned} \quad (14)$$

where, $c_0 = 6.59$; $c_1 = 2.58$; $c_2 = 4.4$; $c_3 = -0.36$; $c_4 = 0.49$; $c_5 = 1.93$; $c_6 = -0.33$; $c_7 = -0.77$; $c_8 = -1.69$; $c_9 = 0.15$; $c_{10} = -1.65$; $c_{11} = -4.68$; $c_{12} = 4.81$; $c_{13} = 0.7$; $c_{14} = -0.85$; $c_{15} = -1.58$; $c_{16} = -0.17$; $c_{17} = -0.79$; $c_{18} = 1.83$; $c_{19} = 0.04$; $c_{20} = 0.18$; $c_{21} = 0.42$; $c_{22} = 0.05$; $c_{23} = 0.32$; $c_{24} = 0.06$. The trained three-layer ANN (PCNN) presented in Equation (9) contains the following connection weights and biases:

$$\mathbf{W}^T = \begin{bmatrix} -0.511 & 0.134 & 0.654 & -1.064 & -0.267 \\ -0.315 & -0.147 & -0.267 & 0.177 & -1.047 \\ -0.060 & -1.266 & 0.759 & -1.248 & -0.331 \\ -0.075 & 0.022 & 0.208 & 0.015 & 0.167 \\ -0.074 & 0.022 & 0.206 & 0.015 & 0.165 \\ 0.103 & -0.177 & 1.253 & -1.045 & 0.535 \\ 0.078 & -0.020 & -0.231 & -0.014 & -0.172 \\ 0.238 & 0.070 & -0.885 & 0.848 & 0.943 \\ 0.123 & 0.456 & -0.387 & 1.547 & -0.017 \\ -0.079 & 0.020 & 0.213 & 0.014 & 0.173 \end{bmatrix}$$

$$\mathbf{W}_H = \begin{bmatrix} 0.869 \\ -0.886 \\ 0.632 \\ -0.291 \\ -0.288 \\ -0.859 \\ 0.299 \\ 0.556 \\ 0.971 \\ -0.299 \end{bmatrix}, \ \mathbf{B}_{\mathrm{H}} = \begin{bmatrix} 0.162 \\ 0.710 \\ 0.319 \\ -0.008 \\ -0.009 \\ -0.570 \\ 0.007 \\ 0.290 \\ -0.373 \\ -0.007 \end{bmatrix}, \ \mathbf{B}_0 = [0.148]$$

Figure 6 presents the performance of the developed models in terms of measured values of dynamic modulus versus the fitted dynamic modulus values. The measured and fitted values are fairly close to the line of equality, indicating that the fitted values are highly correlated with the measured ones.

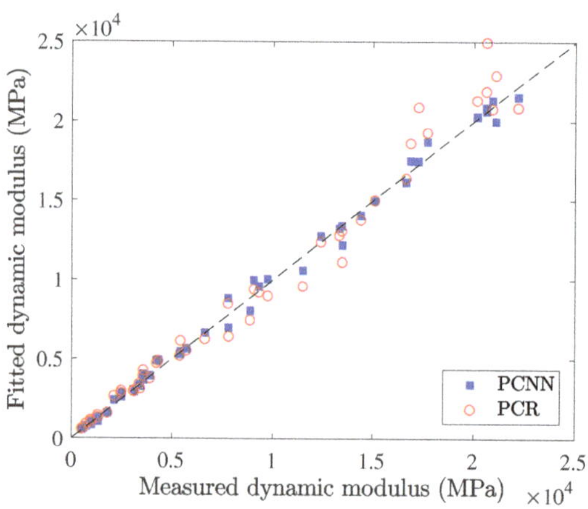

Figure 6. Measured values of dynamic modulus versus fitted values by PCR and PCNN.

Comparisons of PCR and PCNN performance to that of the existing predictive models are conducted based on three statistics: average difference (AD), average absolute difference (AAD), and correlation between measured and fitted values of response (r_{fit}). A summary of the definitions of these statistical components and their formulas is presented in Table 5. In the formulas presented in Table 5, y_i is the ith measured response, $\hat{y}_i$ is the ith fitted response, and n is the number of data points.

The results of the comparison are presented in Table 6. According to the values of r_{fit} in Table 6, the estimated dynamic modulus values obtained form PCR and PCNN models are highly correlated with measured values according to the values, showing that the both PCR and PCNN performed well in terms of modeling the response variable.

Table 5. Statistics which are used to compare model performance.

Statistical Component	Formula	Definition
Average difference (AD)	$AD = \frac{1}{n}\sum_{i=1}^{n}(y_i - \hat{y}_i)$	An estimate of systematic model bias
Average absolute difference (AAD)	$AAD = \frac{1}{n}\sum_{i=1}^{n}\lvert y_i - \hat{y}_i\rvert$	Average closeness of the fitted and measured values of response
r_{fit}	$r_{fit} = \frac{n\sum_{i=1}^{n} y_i\hat{y}_i - (\sum_{i=1}^{n} y_i)(\sum_{i=1}^{n}\hat{y}_i)}{\sqrt{n\sum_{i=1}^{n} y_i^2 - (\sum_{i=1}^{n} y_i)^2}\sqrt{n\sum_{i=1}^{n}\hat{y}_i^2 - (\sum_{i=1}^{n}\hat{y}_i)^2}}$	Correlation of the measured and fitted values of response
Coefficient of determination (R^2)	$R^2 = 1 - \frac{SS_{res}}{SS_{total}}$	Portion of the response variation elucidated by regressors in the fitted model in linear models

Although the corresponding values of r_{fit} for modified Witczak, Hirsch, and Alkhateeb models are 0.93, 0.95, and 0.95, respectively, the average difference and average absolute difference with respect to the measured response are significantly higher than those of PCR and PCNN. This means that the fitted values by the modified Witczak, Hirsch, and Alkhateeb models are not close as those fitted by PCR and PCNN to the response value. In other words, r_{fit}, which reflects the correlation between response and estimated response (if one goes up the other one goes up), could be biased, and in this situations other statistics (AD, and AAD) could be used to evaluate the goodness of fit. The dynamic modulus measured and predicted values are presented in Figure 7 for four asphalt mixtures. According to the presented master curves the current study (PCNN model) provides the closest values of E^* to the measurements for all of three test temperatures, while, the conventional models either overestimate or underestimate the response variable.

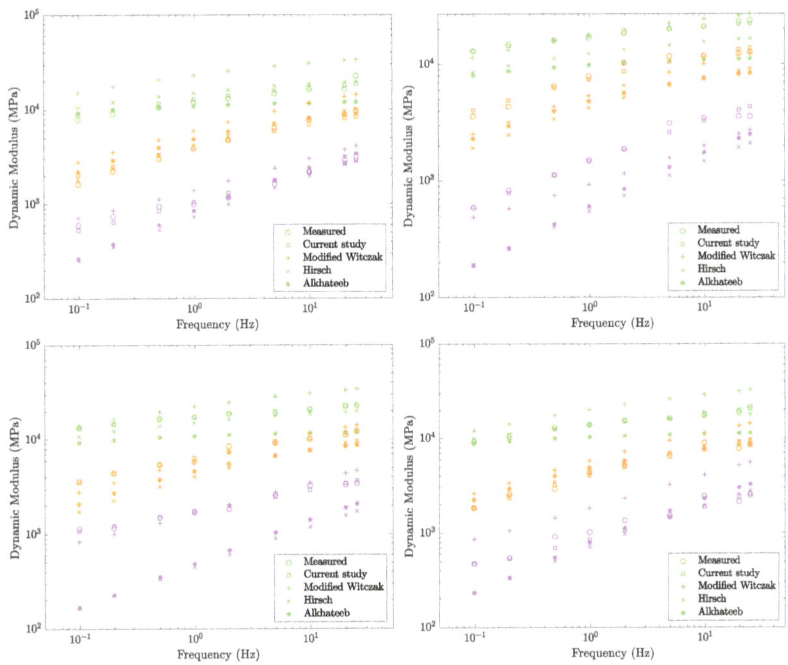

Figure 7. Comparing the measured and predicted (current study and conventional models) dynamic modulus. Results are presented at three temperatures. Green color indicates 0.4 °C, Orange color indicates 17.1 °C, and Purple color indicates 33.8 °C.

Table 6. Performance Comparison of the Developed and Existing Models using Statistical Analysis (na*: not applicable. R^2 is applicable to models with linear parameters and thus it is not applicable to PCNN due to the non-linear nature of its parameters.).

		Average Difference (MPa)	Average Absolute Difference (MPa)	r_{fit}	R^2
PCR	Training	3.9	575.3	0.996	0.99
	Testing	−162.3	718.9	0.995	na
PCNN	Training	13.2	380.7	0.997	na
	Testing	9.7	337.5	0.997	na
Modified Witczak		−2460	3152.1	0.93	0.88
Hirsch		1241.6	1785.7	0.95	0.91
Alkhateeb		2844.5	2984.5	0.95	0.90

A graphical comparison of the PCR and PCNN performance and that of the existing models is presented in the following section.

3.2. Receiver Operating Characteristic Analysis (ROC)

A receiver operating characteristic (ROC) graph is a technique for visualizing, organizing, and selecting classifiers based on their performance. ROC graphs are widely used in medical decision-making as well as in machine learning and data-mining research [35]. True ROC curves plot the false positive rate (probability of false alarm) on the x-axis and the true positive rate (probability of detection) on the y-axis. A classifier is said to perform well if the ROC curve climbs rapidly towards the upper left-hand corner. The more the curve deviates from $y = x$ behavior, the more accurate the prediction is [36]. We can borrow from the concept of ROC curve to obtain a measurement of fit for the competing models and a ROC graph for this study is presented in Figure 8 for this study. As described in Equation (15), the x-axis indicates the standardized residuals ordered from the lowest to the highest (e_i^{**}). Residuals are sorted in ascending order and divided by the largest one that belongs to the Hirsch model.

$$e_i^{**} = \frac{|e_i^*|}{|e_{i_{MAX}}^*|} \tag{15}$$

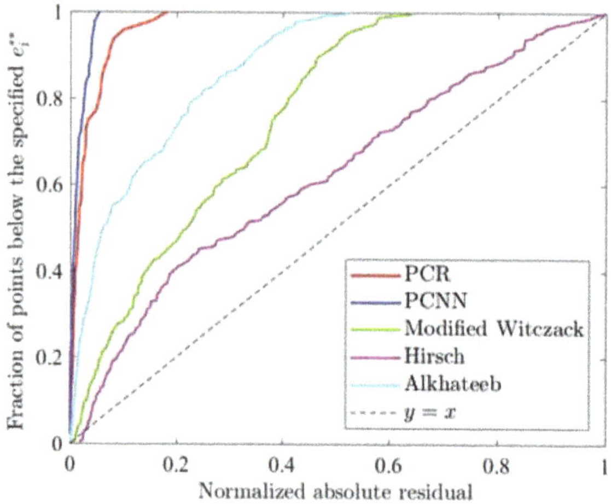

Figure 8. ROC curves for the developed and the existing predictive models for the dynamic modulus. The curves indicate the goodness of the fit provided by the models with regard to a pre-specified residual. The PCNN model showed the highest performance (farthest from the $y = x$ line), and the Hirsch model showed the lowest (closest to the $y = x$ line).

The y-axis indicates the fraction of points whose standardized residuals are less than e_i^{**} [37]. Although the curves obtained for all of the models are monotonically non-decreasing and climb towards the upper left-hand corner (the desired situation shows that the predictive models perform well), PCNN and PCR curves were the highest, proving their better performance to be better than that of existing predictive models.

A convenient global measure of the goodness-of-the-fit is the Area Under the Curve (AUC). To compare classifiers, it is more desirable to reduce ROC performance, to a single scalar value representing expected performance. Since the AUC is a portion of the area of the unit square, its value will always lie between 0 and 1. The AUC values for the PCNN, PCR, Alkhateeb, modified Witczak, and Hirsch models are 0.9864, 0.9717, 0.8746, 0.7609, and 0.6320, respectively. One can use the ROC and AUC analysis results and rank the predictive models according to their performances. In this study, the PCNN model reflected the highest performance in predicting the dynamic modulus value, while the Hirsch model ranked the lowest among all the models.

3.3. Model Validation

The current regression model is presented in the following general form as

$$y_i = f_i(\mathbf{Z}_\mathrm{i}, \boldsymbol{\theta}) + e_i^*. \tag{16}$$

In the above equation f_i is the ith expectation function, $\boldsymbol{\theta}$ is the vector of parameters, and e_i^* is a random deviation of y_i from f_i. This term is assumed to be independent and normally distributed with a mean of zero and unknown variance σ^2 for $i = 1, \cdots, n$, where n is the number of input vectors. If the above assumptions are violated, the results of the analysis could be misleading or erroneous. These assumptions can be testified by examining residuals as defined by

$$e_i^* = y_i - \hat{y}_i. \tag{17}$$

The assumption of independency holds when the residuals plot does not reflect a trivial pattern. The normality assumption is assessed by creating a normal probability plot of the residuals. When the error has a normal distribution, this plot will appear as a straight line [10]. These assumptions were checked for PCR and PCNN, as presented in Figure 9. The assumption of equal variances does not appear to be violated because there are no trivial pattern in this plot. Figure 9 presents the normal probability of the residuals in which it can be seen that the data points are close to the straight line and the normality assumption is validated.

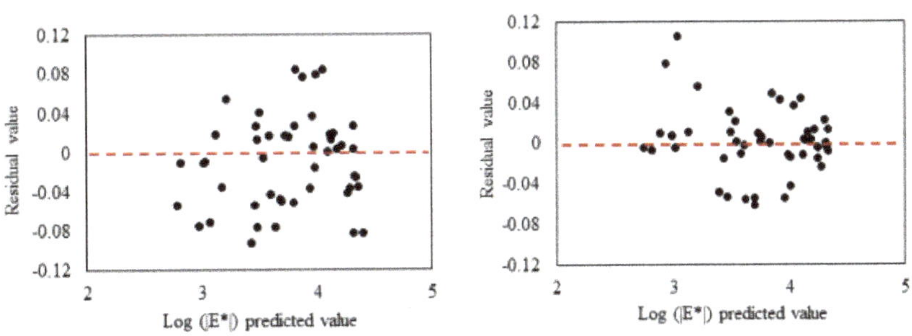

Figure 9. *Cont.*

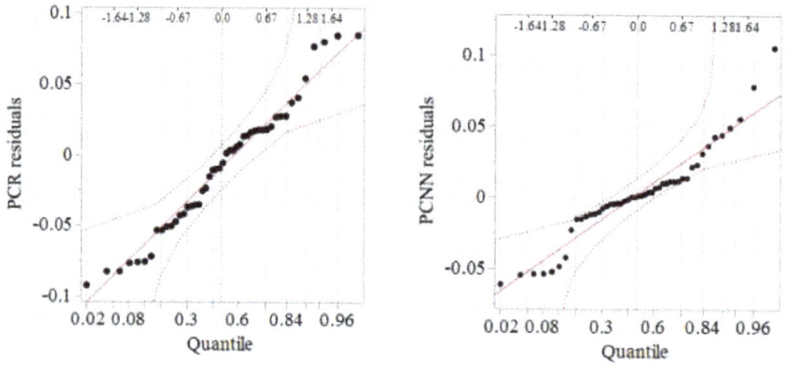

Figure 9. Checking the assumptions of independency using residual plot for PCR (**left**) and PCNN (**right**) models and the normality using normal probability plot of the residuals for PCR (**left**) and PCNN (**right**) models.

4. Application of the Framework: Flexible Pavement Design and Optimization

The above framework is used along with an optimization algorithm to answer the following two central questions:

- what design parameters result in the maximum $|E^*|$?
- what design parameters result in a pre-specified $|E_0^*|$?

One can see that the first item corresponds to the optimal design problem while the second one corresponds to the so-called inverse design.

Since it was shown through multiple statistical measurements that PCNN had the best prediction capability, this model is used in the following section to solve the optimization problems. The ANN used in PCNN is essentially an interconnected nonlinear function, and this necessitates the application of a global optimizer. Moreover, the effective variable space enters the problem as a series of constraints and further restricts the available algorithms. The optimal design problem is formulated as follows:

$$\begin{aligned} \text{maximize} \quad & |E^*| = F_{ANN}(\mathbf{x}) \\ \text{with respect to} \quad & \mathbf{x} = (x_1, \ldots, x_{14}) \\ \text{subject to} \quad & (\mathbf{x} - \mathbf{v})^T A(\mathbf{x} - \mathbf{v}) \leq \mathbf{1}, \\ & (\mathbf{x}_{pca} - \mathbf{v}')^T A'(\mathbf{x}_{pca} - \mathbf{v}') \leq \mathbf{1}, \end{aligned} \tag{18}$$

where the vector of fourteen variables is $\mathbf{x}$, and $(\mathbf{x} - \mathbf{v})^T A(\mathbf{x} - \mathbf{v}) \leq \mathbf{1}$ are the enclosing ellipsoid constraint equations for the original and PCA-based variables. A penalty function approach is used to convert the above constrained problem to an unconstrained one [38]. In this case, when the penalty function is active, it decreases (increases) the objective function when the problem is one of maximization (minimization), and the degree of penalty is based on the closeness of the solution to the corresponding constraint.

Since the inverse design problem aims at finding the specification of a predefined goal, it is defined as a minimization problem as follows:

$$\begin{aligned} \text{minimize} \quad & error = ||E^*| - |E_0^*|| \\ \text{with respect to} \quad & \mathbf{x} = (x_1, \ldots, x_{14}) \\ \text{subject to} \quad & (\mathbf{x} - \mathbf{v})^T A(\mathbf{x} - \mathbf{v}) \leq \mathbf{1}, \\ & (\mathbf{x}_{pca} - \mathbf{v}')^T A'(\mathbf{x}_{pca} - \mathbf{v}') \leq \mathbf{1}, \end{aligned} \tag{19}$$

where $|E_0^*|$ is the desired (goal) dynamic modulus. Although a similar penalization method can also be used to address the constraints in this case, for the above problem the constraints will penalize the objective function when they are active.

Reliable solution of the above problems requires the application of a gradient-free optimization algorithm. Gradient-based optimization algorithms are not applicable in this case because of the network-based nature of the ANNs. Evolutionary-based algorithms are potentially easy-to-use algorithms in the above problems. Novel algorithms have been used to solve complex optimization problems in recent years [39,40], and in this case, Mean-Variance Mapping Optimization (MVMO), an in-house optimization algorithm based on the work by Elrich et al. [41,42], is used. The constraints are handled using the approach described in Aslani et al. [43], in which, the convergence rate of a constrained MVMO was compared to the already-developed methodologies using benchmark structural problems. Authors in [22] indicated that a constrained MVMO is capable accurately identifying an optimal value with a minimum number of simulations. It should be noted that the choice of optimization algorithm is not the principal focus of this study.

Figure 10 (left) depicts the convergence achieved for the first design problem by the constrained MVMO algorithm. The initial data points are random making it heavily penalized, and then the objective function increases as the algorithms evolves. Exploration-exploitation behavior is achieved using adaptive strategies in the course of optimization for MVMO. $\delta = 0.05$ is used as the threshold in Figure 4. Solving the maximization problem resulted in $|E_{max}^*|$= 53,703 MPa. The optimal design parameters are presented in the first column of Table 7.

To find the maximum amount of dynamic modulus one could design for without low temperature failure in the asphalt binder, the maximization problem was solved one more time with an additional constraint of $G^* sin\delta \leq 5000$, resulting in $|E_{max}^*|$= 36,307 MPa. Corresponding design parameters are presented in the second column of Table 7 as the optimal design 2.

Figure 10 (right) shows the convergence of the algorithm for the inverse design problem after starting randomly from three different initial points, with the algorithm is terminated when the error reaches about 10^{-9}. A pre-specified $|E_0^*|$ of 20,417 MPa is considered and the inverse problem of finding the corresponding design parameters is solved. Because of non-linearity of the function, the problem has no unique solution. Three of the possible solutions are presented as designs 1 to 3 in Table 7.

Finally, the five sets of design parameters are compared with current design specification, with the results shown in Table 7. The percentage of aggregate passing by each sieve size is within the acceptable range of the gradation specification. Gradation charts are presented in Figure 11. The obtained percentages of air voids are 4%, which is the target value in the design specification. The obtained values for VMA are slightly less than 14% for a nominal maximum aggregate size (NMAS) of 12.5 mm because the VMA values of the nine mixtures used to train the PCNN are slightly less that 14% (see Table 1). The acceptable range for VFA varies with the amount of traffic load measured in million Equivalent Single Axle Loads (ESALs) as follows:

- traffic loading $< 0.3 \rightarrow 70 <$ VFA < 80
- $0.3 <$ traffic loading $< 3.0 \rightarrow 65 <$ VFA < 78
- traffic loading $> 3.0 \rightarrow 65 <$ VFA < 75

The VFAs obtained for all of the five sets of design are satisfied for all of the traffic categories.

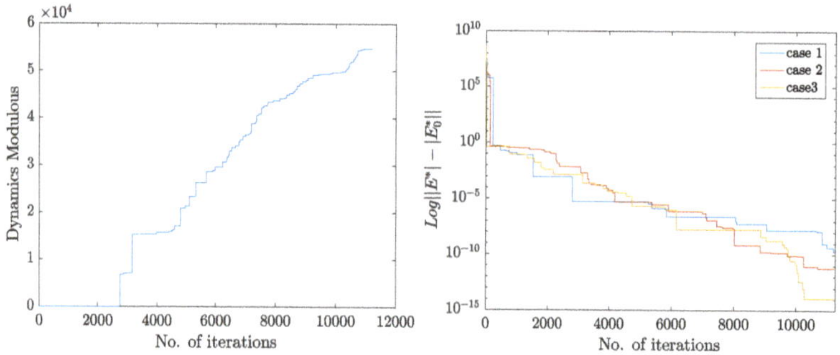

Figure 10. Convergence for the optimal design problem (**left**) and inverse (**right**) design of dynamic modulus.

Table 7. Th Corresponding Design Parameters Obtained from Solving Optimal Design and Inverse Design Problems.

Identity	Optimal Design 1	Optimal Design 2	Design 1	Design 2	Design 3	Design Specification			
						Control Points		Restricted Zone	
						Lower	Upper	Lower	Upper
%Passing from 3/4″	100	100	100	100	100	-	100	-	-
%Passing from 1/2″	93.38	94.03	92.25	91.88	91.80	90	100	-	-
%Passing from 3/8″	81.74	81.72	79.57	79.92	80.70	-	90	-	-
%Passing from #4	53.00	53.90	55.36	55.23	54.39	-	-	-	-
%Passing from #8	39.56	40.51	41.37	41.08	40.92	28	58	39.1	39.1
%Passing from #30	20.75	20.68	21.02	20.87	20.83	-	-	19.1	23.1
%Passing from #50	11.66	11.60	12.08	11.81	12.02	-	-	15.5	15.5
%Passing from #100	6.22	6.21	6.52	6.38	6.40	-	-	-	-
%Passing from #200	4.10	3.85	4.38	4.58	4.56	2	10	-	-
G* (Mpa)	103.13	7.81	133.51	30.20	11.82	-	-	-	-
Phase angle (degree)	35.71	39.60	47.69	47.27	44.77	2	8	-	-
Vbeff%	4.11	4.18	4.02	4.06	4.05	-	-	-	-
VMA	13.47	13.56	13.41	13.45	13.44	-	-	-	-
VFA	70.29	70.50	70.11	70.24	70.24	-	-	-	-
Va%	4.00	4.00	3.99	4.00	4.01	4		-	-

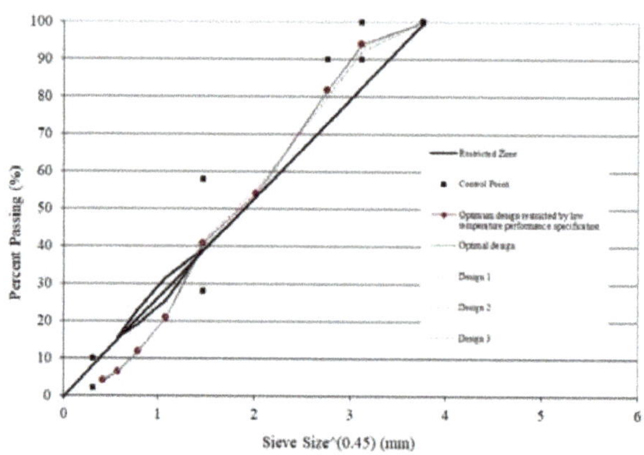

Figure 11. Aggregate gradation graphs with 12.5 mm NMAS particle size distribution obtained from PCNN.

5. Conclusions

This study used the HMA dynamic modulus data and focused to evaluate the quality of predictor variables to be used in a procedure of model development. Correlation analysis is performed to identify cross-correlated input variables, and correlated inputs are replaced by orthogonal pseudo-inputs (PCs) obtained using PCA. Two separate models are developed using multivariate regression and ANN (called PCR and PCNN, respectively). Extrapolation in empirical modeling is addressed by adding the constraint of an n-dimensional enclosing ellipsoid to the modeling problem. Performances of the proposed models were compared to existing predictive models using both statistical analysis and ROC analysis. The models developed satisfactorily estimated the dynamic modulus value, with PCNN indicating remarkably better performance when fitted to the test data than the existing predictive models from the literature. These PCA-based approaches are thus highly recommended as precise modeling strategies in this application. Moreover, these methodologies appear to be capable of modeling other material properties and future investigation in this regard is recommended. To determine this framework's application in pavement design, two optimization problems including optimal design and inverse design have been presented and solved using a mean-variance mapping optimization algorithm. The results for the two problems are in a good agreement with the HMA mix design specification and thus could be a reasonable starting point in solving real-life design problems. Although, the developed models as well as obtained optimal design parameters are based on the empirical database created in this study, the suggested framework has the capability of being re-trained and adjusted to fit new data. For obtaining more reliable and applicable results, a larger empirical database would be required.

Author Contributions: Conceptualization, P.G. and M.A.; methodology, P.G., M.A., and D.K.R.; software, P.G. and M.A.; validation, P.G., M.A., D.K.R., and R.C.W. formal analysis, P.G., M.A., D.K.R., and R.C.W.; investigation, P.G., M.A., D.K.R., and R.C.W.; resources, P.G., M.A., D.K.R., and R.C.W.; data curation, P.G., M.A., D.K.R., and R.C.W.; writing–original draft preparation, P.G., M.A., D.K.R., and R.C.W.; writing–review and editing, P.G., M.A., D.K.R., and R.C.W.; visualization, P.G., M.A., D.K.R., and R.C.W.; supervision, P.G., M.A., D.K.R., and R.C.W.; project administration, R.D.W.

Funding: This research received no external funding.

Acknowledgments: The authors would like to thank anonymous reviewers for their fruitful comments on the manuscript.

Conflicts of Interest: The authors declare no conflict of interest.

References

1. Ghasemi, P. Performance Evaluation of Coarse-Graded Field Mixtures Using Dynamic Modulus Results Gained from Testing in Indirect Tension Mode of Testing. Graduate Theses and Dissertations 16717. Available online: https://lib.dr.iastate.edu/etd/16717 (accessed on 1 August 2019).
2. Birgisson, B.; Roque, R.; Kim, J.; Pham, L.V. *The Use of Complex Modulus to Characterize the Performance of Asphalt Mixtures and Pavements in Florida*; Technical Report; Florida Department of Transportation: Tallahassee, FL, USA, 2004.
3. Arabali, P.; Sakhaeifar, M.S.; Freeman, T.J.; Wilson, B.T.; Borowiec, J.D. Decision-making guideline for preservation of flexible pavements in general aviation airport management. *J. Transp. Eng. Part B Pavements* **2017**, *143*, 04017006. [CrossRef]
4. Bozorgzad, A.; Lee, H.D. Consistent distribution of air voids and asphalt and random orientation of aggregates by flipping specimens during gyratory compaction process. *Constr. Build. Mater.* **2017**, *132*, 376–382. [CrossRef]
5. Guide, N.D. *Guide 1-37A, Guide for Mechanistic-Empirical Design of New and Rehabilitated Pavement Structures, National Cooperative Highway Research Program*; Transportation Research Board, National Research Council: Washington, DC, USA, 2004.
6. AASHTO, A. *Mechanistic-Empirical Pavement Design Guide: A Manual of Practice*; AAoSHaT Officials (American Association of State Highway and Transportation Officials): Washington, DC, USA, 2008.
7. Nobakht, M.; Sakhaeifar, M.S. Dynamic modulus and phase angle prediction of laboratory aged asphalt mixtures. *Constr. Build. Mater.* **2018**, *190*, 740–751. [CrossRef]
8. Peng, C.; Feng, J.; Feiting, S.; Changjun, Z.; Decheng, F. Modified two-phase micromechanical model and generalized self-consistent model for predicting dynamic modulus of asphalt concrete. *Constr. Build. Mater.* **2019**, *201*, 33–41. [CrossRef]

9. Shu, X.; Huang, B. Dynamic modulus prediction of HMA mixtures based on the viscoelastic micromechanical model. *J. Mater. Civ. Eng.* **2008**, *20*, 530–538. [CrossRef]
10. Devore, J.L. *Probability and Statistics for Engineering and the Sciences*; Cengage Learning: Boston, MA, USA, 2011.
11. El-Badawy, S.; Abd El-Hakim, R.; Awed, A. Comparing Artificial Neural Networks with Regression Models for Hot-Mix Asphalt Dynamic Modulus Prediction. *J. Mater. Civ. Eng.* **2018**, *30*, 04018128. [CrossRef]
12. Andrei, D.; Witczak, M.; Mirza, M. Development of a revised predictive model for the dynamic (complex) modulus of asphalt mixtures. In *Development of the 2002 Guide for the Design of New and Rehabilitated Pavement 451 Structures*; NCHRP: Washington, DC, USA, 1999.
13. Bari, J.; Witczak, M. New predictive models for viscosity and complex shear modulus of asphalt binders: For use with mechanistic-empirical pavement design guide. *Transp. Res. Rec. J. Transp. Res. Board* **2007**, *2001*, 9–19. [CrossRef]
14. Christensen, D., Jr.; Pellinen, T.; Bonaquist, R. Hirsch model for estimating the modulus of asphalt concrete. *J. Assoc. Asph. Paving Technol.* **2003**, *72*, 97–121.
15. Jamrah, A.; Kutay, M.E.; Ozturk, H.I. *Characterization of Asphalt Materials Common to Michigan in Support of the Implementation of the Mechanistic-Empirical Pavement Design Guide*; Technical Report; Transportation Research Board: Washington, DC, USA, 2014.
16. Al-Khateeb, G.; Shenoy, A.; Gibson, N.; Harman, T. A new simplistic model for dynamic modulus predictions of asphalt paving mixtures. *J. Assoc. Asph. Paving Technol.* **2006**, *75*, 1254–1293.
17. Sakhaeifar, M.S.; Richard Kim, Y.; Garcia Montano, B.E. Individual temperature based models for nondestructive evaluation of complex moduli in asphalt concrete. *Constr. Build. Mater.* **2017**, *137*, 117–127. [CrossRef]
18. Ghasemi, P.; Aslani, M.; Rollins, D.K.; Williams, R.C.; Schaefer, V.R. Modeling rutting susceptibility of asphalt pavement using principal component pseudo inputs in regression and neural networks. *Int. J. Pavement Res. Technol.* **2018**. [CrossRef]
19. Ren, R.; Han, K.; Zhao, P.; Shi, J.; Zhao, L.; Gao, D.; Zhang, Z.; Yang, Z. Identification of asphalt fingerprints based on ATR-FTIR spectroscopy and principal component-linear discriminant analysis. *Constr. Build. Mater.* **2019**, *198*, 662–668. [CrossRef]
20. Fodor, I.K. *A Survey of Dimension Reduction Techniques*; Technical Report; U.S. Department of Energy: Washington, DC, USA, 2002.
21. Johnson, R.A.; Wichern, D.W. *Applied Multivariate Statistical Analysis*; Prentice-Hall: Upper Saddle River, NJ, USA, 2014; Volume 4.
22. Ghasemi, P.; Aslani, M.; Rollins, D.K.; Williams, R. Principal component analysis-based predictive modeling and optimization of permanent deformation in asphalt pavement: Elimination of correlated inputs and extrapolation in modeling. *Struct. Multidiscip. Optim.* **2018**, *59*, 1335–1353. [CrossRef]
23. Kim, Y.R.; Underwood, B.; Far, M.S.; Jackson, N.; Puccinelli, J. *LTPP Computed Parameter: Dynamic Modulus*; Technical Report; Federal Highway Administration Research and Technology: Washington, DC, USA, 2011.
24. Sakhaeifar, M.S.; Richard Kim, Y.; Kabir, P. New predictive models for the dynamic modulus of hot mix asphalt. *Constr. Build. Mater.* **2015**, *76*, 221–231. [CrossRef]
25. Rollins, D.K.; Zhai, D.; Joe, A.L.; Guidarelli, J.W.; Murarka, A.; Gonzalez, R. A novel data mining method to identify assay-specific signatures in functional genomic studies. *BMC Bioinf.* **2006**, *7*, 377. [CrossRef]
26. Jolliffe, I.T. Principal Component Analysis, Second Edition. *Encycl. Stat. Behav. Sci.* **2002**, *30*, 487,
27. Kuźniar, K.; Waszczyszyn, Z. Neural networks and principal component analysis for identification of building natural periods. *J. Comput. Civ. Eng.* **2006**, *20*, 431–436. [CrossRef]
28. Hua, X.; Ni, Y.; Ko, J.; Wong, K. Modeling of temperature–frequency correlation using combined principal component analysis and support vector regression technique. *J. Comput. Civ. Eng.* **2007**, *21*, 122–135. [CrossRef]
29. Kartam, N. Neural Netwroks in Civil Engineering: Systems and Application. *J. Comput. Civ. Eng.* **1994**, *8*, 149–162.
30. Sanabria, N.; Valentin, V.; Bogus, S.; Zhang, G.; Kalhor, E. *Comparing Neural Networks and Ordered Probit Models for Forecasting Pavement Condition in New Mexico*; Technical Report; Transportation Research Board: Washington, DC, USA, 2017.

31. Gong, H.; Sun, Y.; Mei, Z.; Huang, B. Improving accuracy of rutting prediction for mechanistic-empirical pavement design guide with deep neural networks. *Constr. Build. Mater.* **2018**, *190*, 710–718. [CrossRef]
32. Cheng, B.; Titterington, D. Neural networks: A review from a statistical perspective. *Stat. Sci.* **1994**, *9*, 2–30. [CrossRef]
33. Kutner, M.H.; Nachtsheim, C.; Neter, J. *Applied Linear Regression Models*; McGraw-Hill/Irwin: Chicago, IL, USA, 2004.
34. Todd, M.J.; Yıldırım, E.A. On Khachiyan's algorithm for the computation of minimum-volume enclosing ellipsoids. *Discret. Appl. Math.* **2007**, *155*, 1731–1744. [CrossRef]
35. Fawcett, T. An introduction to ROC analysis. *Pattern Recognit. Lett.* **2006**, *27*, 861–874. [CrossRef]
36. Bi, J.; Bennett, K.P. Regression error characteristic curves. In Proceedings of the 20th International Conference on Machine Learning (ICML-03), Washington, DC, USA, 21–24 August 2003; pp. 43–50.
37. Marti-Vargas, J.R.; Ferri, F.J.; Yepes, V. Prediction of the transfer length of prestressing strands with neural networks. *Comput. Concr.* **2013**, *12*, 187–209. [CrossRef]
38. Rahami, H.; Kaveh, A.; Aslani, M.; Asl, R.N. A hybrid modified genetic-nelder mead simplex algorithm for large-scale truss optimization. *Int. J. Optim. Civ. Eng.* **2011**, *1*, 29–46.
39. He, J.; Yao, X. From an individual to a population: An analysis of the first hitting time of population-based evolutionary algorithms. *IEEE Trans. Evol. Comput.* **2002**, *6*, 495–511.
40. Cai, Z.; Cai, Z.; Wang, Y. A Multiobjective Optimization-Based Evolutionary Algorithm for Constrained Optimization. *IEEE Trans. Evol. Comput.* **2006**, *10*, 658–675. [CrossRef]
41. Rueda, J.L.; Erlich, I. Testing MVMO on learning-based real-parameter single objective benchmark optimization problems. In Proceedings of the 2015 IEEE Congress on Evolutionary Computation (CEC), Sendai, Japan, 25–28 May 2015; pp. 1025–1032.
42. Erlich, I.; Venayagamoorthy, G.K.; Worawat, N. A mean-variance optimization algorithm. In Proceedings of the IEEE Congress on Evolutionary Computation, Barcelona, Spain, 18–23 July 2010; pp. 1–6.
43. Aslani, M.; Ghasemi, P.; Gandomi, A.H. Constrained mean-variance mapping optimization for truss optimization problems. *Struct. Des. Tall Spec. Build.* **2018**, *27*, e1449. [CrossRef]

infrastructures

Article

Synthetic Aggregates for the Production of Innovative Low Impact Porous Layers for Urban Pavements

Piergiorgio Tataranni * and Cesare Sangiorgi

Department of Civil, Chemical, Environmental, and Materials Engineering, University of Bologna, 40136 Bologna, Italy
* Correspondence: piergiorg.tataranni2@unibo.it

Received: 28 June 2019; Accepted: 3 August 2019; Published: 6 August 2019

Abstract: According to the latest estimates, 40% of urban areas are covered by pavements. One of the most remarkable effects on the urban environment is the increase in impermeable surfaces which leads to problems related to water infiltration into the ground and the increase in wash-off volumes. The use of permeable and porous layers in urban applications for cycle lanes, footpaths and parking areas is growing in interest, increasing the potential for control and management of urban runoff. In this paper, a physical and mechanical characterization is proposed of an innovative mixture, prepared with a polymeric transparent binder for semi-porous layers with reduced contribution to the urban heat island effect. Two versions of this mixture are compared, one with just virgin and the one with artificial synthetic aggregates, produced through the alkali-activation of waste basalt powder. Results show suitable properties for both materials if compared to porous asphalt concretes in traditional pavements. Furthermore, the application of synthetic aggregates seems to be a viable solution for the production of innovative and eco-friendly mixtures, allowing the recycling of waste materials.

Keywords: permeable pavements; porous asphalt concrete; polymeric transparent binder; synthetic aggregates; urban pavements; urban runoff

1. Introduction and Objective

According to the latest estimates, the expansion levels imposed by modern society will involve a dramatic growth in the size of cities by the middle of this century [1]. Today, urban areas are formed by multiple paved areas that facilitate people's daily mobility in different ways. From this perspective, pavements can no longer be considered as simple structures to connect places, since they are composed of roads, special lanes, bike lanes, footpaths and squares in which people spend a great part of their life. This rapid development presents challenges in terms of urban and environmental planning, considering that today 40% of urban areas are covered by pavements. This urban land-cover change is affecting the local ecosystems and the underlying surface conditions, resulting in an important increase in the portion of impervious surfaces. The natural hydrological process may be irreversibly changed [2,3], as a consequence of limited rainwater infiltration and the increased surface runoff [4]. Studies have demonstrated a clear correlation between increased flooding hazard registered in urban areas and the changes in the natural drainage network [5,6]. The growth of impervious surfaces has increased the runoff peaks as well as the stormwater runoff accumulation. From an environmental perspective, these impervious areas promote the stagnation of water containing urban pollutants such as road-deposited sediments and potentially toxic metals and metalloids. Studies demonstrated that the presence of contaminants is mainly related to the different land use: high quantities of polycyclic aromatic hydrocarbons were detected in some commercial sites, due to higher vehicle concentrations. Furthermore, heavy metals characterize the stormwater runoff from industrial land-use areas [7,8]. The presence of these pollutants can be considered as one of the primary contributors to water quality depletion in natural water bodies [9]. By 2030 the urban land cover will increase by 1.2 million km^2 and according to recent studies, the global urban land cover is expected to be over 200% if compared to year 2000 [10].

Engineers from all over the world are facing the new challenges in managing the urban planning for a sustainable future. The importance of taking decisive actions to tackle these problems is obvious. From an environmental point of view, Best Management Practices (BMPs) are today widely applied as runoff control system in urban areas in order to reduce the so-called nonpoint source (NPS) pollution [11]. BMPs promote the development of detention basins and extended detention basins, which act as storage and water treatment areas. The latest research has highlighted the performance of innovative "smart" BMPs improved with a real-time active control of the stormwater detention basin outflows, able to achieve up to 90% of pollutant removal and considerably reduce the outflows volumes [12].

As for the increased wash-off volumes in urban areas, some countries have developed a series of guidelines and stormwater management plans and proposed some specific techniques to monitor the urban flooding. To date, the most effective method to control urban runoff is based on the urban design that promotes the development of permeable pavements and surfaces and green areas. This approach is traditionally counted among the infiltration-based technologies to control the urban runoff [13].

Permeable pavements can be considered as a suitable and sustainable alternative to traditional pavements produced with common asphalt or cement concrete. Considering that paved surfaces represent around the 25% of impervious urban surfaces, the possibility of using porous pavements can be effective in controlling the urban runoff [14,15]. Several studies demonstrated that porous pavements are considerably more effective in reducing the wash-off volumes if compared to drainage surfaces [16]. Moreover, the efficiency of these permeable surfaces is highlighted considering their contribution in decreasing the flood peak and its hysteresis, which is generally related to the thickness of the porous structure [17]. Starting from the traditional porous asphalts or concretes, some now materials such as modular elements, paving blocks or plastic grid system are today widely used in urban areas to create permeable areas and structures [18,19].

The porous layers are also effective against the Urban Heat Island (UHI) effect [20]. This phenomenon is another consequence of the dramatic development of urbanization and is evaluated as the overheating of urban temperatures compared to the relatively colder conditions of suburban zones and rural areas [21]. The traditional materials used for pavements and roofs absorb and store most of the solar energy falling on their surface during the day, which is then released in the form of heat during night-time. The albedo is the measure of the sunlight reflection of a surface out of the total solar radiation, and it ranges between zero (corresponding to a full absorption) and 100 (representing a completely reflective surface). The dark surfaces that traditionally distinguish the road pavements and roofs are characterized by a sunlight reflection up to 20% [22]. Several studies and applications have been carried out to face the UHI through the adoption of innovative materials or simply increasing the solar reflection of surfaces [23,24]. Researches verified the mitigation of UHI through the use of porous mixtures and light-colored surfaces [25].

In the case study presented in this paper, a semi-porous mixture prepared with a polymeric transparent binder and a pale limestone aggregate is proposed. This innovative and eco-friendly material has the dual target of reducing urban runoff and UHI, by coupling a porous structure with a light-colored pavement surface. Moreover, the proposed material can be used as surface layer for the construction of a water-retaining pavement, which is a novel type of cool pavement [26]. In this case, the reduction in the pavement temperature is promoted by the evaporation cooling of water stored in the thickness of the layer.

Furthermore, the sustainability impact of this material is optimized by the partial substitution of natural aggregates with artificial synthetic aggregate produced through the alkali-activation of waste basalt powder.

2. Materials and Methods

After reviewing some Italian technical specifications for semi-porous asphalt concretes, two mixtures were studied: one made with 100% pale limestone aggregates (SPT) and one with the replacement by 21% of virgin material with synthetic aggregates (SPS) produced through the alkali-activation of basalt waste powder.

2.1. Polymeric Binder

The polymeric material used as binder is a commercial single-component polyolefin resin sold in solid form (chips). It is used to produce conglomerates for low visual impact pavements to preserve the aesthetic of rural, residential or historical areas. Compared to other colored asphalt concretes, no pigment or additive is added to the original mixture. In this case the binder is transparent, so the final color of the material will be the given by the natural color of the aggregates used for the mix design. This allow to perfectly integrate the pavement in the surrounding landscape.

In operational terms, the binder is added at ambient temperature into the drum mix of a common asphalt plant, where it melts using the aggregates' heat. Once mixed, the material is laid down and compacted with the same technologies and methodologies adopted for common asphalt concretes. The properties of the binder are presented in Table 1 where the technical parameters are assessed in compliance with tests traditionally used for the characterization of bitumen for asphalt concretes.

Table 1. Technical properties of the polymeric binder.

Technical Features	Typical Values	Reference Standard
Dynamic viscosity at 160 °C	>700 mPa·s	EN 13702
Penetration at 25 °C	25–55 dmm	EN 1426
Softening point	>75 °C	EN 1427
Flash point	>220 °C	DIN 51755, ASTM D92,D93
Density at 25 °C	850 kg/m^3	EN ISO 3838

2.2. Synthetic Aggregates

The synthetic artificial aggregates were produced using the alkali-activation process. This is a chemical reaction that is established between two groups of materials, namely precursors and activators, with specific chemical properties. The former are materials generally rich in silica and alumina, the latter are needed to promote the chemical reaction. The result of the synthesis process is the development of a hardened binder based on a combination of hydrous alkali-aluminosilicate and/or alkali-alkali earth-aluminosilicate phases [27–29]. In the work presented in this paper, a specific mix of waste basalt powder and metakaolin was used as the precursor.

The basalt powder is a waste material completely passing the 50 μm sieve resulting from the extraction and production processes in a basalt quarry. The use of basalt is not new for the alkali-activated materials, considering that its chemical properties are suitable for the synthesis reaction [30].

The metakaolin is obtained by heat treatment (around 700 °C) of a natural kaolin and its application to produce alkali-activated materials is a common practice. It is today well-known the improvement in the mechanical properties of the final mixture conferred by the presence of metakaolin [31]. Table 2 summarizes the chemical composition of both basalt and metakaolin powders.

The activators are needed to create a high alkaline ambient to develop the synthesis of the material. In the case under study a liquid compound of Sodium Silicate (SiO_2/Na_2O ratio of 1.99) and Sodium Hydroxide (10 M), mixed according to a specific dosage (SS/SH = 4), was used as the activator. After the mixing phase of precursor and activators, the material was handcrafted to produce granular samples and then cured for 12 hours in an oven at 60 °C. Some mechanical and physical tests were carried out on the synthetic paste and aggregates in order to characterize the material.

As suggested by the well-established literature in the alkali-activated material field, the evaluation of the mechanical properties was based on the assessment of the compressive strength of the alkali-activated mixture. Considering the absence of specific test procedures for the mechanical analysis of this type of synthetic material, the EN 1015-11 [32] standard was taken as reference. According to the standard, the compressive strength of a hardened mortar is evaluated applying a load without shock and increasing it continuously at a rate within the range 50 N/s to 500 N/s so that failure occurs within a period of 30 to 90 second. The maximum load is registered and the compressive strength is than calculated. Furthermore, in order to evaluate the development of the mechanical properties of the mixture during the curing period, the compressive strength tests were repeated after 3, 7, 14, 21 and 28 days after the mixing and casting procedures. In compliance with the standard, tests were carried out on 40 × 40 × 40 mm cubic samples.

Table 2. Chemical properties of basalt and metakaolin.

Compound	Unit	Basalt	Metakaolin
SiO_2	% p/p	45.3	55.2
CaO	% p/p	8.8	0.2
Na_2O	% p/p	1.7	0.6
Al_2O_3	% p/p	21.6	40.3
Fe_2O_3	% p/p	8.5	1.4
SO_3	% p/p	<0.1	0.2
MgO	% p/p	2.0	0.1
P_2O_5	% p/p	0.7	<0.1
TiO_2	% p/p	0.2	1.5
ZnO	% p/p	<0.1	<0.1
K_2O	% p/p	9.7	0.2

Four samples were tested for each test repetition and the average results are presented in Figure 1.

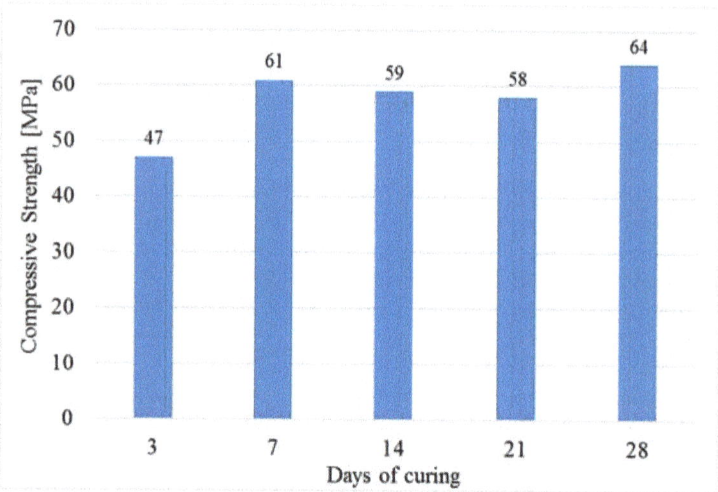

Figure 1. Compressive strength on alkali-activated mixture samples after 3, 7, 14, 21 and 28 days of curing.

Results in Figure 1 highlighted the achievement of a considerable compressive strength (47 MPa) just after 3 days of curing. From day 7 to day 28, there is visible a slight variation in results, ranging from 58 to 65 MPa. This phenomenon is quite common for the alkali-activated materials and it is mainly related to the high influence of the mixing and casting procedures on the final properties of the material. Hence, the curing procedure can be considered ended after 7 days and the compressive strength achieved is considerably high.

2.3. Experimental Program

The experimental program was divided in several laboratory phases, with the final aim of defining the physical, mechanical and functional properties of the two semi-porous mixtures.

Taking as a reference Italian technical specifications for semi-porous asphalt concretes, the physical characterization was based on the evaluation of the air voids content for samples prepared through gyratory compaction in accordance with the EN 12697-31 [33] standard (compaction pressure of 600 kPa and 80 gyrations). The physical analysis was then supported by means of the Indirect Tensile Strength (ITS, EN 12697-23) [34] test carried out at 25 °C. The mechanical characterization was based on evaluation of dynamic behavior using the Indirect Tensile Stiffness Modulus (ITSM, EN 12697-26 annex C) [35] test. This was repeated at 10, 20 and 30 °C to check the thermal sensitivity of the two mixtures. Considering the porous structure of the material, the durability was assessed in terms of reduction of ITS (ITSR, EN 12697-12) [36] after freeze and thaw cycles and in terms of raveling resistance using the Cantabro test (EN 12697-17) [37]. The functional properties of the experimental mixtures were then evaluated as vertical permeability (EN 12697-19) [38] and skid resistance (EN 13036-4) [39].

3. Mixtures Characterization

3.1. Mixture's Details

Two mixtures were designed in compliance with a gradation band suggested by Italian technical specifications for semi-porous layers. The SPT mix was produced with 100% pale limestone aggregates. The SPS mix replaced 21% of the natural aggregate with synthetic aggregates (Figure 2).

(a) (b) (c)

Figure 2. Pale limestone aggregates (**a**), synthetic aggregates (**b**) and transparent polymeric binder (**c**).

The synthetic aggregates were sieved in order to substitute only the natural aggregate particles with the same dimensions (6.3/12.5 mm). Therefore, the gradations of the two mixtures were kept constant in order to have comparable grading curves (Figure 3). After several laboratory trials, the optimum binder content was determined to be 5.5% by weight of the aggregates. In this case, the evaluation of the optimum binder content was carried out in compliance with the Marshall mix design method [40]. The mixing procedure provided for the heating of aggregates at 170 °C and the addition of the polymeric binder chips at ambient temperature directly into the laboratory mixer together with the hot aggregates.

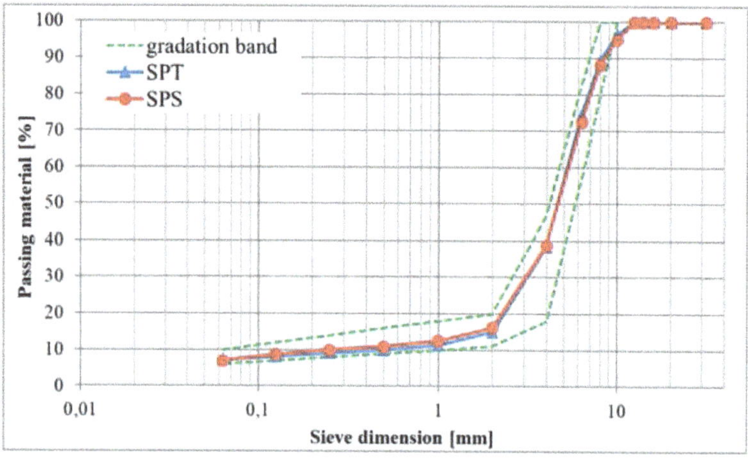

Figure 3. SPT and SPS particles distributions and gradation limits.

As showed in Figure 3, both the grading distributions fit the reference gradation band and there is no remarkable difference between the two mixtures.

The final aspect of the samples after compaction and its surface texture as well as the porous structure are shown in Figure 4. From a visual analysis the synthetic aggregates seem to be well distributed into the mixture, highlighting no issues in terms of workability and segregation during the mixing and compaction process.

Figure 4. Gyratory sample with synthetic aggregates.

3.2. *Air Voids Content*

In order to verify the quality of the mix design, the air voids (Av) content (EN 12697-8) [41] was evaluated for the two experimental mixtures. Four gyratory samples were produced for each mix in compliance with EN 12697-31 standard (80 gyrations). The bulk density of the specimens was calculated using the geometrical procedure, as suggested by the EN 12697-6 [42] standard for open graded bituminous mixtures. The results are presented in Table 3.

Table 3. Air voids content.

Mixture	Av (%)	Mixture	Av (%)
SPT_1	23.7	SPS_1	26.3
SPT_2	23.8	SPS_2	26.5
SPT_3	23.3	SPS_3	26.6
SPT_4	23.6	SPS_4	27.3
Avg.	**23.6**	**Avg.**	**26.7**

Both mixtures show a significant Av content. It is worth noting that the Italian technical specification suggests a lower Av limit of 16% for semi-porous mixtures. The remarkable porosity of the mixture is probably related to the adopted particles size distribution, which was close the lower limit of the gradation band for the material passing the 1 mm sieve and to the upper limit for the retained material at 4 mm sieve. In this case, the obtained aggregates distribution is more open graded.

The higher Av content for the SPS mixture is probably related to the rounded shape of the handcrafted artificial aggregates.

3.3. Mechanical Characterization: ITS and ITSM

The mechanical characterization was based on both static and dynamic tests: Indirect Tensile Strength (ITS) and Indirect Tensile Stiffness Modulus (ITSM).

The ITS test is generally useful for evaluation of the strength of cohesion between aggregates and mastic [43]. According to EN 12697-23, a load with a constant velocity of 50 mm/min is applied diametrically to a cylindrical specimen until failure.

The test was carried out using 3 gyratory specimens (80 gyrations) conditioned at 25 °C for 4 hours. The results are shown in Table 4. Both experimental mixtures comply with the minimum ITS value suggested by the reference technical specification (ITS ≥ 0.50 MPa). The SPS mix shows better performance despite its higher porous structure. In wider terms, a suitable ITS is reached despite the considerable amount of air voids content in the mixtures.

Table 4. Indirect Tensile Strength (ITS) test results.

Mixture	ITS (MPa)	Mixture	ITS (MPa)
SPT_1	0.48	SPS_1	0.55
SPT_2	0.50	SPS_2	0.60
SPT_3	0.51	SPS_3	0.53
Avg.	**0.50**	**Avg.**	**0.56**

These results represent also a further validation of the quality of the mastic formed by the polymeric transparent binder and the finest part of the aggregates distribution.

An advanced mechanical characterization was based on evaluation of mixture behavior under dynamic load using the ITSM test carried out on 3 gyratory samples (80 gyrations) for each mixture at 3 different temperatures: 10, 20 and 30 °C. According to the EN 12697-26 part C standard, the Modulus is determined through a pulse loading with a rise-time of 124 ms, to generate a predefined horizontal deformation of 7±2 μm in the core of the cylindrical specimen. The tests were carried out on 3 samples conditioned at 3 temperatures in order to verify the thermal sensitivity of the mixtures and how the presence of synthetic aggregates could affect this property.

The average results are presented in Figure 5 and summarized in Table 5.

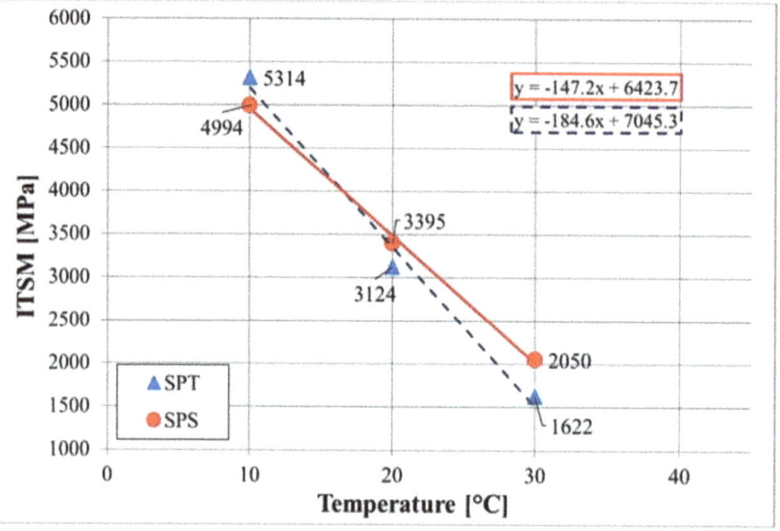

Figure 5. Average Indirect Tensile Stiffness Modulus (ITSM) results at 10, 20 and 30 °C.

Table 5. ITSM test results.

Mixture	ITSM @ 10 °C (MPa)	ITSM @ 20 °C (MPa)	ITSM @ 30 °C (MPa)
SPT_1	5149	3034	1709
SPT_2	5420	3109	1610
SPT_3	5373	3230	1550
Avg. SPT	**5314**	**3124**	**1622**
SPS_1	4990	3416	2095
SPS_2	4865	3368	2036
SPS_3	5127	3402	2005
Avg. SPS	**4994**	**3395**	**2050**

As overall results, both mixtures show a consistent stiffness at each test temperature.

If 20 °C is considered as the reference temperature, there are no substantial differences in stiffness between the two mixtures. The adopted technical specification does not suggest any limitations in terms of stiffness moduli. However, according to the scientific literature and real applications, the achievement of 3000 MPa at 20 °C can be considered as a suitable requirement for porous asphalts, considering their relatively weak structure.

In terms of thermal sensitivity, the two mixtures show a different variation in stiffness in relation to test temperature. The SPT mixture would appear to have a mechanical behavior that is more influenced by test temperature than the SPS mixture. Typically, for asphalt concretes an excessive stiffness at low temperatures and a low mechanical response at high temperatures could result in detrimental issues in terms of durability. In light of the above, the experimental mixture shows a positive increase in stiffness at high temperature and this could have a favorable effect in terms of rutting resistance. Still, the SPS mixture shows a stiffness trend line that corresponds to a reduced thermal sensitivity. This might be a consequence of the partial substitution of natural aggregates with synthetic materials. Future testing will assess the level of thermal transmittance for these artificial aggregates in order to validate this speculation.

3.4. Durability Evaluation: ITSR and Cantabro Tests

Considering the porous structure of the material, the durability evaluation was based on the assessment of the water sensitivity of the mixtures, in terms of reduction in ITS and raveling resistance.

In the first case a set of 3 specimens for each mix were subjected to 10 freeze and thaw cycles, from −20 to 20 °C, for 5 days before being tested. In fact, moisture damage can be considered as one of the main forms of pavement deterioration, which is also promoted by the formation of ice [44]. According to well-established literature, high ITS and ITSR values could guarantee a good resistance to moisture damage [45]. According to the EN 12697-12, the reduction in ITS is calculated as the ratio between results obtained in wet and normal dry conditions (EN 12697-12).

The Cantabro test is typically used in Europe for evaluating the raveling resistance of porous asphalt concretes (EN 12697-17). The test enables the estimation of the abrasiveness of porous asphalt, as these mixtures have high air voids, the contact areas between aggregates, which guarantee cohesion of the asphalt concrete, are limited. It is worth noting that it does not reflect the abrasive effect by studded tires. Thus, the cohesion is evaluated in terms of particles loss (PL) when a set of Marshall samples (EN 12697-30, 50 blows per side) is placed in a Los Angeles machine for 300 revolutions, with a speed of 30 revolutions per minute. Four Marshall samples were tested for each mixture. In compliance with the standard, the specimens were stored for 2 days at a temperature of 25 °C prior to testing. Table 6 summarizes the results for ITSR and Cantabro tests.

Table 6. ITSR and Cantabro test results.

Mixture	ITS_{normal} (MPa)	ITS_{wet} (MPa)	ITSR (%)	Particle Loss (%)
SPT_1	0.49	0.46	/	3.3
SPT_2	0.50	0.46	/	3.5
SPT_3	0.51	0.39	/	3.3
SPT_4	/	/	/	3.5
Avg. SPT	**0.50**	**0.44**	**89**	**3.3**
SPS_1	0.55	0.45	/	3.9
SPS_2	0.60	0.44	/	3.8
SPS_3	0.53	0.39	/	3.4
SPS_4	/	/	/	3.4
Avg. SPS	**0.56**	**0.43**	**77**	**3.6**

The technical specifications generally suggest a minimum ITSR value equal to 75%. Both experimental mixtures exceed this threshold value. It is worth noting that the reduction in ITS is generally evaluated for samples kept in a water bath (40 °C) for 72 hours prior testing. In the case under study, the ITSR results are in line with the suggested lower limit even if the mixtures have been subject to a considerable higher deterioration. The lower ITSR results for SPS are probably related to the higher air voids content that might had a detrimental effect during the freeze and thaw cycles.

In terms of particle loss, there is no significant difference between the two mixtures. The most common Italian technical specification suggests a maximum particle loss (PL) value equal to 20%, for porous asphalt, which is substantially higher if compared to test results. This is a further validation of the quality of the cohesion between particles guaranteed by the polymeric transparent binder considering the remarkable porosity of the two mixtures.

3.5. Functional Properties: Skid Resistance and Vertical Permeability Test

Surface friction and vertical permeability are important functional properties for porous layers.

The friction between tires and road pavement involves two components: adhesion and hysteresis. The first phenomenon is connected to the microtexture of the pavement, which is generally evaluated by means of the skid resistance test [46]. The most common measure of the skid resistance is given in terms of Pendulum Test Value (PTV, EN 13036-4) using the British Portable Pendulum. According to the standard, the frictional force is the force acting tangentially in the contact area and it is measured as the loss of energy of a standard rubber slider that slides across the test surface. The PTV value in given by the average result of five repetitions for each single test point. The final result is adjusted with specific factors depending on the surface temperature. It is worth mentioning that the test surface must be wetted prior to testing.

As for the permeability, it is the most important property for a porous asphalt. According to the EN 12697-19, it can be evaluated in lab in terms of vertical and horizontal permeability: in this research, the former property was assessed. The vertical permeability is considered as the water flowing in a vertical direction through the specimen thickness. The test procedure imposes that a water column of constant height is kept on the surface of the porous sample and the vertical permeability is evaluated in terms of the amount of water flowing through the sample in a specific range of time using the Darcy's Law.

The results of both tests are shown in Table 7.

Table 7. Average Pendulum Test Value (PTV) and vertical permeability results.

Mixture	PTV	Vertical Permeability (10^{-3} m/s)
SPT_1	53	0.50
SPT_2	48	0.52
SPT_3	49	0.63
Avg. SPT	**50**	**0.55**
SPS_1	55	0.54
SPS_2	55	0.54
SPS_3	54	0.61
Avg. SPS	**55**	**0.56**

In terms of skid resistance, a small difference was found between the two mixtures. The SPS mixture has a higher friction, possibly related to the different surface texture affected by the higher porosity and the different micro and macro texture of the synthetic aggregates. Both PTV values are acceptable but the surface texture needs to be improved if compared to values generally suggested by technical specifications for asphalt pavements (PTV ≥ 50). However, an increase in PTV is expected after a primary polishing of the binder film that covers the aggregates by the traffic. Nevertheless, an optimization of the grading distribution might improve the macro-texture of the material.

No significant difference was found in terms of vertical permeability for the two mixtures. It is worth noting that the reference standard suggests a minimum value of $0.5 \cdot 10^{-3}$ m/s for traditional porous layers. The obtained values are remarkable, considering that the adopted gradation band is suggested for semi-porous layers. Nevertheless, taking into account the significant porosity of the two mixtures, the air voids interconnection must be improved. Future imaging tests with e.g. Nuclear Magnetic Resonance (NMR) technology will evaluate the inner structure of the samples in order to verify the interconnectivity and tortuosity of air voids.

4. Conclusions

In the present research, a low impact semi-porous concrete produced with transparent polymeric binder and pale limestone aggregates is proposed. To improve the sustainability of the material, an experimental mixture was produced with the partial substitution of natural aggregates with artificial ones obtained through the alkali-activation of waste basalt powder. The research program provided for a physical and mechanical laboratory characterization.

On the basis of the presented results, the following conclusions can be drawn:

- The adopted particles size distribution and polymeric binder amount allow the achievement of good workability and a higher porosity if compared to traditional semi-porous asphalt concretes. The presence of synthetic aggregates did not affect the mixing procedure and the workability properties of the mixture. The visual analysis of the samples and their inner structure highlighted a correct distribution of the artificial aggregates within the mixture.

- The increased air voids content for the SPS mixture is more likely related to the rounded shape of the handcrafted particles, which limited the aggregates interlocking resulting in a more open structure.
- The mechanical characterization and durability evaluation highlighted how both mixtures exceed the threshold limits imposed by the technical specifications for semi-porous asphalt layer, despite the significant porosity of the materials. Consistent properties were registered in terms of water susceptibility considering the severe conditioning given by the freeze and thaw cycles. Furthermore, both mixtures revealed a ravelling resistance considerably lower than the threshold limit for porous asphalts, as a further confirmation of the good cohesion between particles guaranteed by the polymeric binder.
- The most relevant difference between the mixtures is related to the low thermal sensitivity of SPS. The experimental mixture showed a favourable increase in stiffness at high temperature and a slight decrease at the lowest ones. This phenomenon might be attributed to the presence of artificial aggregates. Future tests will assess the thermal transmittance of the synthetic aggregates in order to validate this conclusion.
- The two experimental mixtures showed a good vertical permeability which is in line with the lowest values suggested for porous layers. However, considering the high porosity of the mixture, most of the air voids may be not fully connected. In terms of skid resistance, both mixtures should be optimized to improve this parameter in their early life, despite the adoption of synthetic aggregates led to higher PTVs. As a general statement, the optimization of the mix design might improve the texture properties of the material as well as the interconnection of air voids.

In the light of the above, the use of transparent polymeric binder seems to be a viable solution for the production of low impact semi-porous layers for use in urban areas. Future studies will investigate the possible substitution of higher quantities of natural aggregate with synthetic material aiming to the production of 100% synthetic mixtures. Furthermore, the use of a centrifugal granulator can improve the quality of the synthetic aggregates and convert the production from the laboratory to the industrial scale.

Author Contributions: Conceptualization, P.T. and C.S.; investigation, P.T.; data curation, P.T. and C.S.; writing—original draft preparation, P.T. and C.S.; writing—review and editing, P.T. and C.S.; supervision, C.S.

Funding: This research received no external funding.

Acknowledgments: Authors are grateful to CORECOM s.r.l. that supplied the polymeric transparent binder during the whole research.

Conflicts of Interest: The authors declare no conflict of interest.

References

1. Artmann, M.; Inostroza, L.; Fan, P. Urban sprawl, compact urban development and green cities. How much do we know, how much do we agree? *Ecol. Indic.* **2019**, *96*, 3–9. [CrossRef]
2. Li, C.; Liu, M.; Hu, Y.; Shi, T.; Qu, X.; Walter, M.T. Effects of urbanization on direct runoff characteristics in urban functional zones. *Sci. Total. Environ.* **2018**, *643*, 301–311. [CrossRef] [PubMed]
3. Leopold, L.B. Hydrology for urban land planning - A guidebook on the hydrologic effects of urban land use. *Circular* **1968**.
4. Whitford, V.; Ennos, A.R.; Handley, J.F. City form and natural process indicators for the ecological performance of urban areas and their alpplication to merseyside, uk.pdf. *Landsc. Urban Plan.* **2001**, *57*, 91–103. [CrossRef]
5. Huang, H.-J.; Cheng, S.-J.; Wen, J.-C.; Lee, J.-H. Effect of growing watershed imperviousness on hydrograph parameters and peak discharge. *Hydrol. Process.* **2008**, *22*, 2075–2085. [CrossRef]
6. Isik, S.; Kalin, L.; Schoonover, J.E.; Srivastava, P.; Lockaby, B.G. Modeling effects of changing land use/cover on daily streamflow: An Artificial Neural Network and curve number based hybrid approach. *J. Hydrol.* **2013**, *485*, 103–112. [CrossRef]

7. Burant, A.; Selbig, W.; Furlong, E.T.; Higgins, C.P. Trace organic contaminants in urban runoff: Associations with urban land-use. *Environ. Pollut.* **2018**, *242*, 2068–2077. [CrossRef] [PubMed]
8. Brown, J.N.; Peake, B.M. Sources of heavy metals and polycyclic aromatic hydrocarbons in urban stormwater runoff. *Sci. Total. Environ.* **2006**, *359*, 145–155. [CrossRef] [PubMed]
9. Zhang, P.; Cai, Y.; Wang, J. A simulation-based real-time control system for reducing urban runoff pollution through a stormwater storage tank. *J. Clean. Prod.* **2018**, *183*, 641–652. [CrossRef]
10. Seto, K.C.; Güneralp, B.; Hutyra, L.R. Global forecasts of urban expansion to 2030 and direct impacts on biodiversity and carbon pools. *Proc. Natl. Acad. Sci. USA* **2012**, *109*, 16083–16088. [CrossRef]
11. Park, D.; Kang, H.; Jung, S.H.; Roesner, L.A. Reliability analysis for evaluation of factors affecting pollutant load reduction in urban stormwater BMP systems. *Environ. Model. Softw.* **2015**, *74*, 130–139. [CrossRef]
12. Sharior, S.; McDonald, W.; Parolari, A.J. Improved reliability of stormwater detention basin performance through water quality data-informed real-time control. *J. Hydrol.* **2019**, *573*, 422–431. [CrossRef]
13. Guan, M.; Sillanpää, N.; Koivusalo, H. Assessment of LID practices for restoring pre-development runoff regime in an urbanized catchment in southern Finland. *Water Sci. Technol.* **2015**, *71*, 1485. [CrossRef] [PubMed]
14. Mullaney, J.; Lucke, T. Practical review of pervious pavement designs. *Clean Soil Air Water* **2014**, *42*, 111–124. [CrossRef]
15. Ortega-Villar, R.; Lizárraga-Mendiola, L.; Coronel-Olivares, C.; López-León, L.D.; Bigurra-Alzati, C.A.; Vázquez-Rodríguez, G.A. Effect of photocatalytic Fe2O3 nanoparticles on urban runoff pollutant removal by permeable concrete. *J. Environ. Manag.* **2019**, *242*, 487–495. [CrossRef]
16. Zhenci, X.; Guo, Y. Simulation test of runoff on different underlying surfaces in urban area. *South North Water Transf. Water Sci. Technol.* **2017**, *10*, 64–66.
17. Zhu, H.; Yu, M.; Zhu, J.; Lu, H.; Cao, R.; Zhang, L. Simulation study on effect of permeable pavement on reducing flood risk of urban runoff. *Int. J. Transp. Sci. Technol.* **2018**. [CrossRef]
18. Saadeh, S.; Ralla, A.; Al-Zubi, Y.; Wu, R.; Harvey, J. Application of fully permeable pavements as a sustainable approach for mitigation of stormwater runoff. *Int. J. Transp. Sci. Technol.* **2019**. [CrossRef]
19. Kamali, M.; Delkash, M.; Tajrishy, M. Evaluation of permeable pavement responses to urban surface runoff. *J. Environ. Manag.* **2017**, *187*, 43–53. [CrossRef]
20. Higashiyama, H.; Sano, M.; Nakanishi, F.; Takahashi, O.; Tsukuma, S. Field measurements of road surface temperature of several asphalt pavements with temperature rise reducing function. *Case Stud. Constr. Mater.* **2016**, *4*, 73–80. [CrossRef]
21. Pisello, A.L. State of the art on the development of cool coatings for buildings and cities. *Sol. Energy* **2017**, *144*, 660–680. [CrossRef]
22. Praticò, F.G.; Giunta, M.; Marino, C.; Attinà, A. Pavement albedo and sustainability: An experimental investigation. In Proceedings of the 7th International Conference on Maintenance and Rehabilitation of Pavements and Technological Control, Auckland, New Zealand, 28–30 August 2012.
23. Starke, P.; Göbel, P.; Coldewey, W.G. Urban evaporation rates for water-permeable pavements. *Water Sci. Technol.* **2010**, *62*, 1161–1169. [CrossRef] [PubMed]
24. Takebayashi, H.; Moriyama, M. Study on Surface Heat Budget of Various Pavements for Urban Heat Island Mitigation. *Adv. Mater. Sci. Eng.* **2012**, *2012*, 1–11. [CrossRef]
25. Santamouris, M. Using cool pavements as a mitigation strategy to fight urban heat island—A review of the actual developments. *Renew. Sustain. Energy Rev.* **2013**, *26*, 224–240. [CrossRef]
26. Bao, T.; Liu, Z.; Zhang, X.; He, Y. A drainable water-retaining paver block for runoff reduction and evaporation cooling. *J. Clean. Prod.* **2019**, *228*, 418–424. [CrossRef]
27. Provis, J.L. Alkali-activated materials. *Cem. Concr. Res.* **2018**, *114*, 40–48. [CrossRef]
28. Duxson, P.; Fernández-Jiménez, A.; Provis, J.L.; Lukey, G.C.; Palomo, A.; Deventer, J.S.J. Geopolymer technology: The current state of the art. *J. Mater. Sci.* **2007**, *42*, 2917–2933. [CrossRef]
29. Tang, Z.; Li, W.; Hu, Y.; Zhou, J.L.; Tam, V.W. Review on designs and properties of multifunctional alkali-activated materials (AAMs). *Constr. Build. Mater.* **2019**, *200*, 474–489. [CrossRef]
30. Saraya, M.E.-S.I.; El-Fadaly, E. Preliminary Study of Alkali Activation of Basalt: Effect of NaOH Concentration on Geopolymerization of Basalt. *J. Mater. Sci. Chem. Eng.* **2017**, *5*, 58–76.

31. Tataranni, P.; Besemer, G.M.; Bortolotti, V.; Sangiorgi, C. Preliminary Research on the Physical and Mechanical Properties of Alternative Lightweight Aggregates Produced by Alkali-Activation of Waste Powders. *Materials* **2018**, *11*, 1255. [CrossRef]
32. *EN 1015-11, Methods of Test for Mortar for Masonry - Part 11: Determination of Flexural and Compressive Strength of Hardened Mortar*; CEN (European Committee for Standardization): Brussels, Belgium, 1999.
33. *EN 12697-31, Bituminous Mixtures – Test Methods for Hot Mix Asphalt. Part 31: Specimen Preparation by Gyratory Compactor*; CEN (European Committee for Standardization): Brussels, Belgium, 2007.
34. *EN 12697-23, Bituminous Mixtures – Test Methods for hot Mix Asphalt. Part 23: Determination of the Indirect Tensile Strength of Bituminous Specimens*; CEN (European Committee for Standardization): Brussels, Belgium, 2018.
35. *EN 12697-26, Bituminous Mixtures – Test Methods for Hot Mix Asphalt. Part 26: Stiffness*; CEN (European Committee for Standardization): Brussels, Belgium, 2012.
36. *EN 12697-12, Bituminous Mixtures – Test Methods for Hot Mix Asphalt. Part 12: Determination of the Water Sensitivity of Bituminous Specimens*; CEN (European Committee for Standardization): Brussels, Belgium, 2018.
37. *EN 12697-17, Bituminous Mixtures – Test Methods for Hot Mix Asphalt, Part 17: Particle Loss of Porous Asphalt Specimens*; CEN (European Committee for Standardization): Brussels, Belgium, 2017.
38. *EN 12697-19, Bituminous Mixtures – Test Methods for Hot Mix Asphalt, Part 19: Permeability of Specimens*; CEN (European Committee for Standardization): Brussels, Belgium, 2012.
39. *EN 13036-4, Road and Airfield Surface Characteristics – Test Methods. Part 4: Method for Measurement of Slip/Skid Resistance of a Surface: The Pendulum Test*; CEN (European Committee for Standardization): Brussels, Belgium, 2011.
40. *EN 12697-30, Bituminous Mixtures—Test Methods for Hot Mix Asphalt—Part 30: Specimen Preparation By Impact Compactor*; CEN (European Committee for Standardization): Brussels, Belgium, 2012.
41. *EN 12697-8, Bituminous Mixtures – Test Methods for Hot Mix Asphalt. Part 8: Determination of Void Characteristics of Bituminous Specimens*; CEN (European Committee for Standardization): Brussels, Belgium, 2019.
42. *EN 12697-6, Bituminous Mixtures – Test Methods for Hot Mix Asphalt. Part 6: Determination of Bulk Density of Bituminous Specimens*; CEN (European Committee for Standardization): Brussels, Belgium, 2012.
43. Sangiorgi, C.; Tataranni, P.; Simone, A.; Vignali, V.; Lantieri, C.; Dondi, G. Waste bleaching clays as fillers in hot bituminous mixtures. *Constr. Build. Mater.* **2014**, *73*, 320–325. [CrossRef]
44. Sangiorgi, C.; Eskandarsefat, S.; Tataranni, P.; Simone, A.; Vignali, V.; Lantieri, C.; Dondi, G. A complete laboratory assessment of crumb rubber porous asphalt. *Constr. Build. Mater.* **2017**, *132*, 500–507. [CrossRef]
45. Chen, J.S.; Chen, S.F.; Liao, M.C. Laboratory and field evaluation of porous Asphalt concrete. *Asian Transp. Stud.* **2015**, *3*, 298–311.
46. Praticò, F.; Vaiana, R. A study on the relationship between mean texture depth and mean profile depth of asphalt pavements. *Constr. Build. Mater.* **2015**, *101*, 72–79. [CrossRef]

infrastructures

Review

Fiber Optics Sensors in Asphalt Pavement: State-of-the-Art Review

Patricia Kara De Maeijer [1,*], Geert Luyckx [2], Cedric Vuye [1], Eli Voet [2], Wim Van den bergh [1], Steve Vanlanduit [3], Johan Braspenninckx [4], Nele Stevens [4] and Jurgen De Wolf [5]

1 EMIB Research Group, Faculty of Applied Engineering, University of Antwerp, 2020 Antwerp, Belgium; cedric.vuye@uantwerpen.be (C.V.); wim.vandenbergh@uantwerpen.be (W.V.d.b.)
2 Com&Sens, 9810 Nazareth, Belgium; gluyckx@com-sens.eu (G.L.); evoet@com-sens.eu (E.V.)
3 Op3Mech Research Group, Faculty of Applied Engineering, University of Antwerp, Groenenborgerlaan 171, 2020 Antwerp, Belgium; steve.vanlanduit@uantwerpen.be
4 Port of Antwerp, Zaha Hadidplein 1, 2030 Antwerp, Belgium; johan.braspenninckx@portofantwerp.com (J.B.); nele.stevens@portofantwerp.com (N.S.)
5 Agency for Roads and Traffic, 1140 Evere, Belgium; jurgen.dewolf@mow.vlaanderen.be
* Correspondence: patricija.karademaeijer@uantwerpen.be; Tel.: +32-3-265-8851

Received: 15 May 2019; Accepted: 14 June 2019; Published: 20 June 2019

Abstract: Pavement design is essentially and usually a structural long-term evaluation process which is needed to ensure that traffic loads are efficiently distributed at all levels of the total road structure. Furthermore, to get a complete analysis of its durability behavior, long-term monitoring should be facilitated, not only from the top by falling weight deflectometer (FWD) or core drilling but preferably from inside the structure and at exactly the same positions during a long-time interval. Considering that it is very hard to devise an efficient method to determine realistic in-situ mechanical properties of pavements, the determination of strain at the bottom of asphalt pavement layers through non-destructive tests is of a great interest. As it is known, fiber Bragg grating (FBG) sensors are the most promising candidates to effectively replace conventional strain gauges for a long-term monitoring application in a harsh environment. The main goals of this paper are to compile an overview of the recent developments worldwide in the application of fiber optics sensors (FOS) in asphalt pavement monitoring systems; to find out if those systems provide repeatable and suitable results for a long-term monitoring; if there are certain solutions to validate an inverse modelling approach based on the results of FWD and FOS.

Keywords: asphalt; fiber optics sensors (FOS); fiber Bragg grating (FBG) sensors; falling weight deflectometer (FWD)

1. Introduction

Pavement design is essentially and usually a structural long-term evaluation process which is needed to ensure that traffic loads are efficiently distributed at all levels of the total road structure. The total road structure consists of a top layer, one or more base layers, subbase and subgrade. The stresses and strains developed at those levels should be within the capabilities of the materials used. The objective of the pavement design is to produce an engineering structure that will distribute traffic loads efficiently within the selected burden parameters by minimizing the whole-life cost of the pavement including the work costs (materials, construction, maintenance and residual value); the user costs (traffic delays, accidents at roadworks, skidding accidents, fuel consumption/tire wear and residual allowance); environmental impact and so forth. Mechanical pavement design usually involves the selection of the materials/mixtures for the different layers of the pavement structure and the calculation of the required thickness of the designed pavement. Stiffness is a fundamental and important parameter that must be fully understood in the selection of the materials. Usually pavement layers have a higher stiffness while the layers underneath such as base and subbase have a greater thickness with a lower stiffness to get the same pressure on the underside. The thickness of the pavement structure layers can be reduced by increasing the stiffness of the layers and introducing multilayered pavement structures which are commonly used in, for example, Belgium, France and The Netherlands. One of the important facets of material behavior is the consideration of the situation where the layers are susceptible to moisture. If the area becomes saturated, the stiffness is reduced (normally imposed stresses taken by the dry material cause the layer to fail). For sure it can be avoided if drainage is installed in such a way that groundwater never reaches the pavement layers. Since moisture may affect the subgrade and the sub-base (and, also the base if it is unbound). Another one is a temperature monitoring, as temperature affects the bitumen-bound layers. It is essential that the design process takes the climatic conditions into account [1,2]. Also, the pavement compaction quality is a very important facet.

In the last 50 years, pavement research and related pavement techniques have grown. These theories, principles, and/or procedures which were based on the knowledge and research achievements at that time, have helped the pavement professionals to make specific analysis, design, construction and maintenance on pavements. Therefore, they have created a far-reaching influence on later pavement technology [3]. On the other hand, during the same period, a lot of new pavement materials were invented and widely used. The properties of these new materials are considerably different compared to the conventional materials. Neither traditional pavement analysis methods nor existing design principles can provide a direct way to consider these differences. This leads to the necessity and the difficulty of modifying the pavement design methods used so far. Currently, there are several important challenges to pavement research related to asphalt pavement analysis and design that include: how to deal with more and more common heavy traffic loads (the allowed max. weight for 5 axles to be 44 tons in Belgium and France and 50 tons in The Netherlands) [4], or even overloads and with increasing traffic volumes; how to consider emerging pavement materials; how to incorporate new materials, new techniques and new design concepts into pavement analysis and design; and how to consider ageing and healing effects.

The assessment of pavement mechanical state and service life is very important for design evaluation and road maintenance. However, this job seems to be a mission impossible, because it is unimaginable to learn how exactly a pavement works inside [5]. Conventional monitoring and investigation methods adopted by researchers are core drilling, pavement cutting, Benkelman beam, falling weight deflector (FWD), automatic deflectometer, surface-curvature apparatus and so forth. They are either destructive or with low-precision or low-frequency and most important is that all these methods are discontinuous and short-term. A pavement which is exposed for a long time to a natural environment, is deteriorating at the coupled effect of load, temperature, water and ultraviolet light. It is very difficult to completely understand the mechanical response of a pavement structure in an actual environment by regular methods [5]. In fact, pavement monitoring is a rather complicated process and each pavement is another case to survey. Furthermore, it cannot be monitored only for a short-term period to get a complete analysis of the pavement behavior.

The past two decades some major technological breakthroughs produced by the fusion of different disciplines have been witnessed. This trend is likely to develop in the future due to the recent significant advancements of fiber optics communications, photonics, biomedical and nanotechnologies worldwide. In parallel with the communications and information technology revolution, fiber and waveguide optics sensor and imaging technologies have enjoyed an unseen technological maturity and revealed enormous potentials for a broad variety of new applications [6].

Asphalt material is often considered to behave in a linearly viscoelastic-plastic manner; thus, its mechanical response is a continuous function of time and temperature. Considering the stiffness of the material, its behavior at lower temperatures is equivalent to a higher strain rate, such as the strain on pavement due to fast moving traffic. In the case of high stiffness, the strain on asphalt should ideally be measured directly for the greatest accuracy; however, instruments capable of making such measurements are not generally available. Electrical strain gauges are often assumed to have negligible stiffness and the stress transferred from the asphalt to an embedded sensor decreases drastically, thereby reducing the sensitivity of the sensor reading [7]. The sensors used for pavement instrumentation must be as much compatible with the heterogeneous nature and mechanical properties of pavement materials. First, the sensors should be as small as possible so that they are not too intrusive in the bituminous layers. Secondly for strain measurements, the stiffness of the sensors has to match that of the asphalt mixture in order to correctly measure the mechanical properties of the pavement. More-over, the embedded sensors must withstand the highest stresses experienced during the pavement construction process (high temperature and compression). After that, if a long-term monitoring is considered, the sensor should be resistant to corrosion and to thermo-mechanical fatigue conditions [8].

Several fiber optic sensor technologies (fiber Bragg grating (FBG) and Fabry-Perot (FP) interferometry) have already been used for the experimental investigation of pavement behavior and pavement monitoring with positive results. FBG is a small portion of an optical fiber several millimeters long in which a diffraction grating is written by ultraviolet (UV) exposure. The optical property of this grating is to reflect a narrow optical band (around a center wavelength called Bragg wavelength) of the incident spectrum. FBG have the intrinsic quality to be very sensitive to thermal and mechanical stimuli. The Bragg wavelength is proportional to the temperature and/or strain variation. Since, this sensor is very brittle, it needs to be packaged. The fiber FP sensor is essentially an optical cavity that is defined by two semi-reflecting parallel mirrors. The FP cavity (at least in a bulk optic form) has highly reflecting mirrors of reflectivity such that the device has a high finesse and consequently its reflection/transmission is spectrally selective and serves as an interference filter element. In its use with optical fibers, the cavity is formed in a short length (1–30 mm) of optical fiber that has partially reflecting coated ends which is then fusion-spliced onto the end of the connecting fiber [9]. Those technologies allow to perform dynamic measurements at a sampling rate of at least of 0.5–1 kHz (for the standard interrogators), they are investigated particularly for the development of traffic classification and weigh-in-motion systems. In particular, FBG sensors have been widely applied in different sensing fields, where they are used as strain and temperature sensors [10–13], for deformation investigation [14–16], rutting performance [17,18], response of asphalt concrete [19,20] and weigh-in-motion [21]. Compared with conventional sensors, FBG sensors are the most promising candidates to effectively replace conventional strain gauges for long-term real-time monitoring applications in a harsh environment. They exhibit several advantages: flexibility, embeddability, high frequency, electromagnetic interference immunity and so forth. FBG and FP technologies deliver a strain local measurement like an electrical strain gauge. Despite their high sensitivity and accuracy, they are not suitable for detection of cracks or damage. Due to their relatively small dimensions compared to those of a pavement, a crack can be detected only if it propagates in the vicinity of the sensor, by means of fiber optic sensing techniques based on the Brillouin scattering or the Rayleigh scattering [8].

The FBG monitoring system prototype installed at UAntwerp—a three-layered pavement test track (width—4 m and length—96 m), the CyPaTs bicycle path—in September 2017, proved that the FBG technique can be successfully applicable for in-situ strain and temperature measurements under real heavy loaded traffic (e.g., truck, paver, roller) in the asphalt pavement structures during the asphalt paving process [22–26]. It also proved that an FBG monitoring system can be functional for long-term pavement monitoring as it is still operational. Nowadays, the determination of strain at the bottom of asphalt pavement layers through non-destructive tests is of a great interest. Therefore, the application of FBG in an asphalt pavement structure can be considered as an advanced research method for a long-term and real-time process.

The main goals of this review are to compile an overview of the recent developments worldwide in the application of fiber optics sensors (FOS) in asphalt pavement monitoring systems; to find out if those systems provide repeatable and suitable results for a long-term monitoring; if there are certain solutions to validate an inverse modelling approach based on the results of FWD and FOS. Section two gives an overview of the standardized FWD methodology compared to several test cases using fiber optics sensors in asphalt pavements. The results of the inverse modeling approach performed by several researchers using FOS and FWD are described in section three. A concluding summary on the application of FOS to determine tensile strain at the bottom of an asphalt layer is given in Section 4.

2. Testing Methods and Technologies

This part of the review describes the standardized FWD methodology and several cases of application of FOS in asphalt pavement as an alternative method to FWD to determine tensile strains at the bottom of the asphalt layer.

2.1. Falling Weight Deflectometer

FWD is by now one of the most common non-destructive testing methods to assess bearing capacity on major road infrastructures by using deflection data generated from a loading device to quantify the response of a pavement structure to known load drops [27,28]. This non-destructive technique allows measuring the deflection response of the pavement at several positions under a given load (Figure 1). The surface deflections obtained from FWD testing are used to back-calculate in situ material properties using software which has been developed in the eighties [29]. These properties are considered representative for the pavement response to a load and can be used to assess stresses and strains that are valid for pavement structural and fatigue analysis and design. However, the accuracy of the results (stresses and strains at critical locations in the pavement) depends upon the assumptions used for the analysis [30].

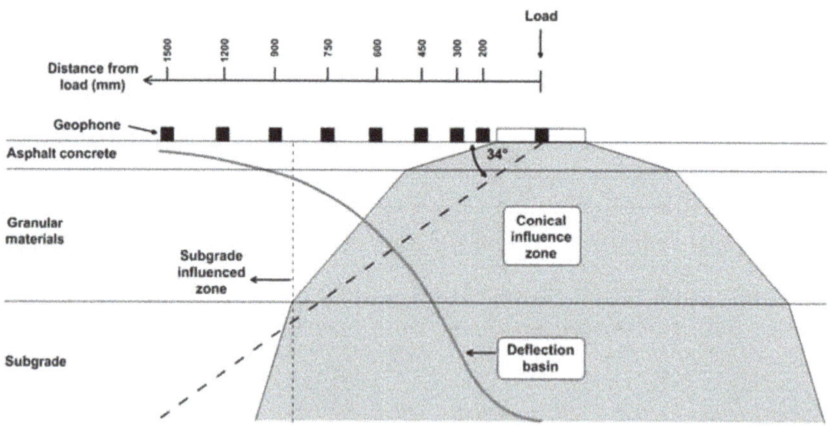

Figure 1. Schematics of the falling weight deflectometer (FWD) test (figure is reproduced from Bilodeau & Doré [31]).

2.2. Fiber Optics Sensors

Pavement monitoring is an essential part of pavement research and plays an important role in pavement management systems [32]. It is known that it is very hard to devise an efficient method to determine realistic mechanical properties of the asphalt pavement. Several gauges were developed to instrument the pavement for its monitoring but considering their high sensitivity to a lot of parameters, not many succeeded to obtain real-time strain data [33] as an alternative method for traditional FWD. FOS gained popularity in the last two decades when an attempt was made to instrument FOS inside the pavement for monitoring purposes. FOS is not commonly used in asphalt technology due to its application restrictions during rough construction processes, which require the sensors to endure high temperatures, moisture, high compaction force, repeated heavy loading and so forth. The accurate measurement of the pavement responses (strain and stress) distributions in pavement structure, combined with temperature, is critical for the understanding of pavement behavior and the modeling of pavement failure [32]. It was noted by Papavasiliou & Loizos [28] that implementing a FOS system is a time consuming and delicate procedure but it has proved to be a useful and promising tool for in-situ strain measurements under real traffic loading; however, due to the deviations between calculated strains (pavement design, FWD data analysis) and FOS signal records (measured strains) more experimental work was needed.

Below several cases of application of FOS in asphalt pavements are discussed in further detail.

2.2.1. Fiber-Optic Strain Gauge (2007) (Laval University, Canada)

The proposed gauge/instrumented core relates to horizontal strain measurements at the bottom of bound surfacing layers of pavements [34,35]. The gauge is designed to be retrofitted in existing pavement surfacing layers (asphalt concrete, Portland cement concrete or other bound material) through a small diameter core hole to minimize perturbation to the pavement layer to be instrumented. Fiber optic strain sensors are imbedded in the polymeric proof body. Even though electric strain gauges can be used for the proposed application, fiber optic sensors are preferred due to their insensitivity to water, frost action and electric fields. Figure 2 shows the schematic design of OpSens interferometric fiber optic strain transducer.

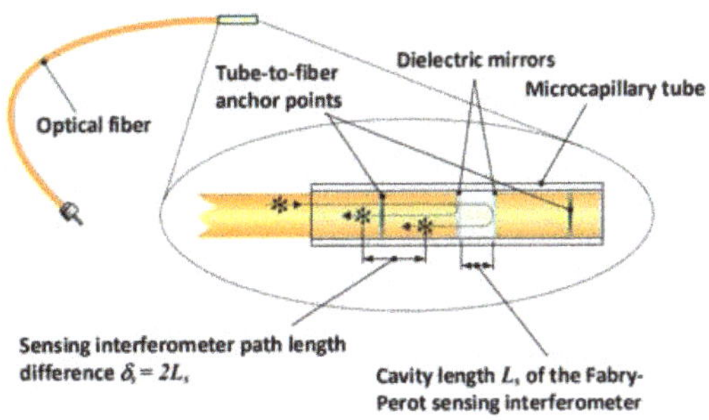

Figure 2. Strain sensor based on the Fabry-Perot (FP) interferometer (figure is reproduced from OpSens Solutions [36]).

The sensor is made of two optical fibers that are precisely aligned inside a microcapillary tube to form an optical FP sensing interferometer. This makes the strain gauge completely immune to any electromagnetic interference and completely insensitive to transverse strains and temperature, as opposed to fiber optic Bragg gratings sensors.

Figure 3 shows a schematic illustration of an instrumented core which includes two measurement levels. Inside each polymeric proof body, two orthogonal fiber optic sensors are inserted (dash lines) for measuring the strain along the longitudinal and the transverse directions of the road. The polymeric proof body is made of a plastic composite having an elastic modulus and a thermal coefficient of contraction like asphalt concrete, allowing both materials to be mechanically compatible. The material is sufficiently robust to protect the gauge when the core is subjected to heavy traffic loads and severe climatic conditions.

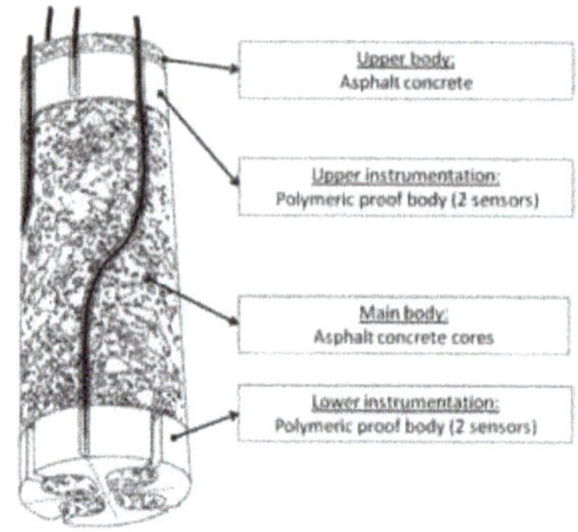

Figure 3. Laval University fiber-optic strain gauge (figure is reproduced from Bilodeau & Doré [31]).

2.2.2. Fiber-Optic Strain Gauge (2007) (National Technical University of Athens, Greece)

The system is based on the FP sensor type, which consists of two semi-reflective mirrors facing each other [37–39]. The mirrors are placed on the tips of multimode optical fibers, which are spot-fused into a capillary. The air gap between the mirrors defines the FP cavity to minimize disturbances to the stress fields around the gauge when the pavement was subjected to loading by heavy vehicles. It was also designed to allow for the installation of two orthogonal strain sensors allowing for the measurement of longitudinal and transversal strains at the bottom of the bound layer. The proof body and consequently the sensor were instrumented in the laboratory at the critical fatigue locations of the cores previously drilled from the test sites. The gauge was installed in the pavement by placing it and gluing it at the desired location in the core hole. The diameter of the cores was less than 10 cm, to minimize any disturbances to the pavement layers being instrumented.

2.2.3. Telecom Fiber Optic Cable as Distributed Sensor (2014) (IFSTTAR, France)

A new non-destructive technique for the detection of cracking in asphalt pavements was tested at a real scale on the pavement fatigue carousel of French Institute of Science and Technology for Transport, Development and Networks (IFSTTAR), which is a large scale circular outdoor facility, unique in Europe by its size (120 m length, 6 m width) and loading capabilities (Figure 4) [8,40]. It was based on the use of a telecommunication fiber optic cable embedded in the pavement layer as distributed sensor. An optical interrogator based on the Rayleigh scattering was used to measure strain profiles. The main advantage of this technique is that it allows to measure strains over a long length of fiber optic with a high spatial resolution, less than 1 cm. By comparing strain profiles measured at different times, an attempt was made to link strain changes with the appearance of damage (cracking) in the pavement. A significant increase of strains was detected by the optical fibers at different points in the pavement structure, before any damage was visible [8].

Figure 4. Installation of fiber optic cable at IFSTTAR accelerated pavement testing facility—pavement fatigue carousel (figures are reproduced from Chapeleau et al. [40]).

However, it must be mentioned here, that the main disadvantage of telecommunication fibers application is its limitation (up to few Hz) in sampling frequency. It is not possible to use the sensing system if determination of eigenfrequencies or performance of dynamic testing on a structure is needed. It is only useful to measure static long-term effects. In regard to FBG sensing systems, it is possible to perform low sampling frequencies for a long-term effect (e.g., creep, permanent deformation) and it is also possible momentarily to go higher in read out frequency to perform dynamic testing or investigate local fast relaxation effects and so forth. Standard telecommunication fibers are very fragile and are not suitable for direct embedding in the asphalt. As such cable design becomes very important, moreover also with respect to strain transfer and intrusiveness in the case cable diameters exceed 2 mm.

2.2.4. Fiber-Optic Strain Plate (2014) (Federal Aviation Administration, USA)

To measure the strain response under traffic load, four test pavement sections were instrumented with an innovative polymeric plate technology [41–44] at the Federal Aviation Administration (FAA) National Airport Pavement Test Facility (NAPTF) in Atlantic City, New Jersey, USA in 2014 (Figure 5) [45]. Each test section was equipped with one instrumented plate positioned perpendicularly to traffic direction. Each plate consisted of a polyphenylene sulphide (PPS) with a rectangular thin body in which 24 fiber optic strain gauges were embedded and bonded with epoxy. The fiber-optic strain gauges working principle is based on the White Light Polarization Interferometry (WLPI) technology. WLPI uses a signal conditioner to sense the path length difference inside a FP interferometer of a known cavity length and delimited by two dielectric mirrors [34].

Figure 5. Instrumented plate installation at the Federal Aviation Administration (FAA) National Airport Pavement Test Facility (NAPTF) in Atlantic City, New Jersey, USA (figures are reproduced from Garg et al. [45]).

With proper calibration, the path length difference can be related to engineering values, such as displacement, stress and strain. Three signal conditioners equipped with 8 channels were used to collect the data of the 24 gauges on one plate. The signal conditioner sends and receives the light and the software interprets and transforms the received signal into physical quantitative values. These values are compiled in a text file at a specified frequency (a 500 Hz data collection frequency was used).

2.2.5. FOS Monitoring System (2014) (Japan)

A full-sized field test was conducted on a slope of an under-construction asphalt-faced dam in 2014 [7]. Four types of sensor installation techniques were executed to embed flexible sensors which were simply coated with polyethylene for direct embedding in an asphalt structure on the asphalt slope (see Figure 6); each sensor had two lines along the slope due to the turning back at the bottom. All four sensor types were placed on the designated location, fixed with a solvent-type primer and covered with an asphalt mortar or sealant to be protected from a supply cart which carried the asphalt mixture (approximately 20 tons in total) over the sensors. Since the used sensors did not have a tension member, which is commonly used as reinforcement in conventional optical communication cables, FOS was expected to keep its sensitivity even if the asphalt exhibits low stiffness. Once the tension member reinforces the sensor itself, the sensor's behavior would not comply with the flexible asphalt structure. In such case, the fact that the transferred stress from the asphalt to the sensor is drastically decreased accordingly leads to low sensor sensitivity. With the aim of flexibility and protection, a thermoplastic polyester elastomer-jacketed optical fiber with a diameter of 0.9 mm was directly coated to a diameter of 5 mm with adhesive polyethylene. Hence, the inner sensing element that is in the center of the fiber with a diameter of 0.25 mm behaved in the same way as the surrounding asphalt through the coating and jacket. A preliminary installation test on the hot mix asphalt fabrication, that is, compacted with aggregates at over 170°C, confirmed that the embedded FOS successfully survived. Also, FOS positioned in a trench of a lower layer was found to survive a heavy machinery construction. Though, only three sensor types out of four have survived the asphalt construction.

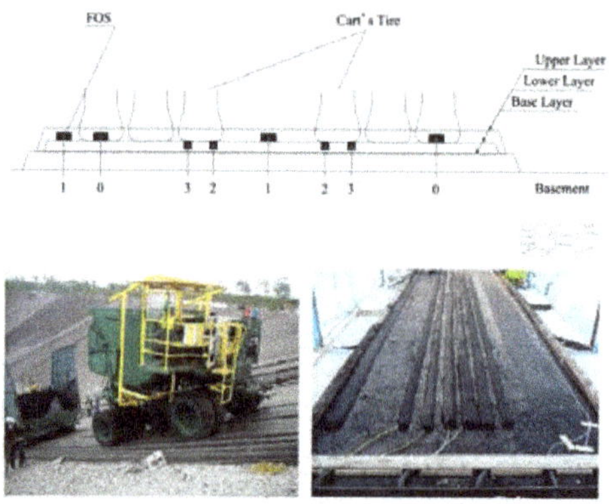

Figure 6. FOS installation on lower asphalt layer (figures are reproduced from Imai et al. [7]).

2.2.6. Asphalt Pavement Structural Health Monitoring with FBG Sensors (2012–2014) (China)

Optical fiber Bragg grating (OFBG) sensors to monitor the 3D strain of an asphalt layer were installed in a highway system in China [46]. Figure 7 shows the proposed sensors which have been assembled in three-dimension by a fiber reinforced polymer (FRP) connection, which has one inset hole for the vertical sensor and two spiral outshoots for the installation of the two transverse sensors. The short-gauged sensor is intended to monitor the vertical strain and the long-gauged ones are used to monitor the horizontal strains of the host structure. Reflective optical fiber Bragg grating signals from all of the three strain sensors are monitored in real time by an optical signal analyzer (OSA) and recorded by the computers for post data processing.

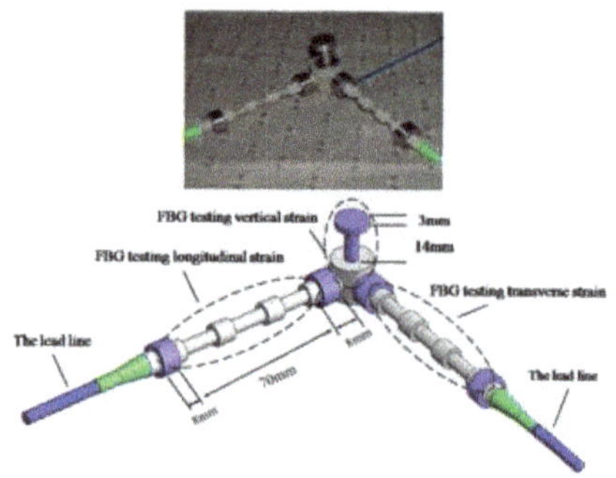

Figure 7. Fabricated 3D strain sensor assembly for 3D structural strain monitoring (figure is reproduced from Zhou et al. [46]).

The project was elaborated in Jilin (China) with the aim: to measure the vertical compressive strain at the top of subgrade, the transverse tensile strain at the bottom of asphalt layer and the vertical compressive strain at the middle of asphalt layer based on key mechanical information and position in flexible pavement [5]. During the FBG sensors installation all cables were bound to avoid loss of signal for being pulled up. Static compaction (without any vibration applied for first two times) was applied to protect sensors from a heavy-duty loading. This study gave an idea about possible pavement monitoring by means of FBG sensors.

Monitoring of the in-situ compaction of an asphalt pavement by means of FBG sensing technology was performed during road construction in China in 2014 [47]. Three measuring points about three kilometers away from each other were chosen. FBG sensors were embedded near the curb of the pavement which is usually not easy to compact. To assure the survival rate of FBG sensors, sensors were embedded in grooves with similar configuration like in research study of Dong et al. [5]. During the monitoring process, rollers were operated at a constant speed of 4.8 km/h; strain and compactness were tested and calculated to identify how many roller passes are necessary for the compaction procedure. To verify the accuracy of compaction quality, the density of the pavement materials was tested by drilling cores from the pavement each 30 m away from the measuring points. When the temperature of pavement dissipated to normal temperature, the response of the pavement was evaluated using a test truck with known weight. Results showed that a lager compactness of HMA would result in a smaller vertical deformation of the pavement under dynamic load. Results also showed that the elastic recovery of the material would be obviously seen in the deformation curve, which indicated that FBG sensing technology has a sufficient precision for monitoring the deformation of an asphalt pavement. FBG sensors embedded in asphalt pavement can also be used for long-term monitoring of pavement structural behavior and provide the basis data for timely maintenance of asphalt pavement.

2.2.7. Installation of FBG Sensors in Asphalt (2016) (ASPARi, The Netherlands)

In the ASPARi project (a research project lead by the University of Twente (The Netherlands) in cooperation with several road construction companies throughout The Netherlands (for more information—https://www.utwente.nl/en/et/trc/projects/aspari/) several experimental programs were carried out where FBG sensors were applied in asphalt pavement [48]. Some of the items that were investigated in this project included: (i) a practical way to install and protect the sensors; (ii) which parameters influence the values delivered by the FBG-sensors; (iii) whether thermal values could be compared with thermocouples; and (iv) the output of the FBG-sensors was investigated to be suitable for models which could be of added value for contractors (long-term performance, vehicle induced loads after the compaction process and the amount of energy used during the asphalting process). It was concluded that it is possible to install the sensors on the desired position in the asphalt. However, protection of the sensors is a more complicated task; some of the sensors were broken because of the experiments. Temperature sensors values deviated from the thermocouple's ones, since some of the sensors were placed in a steel tube for protection. It was concluded that the FBG, partly due to its versatility, still proves to be a promising technology. However, there is still a lack of specific knowledge about deployment technology for the early stage of the asphalting process. Therefore, it was required that more research is conducted on the possible adaptation of the FBG by the road construction industry. As a result, the models to be developed in this research project had to wait until the input, which consisted of the data acquired from the FBG-sensors, could be more reliable, accurate and robust.

2.2.8. FBG Monitoring System (2017) (UAntwerp, Belgium)

Two new approaches to FBG sensors installation in three asphalt pavement layers were implemented for the first time in Belgium: (i) the installation of FBG sensors in prefabricated asphalt specimens with dimensions 50 × 15 × 500 mm with a 2 mm deep groove at the bottom of the specimen in the base layer, directly towards the base and (ii) the installation of FBG sensors at the surface of the previously constructed asphalt layer in 2 mm deep grooves. Both innovative approaches allowed the implementation of FBG sensors without sawing the whole layer into two parts. The installation of the FBG monitoring system prototype was a part of the project–CyPaTs, in which a bicycle path (length—96 m and width—4 m) was accomplished at UAntwerp in 2017 [49]. The installed FBG sensors were commercially available, organic modulated, ceramic-coated Draw Tower Gratings (DTG®) [50] with outer diameter of 0.2 mm, embedded in a glass fiber reinforced plastic (GFRP) round profile with an outer diameter of 1 mm and protected with an additional high-density polyethylene (HDPE) coating with outer diameter of 0.5 mm. The installed FBG monitoring system prototype consisted of several FBG chains: 2 fibers with 30 DTG® (spacing between sensors 10 cm) and 4 fibers with 5 DTG® (spacing between sensors 80 cm) and two temperature sensors (FBG based ~40 mm SS housing and ~1 mm diameter) embedded in three asphalt layers with a cross section configuration (width—4 m and length—3.2 m). The strain and temperature data were obtained using an interrogator FBG-SCAN 808D with 8 channels (1507–1593 nm wavelength range, 250 Hz measurement frequency for all channels). The FBG sensors configuration embedded in the three asphalt layers in both transverse and longitudinal directions at the bottom of each layer can be seen in Figure 8. All FBG sensors in all three asphalt layers survived during pavement construction. It was possible to learn how exactly pavement works inside during the pavement construction [22–26]. Fiber egress points were designed as such to come out at side of the pavement. Redundancy was built-in by the option to measure the strain wires from both sides. Monitoring of the FBG system was performed since the construction of the pavement. All sensing fibers were connected to a single mode multifiber (SMF) backbone cable to enable continuous monitoring from inside the building. This FBG system's appliance as long-term monitoring system is possible and it can be installed, for example, in heavy-duty roads during their construction.

Figure 8. UAntwerp FBG monitoring system sensors configuration in the three asphalt layers with sensors embedded in prefab specimens and grooves at CyPaTs bicycle path [22–26].

The same advanced UAntwerp FBG monitoring system—in collaboration with Port of Antwerp (Belgium) and Com&Sens (Belgium)—will be installed this summer in the Port of Antwerp to monitor a heavy-duty pavement in real-time and over the long-term.

3. Discussion

Pavement design is essentially and usually a structural long-term evaluation process which is needed to ensure that traffic loads are efficiently distributed at all levels of the total road structure.

Furthermore, it cannot be monitored only for a short-term period to get a complete analysis of the pavement behavior. As shown in the selected literature, FBG sensors are the most promising candidates to effectively replace conventional strain gauges for a long-term monitoring application in a harsh environment.

This part of the review describes the outcome of the results if those systems provide repeatable and suitable results for a long-term monitoring and of the inverse modeling approach performed by several researchers from the above-mentioned research studies using FOS, strain gauges and FWD. It must be noted that only few studies made the comparison between the FOS strain results and the FWD measurements. Most of the studies were focused on the feasibility of applying FOS in pavement structures.

In the research group of NTUA, Greece [39] a FOS system was used for horizontal (tensile) strain measurements at the top and the bottom of a foamed asphalt (FA) layer. These locations were selected, considering the results of the strain response analysis based on back-calculated moduli at characteristic locations within the body of the recycled layer in a similar pavement structure. Tensile strains at these locations were critical in terms of possible fatigue failure and directly related to the performance of the pavement structure. Strain measurements were conducted with the FOS system during FWD loading (40 kN). In-depth temperature measurements were conducted with thermometers into drilled holes and using a single drop of glycerol, according to COST 336. Back-calculation of the recycled pavement layer moduli was carried out using MODCOMP© software (US). The horizontal (tensile) strains at the top as well as at the bottom of the foamed asphalt layer were calculated using finite-element linear analysis (FEA), using the ABAQUS© software (US). Compared to a much simpler multilayer elastic system analysis, the three-dimensional (3D) FEA was considered as more precise for the simulation of the FWD loading. The back-calculated moduli of the pavement's layers were used as input for the forward FEA. Because the target of the analysis is the comparison of the calculated strains with the measured values, no adjustment of the back-calculated foamed asphalt moduli was conducted. The maximum measured tensile strains at the bottom of the FA layer, transversely or longitudinally, ranged from 6 to 15 $\mu\varepsilon$ when applying FWD loading of 40 kN. The maximum calculated tensile strains (FEA) ranged from 19 to 34 $\mu\varepsilon$. The maximum measured tensile strains at the top of the foamed asphalt layer (respectively bottom of the asphalt layers), transversely or longitudinally, ranged from −10 to 7 $\mu\varepsilon$ when applying an FWD loading of 40 kN. The maximum calculated tensile strains (FEA) ranged from 3 to 19 $\mu\varepsilon$. In all cases, the measured strains were lower than the relative calculated strains, indicating that the FEA overestimates the tensile strains, especially at the bottom of the FA layer. This discrepancy can be attributed to various sources of uncertainty during the calculations. There may also be errors and limitations in instrumentation design and installation that contribute to the mentioned differences [39]. It can be noted here that the strain measurements could also be underestimated or that the sensing system itself had influence on the material behavior. It was concluded that FWD can only be used as excitation to obtain tensile strain values of FA pavement materials by means of FOS. This outcome is promising towards the possibility of potential use of FWD for simulating the strain induced by a moving tire. However, as it was noted by Loizos et al. [39], more in situ data (measured strains) could lead to more reliable conclusions.

In the research group of University of Laval, Canada, it was noted by Bilodeau & Doré [31] that an existing alternative approach for determining the tensile strain occurring at the bottom of the asphalt concrete layer consists of bypassing the necessity of back-calculating the modulus of each layer by direct strain estimation from the deflection basin. In most cases, this approach overestimates the tensile strains occurring at the bottom of asphalt/concrete layers, leading to an underestimation of the pavement's fatigue life. A model has been developed based on a theoretical analysis and a field calibration using data obtained at the Laval University experimental pavement site which allowed to compute tensile strain values that are in good agreement with tensile strain values obtained using theoretical models and, also with field measurements. The proposed approach to determine the tensile strain at the bottom of asphalt concrete layers using the deflection basin is based on Equations (1)–(3)

to estimate the modulus of the mechanically involved layers, which are the asphalt concrete and base layers:

$$\varepsilon_t = \frac{H_{AC}(0.199 + 3.868 \times 10^{-2} \log(E_{AC}) - 1.122 \times 10^{-4} E_{base} - 8.627 \times 10^{-4} H_{AC}}{2R} \quad (1)$$

$$\log|E^*| = 0.95 + \frac{3.27}{1 + e^{(-2.67 - 0.51 \log f + 0.07T)}} \quad (2)$$

$$E_{base} = [41.333 \ln(E_{AC}) - 438.43][\ln(H_{AC}) - 5.9877] + 50.683 \quad (3)$$

in which H_{AC} is the thickness (mm), E_{AC} is modulus of the asphalt concrete layer (MPa), E_{base} is the modulus of the aggregate base layer (MPa), $|E^*|$ is the dynamic modulus (MPa), f is the loading frequency (Hz) and T is the temperature (°C). To use the proposed model, the following values have to be obtained: temperature of the asphalt concrete, loading frequency of the FWD, deflection basin (d_0 and d200 in mm), loading plate radius (mm) and thickness of asphalt concrete (mm). Therefore, the model is based on easily obtainable parameters and has a good prediction capacity. The tensile strains induced at the bottom of the asphalt concrete layer were measured with two different fiber optic strain gauges (asphalt concrete cores and plate). The FWD drops (40 kN) were applied directly on top of the sensors. The temperature was measured using a sensor located inside the asphalt concrete layer. Between the FWD tests, the temperature of the instrumented pavement zone was also controlled with a thermal blanket connected to two thermal baths controlling the circulated fluid temperature. An average loading frequency of 34 Hz was found for the FWD tests. A calibration factor of 1.87 is found between the model and the field conditions. This difference may be attributed to numerous factors, such as the difference between idealized modelling conditions, which were set for simplification reasons and non-ideal field conditions. As part of the field validation and calibration process, the objective was to identify a coefficient such as the one that was presented to consider the main differences between the model and the field conditions. Equation (1) allows the tensile strain to be determined at the bottom of the asphalt concrete layer as determined with finite element modelling using a combination of asphalt concrete thickness and modulus, base layer modulus, as well as using the deflection basin through the radius of curvature [31].

In the study of Primusz et al. [51], an approach was presented to define the modulus of the examined pavement layers, knowing the deflection curve and the thickness of the bound layer, without using further back-calculation. According to the performed research, there is a very strong correspondence between the E modulus of the bottom layer, the vertical deflection interpreted at the load axis and the radius of curvature of the pavement:

$$E = 1224.45 \cdot D_0^{-1.623} \cdot R_0^{-0.629} \quad (4)$$

where D_0 is the measured vertical deflection (mm), R_0 is radius of curvature (m).

The objective of the study of Grellet et al. [44], as a part of a collaborative project between Laval University (Quebec City, Canada) and the IFSTTAR (Nantes, France), was to integrate viscoelastic properties in an asphalt pavement model in order to understand and predict the two types of cracking mechanisms: (i) initiated at the bottom of the asphalt layer and propagating toward the surface (bottom-up cracks); (ii) the second is initiated near the surface of the pavement and propagates downward through the bound layers (top-down cracks). The field tests have been conducted at the IFSTTAR's accelerated pavement testing facility and at the SERUL (Laval University Road Experimental Site). Results from these studies showed that fiber optic sensors allowed adequately characterizing the strains occurring within the layer and evaluating the effects of the load configurations [43]. It was concluded in Grellet et al. [44] that a better pavement modeling is obtained using viscoelastic properties for the mechanical behavior of the asphalt layers and for the interface. Modeling the tack coat with a viscoelastic layer modifies the stresses and strain distribution through the layers and alters the prediction of pavement performance. Significant tensile stresses appear near the surface and could produce top-down cracking. The integration of the viscoelastic interface imposes a redistribution of the stresses through the layer. Tensile stresses increase near the surface and near the interface but decrease at the bottom of the layers. However, strains are higher considering the interface. The high extension strains (more than 250 με) have been measured at the bottom of the bituminous wearing course of a thin pavement structure at a high temperature (30°C) with a slow return of the strains to zero after loading.

The high tensile strains observed at high temperature at the bottom of the wearing course suggest that at high temperature, the interface with the lower layers cannot be considered as fully bonded and that some sliding between the wearing course and the base course occurs, generating these tensile strains. In other words, this means that the degree of bonding of the interface seems to change with temperature. This could be explained by the behavior of the tack coat (bitumen emulsion) at the interface; at high temperature, this emulsion presents a low stiffness, which reduces the shear resistance of the interface [51] It was stated that several temperature and load parameters must be evaluated to determine the most critical conditions.

An interesting outcome can be found in the study of Duong et al. [52], showing the results of the monitoring by means of strain gauges of an experimental pavement section recently re-constructed on a French motorway. The measurements showed that at high temperatures (above 30°C), high strain levels (150 με) are measured at the bottom of the bituminous layers. These strains exceed the limit fatigue strain, leading to failure for 1 million load cycles, determined using standard two-point bending fatigue tests, performed at 10°C and 25 Hz. Similar tests results had been obtained previously at IFSTTAR in accelerated pavement tests. Calculations were performed with the ALIZE-LCPC and Viscoroute programs (France), to fit the experimental strains, using elastic and viscoelastic pavement models. These calculations have shown that the pavement interfaces cannot be considered as fully bonded and that their level of bonding clearly changes with temperature. Different modelling cases have been tested and the best predictions have been obtained when modelling the interfaces as thin elastic layers (2 mm thick), with a low elastic modulus (in the range of 120 to 20 MPa for the range of low to high temperatures obtained on site). These interface layer moduli decreased when the temperature increased, and a particularly significant drop was observed between 25°C and 30°C with both ALIZE-LCPC and Viscoroute calculations. These results stress the fact that in pavement calculations, great attention should be paid to the modelling of pavement layer interfaces. These interfaces cannot be considered as fully bonded in all cases and in particular, their degree of bonding may decrease at high temperatures (above 30°C), where the stiffness of the tack coat at the interface becomes very low. This can lead to higher tensile strains at the bottom of the bituminous layers than predicted by standard design calculations, with bonded layers and thus to higher fatigue damage than predicted.

4. Conclusions

The most significant cases/attempts to perform experimental measurements with optical fibers in asphalt pavements in the last two decades have been included in this paper. Some of the main conclusions can be summarized as follows: the available technical information described mostly the attempts to install the FOS in the pavement; only a few studies provided technical details on the FOS installation; only a few cases could be referred to as long-term pavement monitoring; most of the cases envisaged more experimental measurements than monitoring of the pavement itself. Some suggested solutions were given to validate an inverse modelling approach based on the results of FWD and FOS. It can be concluded that the application of FOS in the asphalt pavement: (i) has proved to be a useful and a promising tool for in-situ strain measurements under real traffic loading at the bottom of the asphalt layers; (ii) it also proved that interfaces of the pavement structure cannot be considered as fully bonded, in particular, at high temperatures, although when using the standardized FWD methodology it is always considered that interfaces are fully bonded; (iii) instrumentation design and installation of FOS contributes to the differences in calculated and measured tensile strain values; (iv) due to the deviations between calculated strains (pavement design, FWD data analysis) and FOS signal records (measured strains) more experimental work is needed to define a calibration factor.

Author Contributions: Writing—original draft preparation, P.K.D.M.; Writing—editing, G.L., C.V., E.V. and W.V.d.b.; Writing—review, S.V., J.B., N.S. and J.D.W.

Funding: This research was funded by the Port of Antwerp (within the project "Duurzame Asfaltverhardingen voor zwaar belaste wegdekken") and by the University Research Fund at the University of Antwerp through the project BOF/STIMPRO/36539 "Development of a novel optical signal processing method for analyzing data of the deformations of the asphalt construction by using Fiber Bragg technology to design new asphalt model," supported by both the Road Engineering Research Section (EMIB) and Op3Mech research group.

Conflicts of Interest: The authors declare no conflict of interest.

References

1. Read, J.; Whiteoak, D.; Hunter, R.N. *The Shell Bitumen Handbook*; Thomas Telford: London, UK, 2003; p. 460.
2. Van Gurp, C.A.P.M. Characterization of Seasonal Influences on Asphalt Pavements with the Use of Falling Weight Deflectometers. Ph.D. Thesis, TU Delft, Delft, The Netherland, 1995. Available online: https://repository.tudelft.nl/islandora/object/uuid:596d5b28-308d-4c74-a3b8-cad632a93619?collection=research (accessed on 12 May 2019).
3. Sun, L. *Structural Behavior of Asphalt Pavements*; Butterworth-Heinemann: Oxford, UK, 2016; p. 1070, ISBN 978-0-1280-2893-3.
4. Permissible Maximum Weights of Lorries in Europe, International Transport Forum. 2015. Available online: https://www.itf-oecd.org/sites/default/files/docs/weights_0.pdf (accessed on 12 May 2019).
5. Dong, Z.; Li, S.; Wen, J.; Chen, H. Asphalt pavement structural health monitoring utilizing FBG sensors. *Adv. Eng. Forum* **2012**, *5*, 339–344. [CrossRef]
6. Bock, W.J.; Gannot, I.; Tanev, S. *Optical Waveguide Sensing and Imaging, NATO Science for Peace and Security Series-B: Physics and Biophysics*; Springer: Dordrecht, The Netherlands, 2008; p. 269. [CrossRef]
7. Imai, M.; Igarashi, Y.; Shibata, M.; Miura, S. Experimental study on strain and deformation monitoring of asphalt structures using embedded fiber optic sensor. *J. Civ. Struct. Health Monit.* **2014**, *4*, 209–220. [CrossRef]
8. Chapeleau, X.; Blanc, J.; Hornych, P.; Gautier, J.-L.; Carroget, J. Assessment of cracks detection in pavement by a distributed fiber optic sensing technology. *J. Civ. Struct. Health Monit.* **2017**, *7*, 459–470. [CrossRef]
9. Meggitt, B.T. Chapter 17-Fiber Optics in Sensor Instrumentation. In *Instrumentation Reference Book*; Boyes, W., Ed.; Butterworth-Heinemann: Oxford, UK, 2010; pp. 191–216. [CrossRef]
10. Wang, J.-N.; Tang, J.-L.; Chang, H.-P. Fiber Bragg grating sensors for use in pavement structural strain-temperature monitoring. *Proc. SPIE* **2006**, *6174*, 61743S. [CrossRef]
11. Luyckx, G.; Voet, E.; Lammens, N.; Degrieck, J. Strain measurements of composite laminates with embedded fibre Bragg gratings: Criticism and opportunities for research. *Sensors* **2011**, *11*, 384–408. [CrossRef] [PubMed]
12. Geernaert, T.; Sulejmani, S.; Sonnenfeld, C.; Luyckx, G.; Chah, K.; Areias, L.; Mergo, P.; Urbanczyk, W.; Van Marcke, P.; Coppens, E.; et al. Microstructured optical fiber Bragg grating-based strain and temperature sensing in the concrete buffer of the Belgian supercontainer concept. *Proc. SPIE* **2014**, *9157*, 915777. [CrossRef]
13. Luyckx, G.; Voet, E.; Grefhorst, R.; Peeters, J.; Degrieck, J. Long term monitoring of an all-composite water lock using fibre optics. In Proceedings of the SAMPE EUROPE 2016, Liege, Belgium, 13–15 September 2016.
14. Liu, W.; Liu, X.; Wang, Z.; Zhi, Z. High temperature deformation investigation of asphalt mixture with nanosized volcanic ash fillers using optical fiber sensor. *Measurement* **2019**, *140*, 171–181. [CrossRef]
15. Li, L.; Huang, X.; Wang, L.; Li, C. Integrated experimental and numerical study on permanent deformation of asphalt pavement at intersections. *J. Mater. Civ. Eng.* **2013**, *25*, 907–912. [CrossRef]
16. Abe, N.; Mizukami, J.; Kimura, M. Influence of moving load to extend the deformation of asphalt pavement. In Proceedings of the 11th ISAP conference on Asphalt Pavements, Nagoya, Japan, 1–6 August 2010; pp. 1054–1063.
17. Xie, J.; Li, H.; Gao, L.; Liu, M. Laboratory investigation of rutting performance for multilayer pavement with fiber Bragg gratings. *Constr. Build. Mater.* **2017**, *154*, 331–339. [CrossRef]
18. Dong, Z.; Tan, Y.; Cao, L.; Li, S. Rutting mechanism analysis of heavy-duty asphalt pavement based on pavement survey, finite element simulation and instrumentation. *J. Test. Eval.* **2012**, *40*, 1228–1237. [CrossRef]
19. Li, K.; Xie, J. Experiment and research of using fiber Bragg grating to monitor the dynamic response of asphalt concrete. *Appl. Mech. Mater.* **2011**, *97–98*, 301–304. [CrossRef]

20. Li, Q.; Cary, C.; Combs, S.; Garg, N. Evaluation of asphalt concrete layer response using asphalt strain gauges and fiber optic strain gauges. In Proceedings of the International Conference on Transportation and Development, Houston, TX, USA, 26–29 June 2016; pp. 42–53. [CrossRef]
21. Chen, S.-Z.; Wu, G.; Feng, D.-C.; Zhang, L. Development of a bridge weigh-in-motion system based on long-gauge fiber Bragg grating sensors. *J. Bridge Eng.* **2018**, *23*, 04018063. [CrossRef]
22. Kara De Maeijer, P.; Van den bergh, W.; Vuye, C. Case study on the technique of installation of fiber Bragg gratings sensors in three asphalt layers. In Proceedings of the 13th ISAP Conference on Asphalt Pavements, Fortaleza, Brazil, 19–21 June 2018; pp. 1–7.
23. Kara De Maeijer, P.; Van den bergh, W.; Vuye, C. Fiber Bragg gratings sensors in three asphalt pavement layers. *Infrastructures* **2018**, *3*, 16. [CrossRef]
24. Kara De Maeijer, P.; Van den bergh, W.; Vuye, C. Case study on strain and temperature real-time monitoring by using fiber Bragg grating sensors embedded in three asphalt layers. In Proceedings of the 4th International Conference on Service Life Design for Infrastructures, Delft, The Netherlands, 27–30 August 2018; Ye, G., Yuan, Y., Romero Rodriguez, C., Zhang, H., Šavija, B., Eds.; RILEM PRO 125. RILEM Publications: Paris, France; pp. 937–940.
25. Kara De Maeijer, P. The Effectiveness of Application of Fiber Bragg Grating Sensors in Pavement Engineering. Presented at the Asphalt Innovatie Symposium (AIS2018), Antwerp, Belgium, 13 December 2018; Available online: https://www.uantwerpen.be/en/research-groups/emib/rers/activities/ais2018 (accessed on 12 May 2019).
26. Kara De Maeijer, P.; Van den bergh, W.; Vuye, C.; Vanlanduit, S.; Braspenninckx, J.; Stevens, N.; Voet, E.; Luyckx, G.; De Wolf, J. Inverse modelling approach-fiber Bragg grating (FBG) measurements in comparison to falling weight deflectometer (FWD) measurements: Review. Presented at the 7th International Conference Bituminous Mixtures and Pavements, Thessaloniki, Greece, 12–14 June 2019; p. 211.
27. Lenngren, C. Advanced backcalculation of FWD data on asphalt pavements. In Proceedings of the 13th ISAP Conference on Asphalt Pavements, Fortaleza, Brazil, 19–21 June 2018.
28. Papavasiliou, V.; Loizos, A. Assessment of the bearing capacity of pavements using fiber optic sensors. In Proceedings of the 10th International Conference on the Bearing Capacity of Roads, Railways and Airfields (BCRRA 2017), Athens, Greece, 28–30 June 2017; Taylor & Francis Group: London, UK, 2017; pp. 653–660, ISBN 978-1-138-29595-7.
29. Kim, Y.R.; Park, H. *Use of Falling Weight Deflectometer Multi-Load Data for Pavement Strength Estimation*; Final Report, Research Project No HWY-00-4; Department of Civil Engineering, North Carolina State University Raleigh: Raleigh, NC, USA, 2002.
30. Chen, D.; Scullion, T. Forensic investigations of roadway pavement failures. *J. Perform. Constr. Facil.* **2008**, *22*, 35–44. [CrossRef]
31. Bilodeau, J.-P.; Doré, G. Estimation of tensile strains at the bottom of asphalt concrete layers under wheel loading using deflection basins from falling weight deflectometer tests. *Can. J. Civ. Eng.* **2012**, *39*, 771–778. [CrossRef]
32. Xue, W.; Wang, D.; Wang, L. A review and perspective about pavement monitoring. *Int. J. Pavement Res. Technol.* **2012**, *5*, 295–302.
33. Bueche, N.; Rychen, P.; Dumont, A.G. Optical fiber feasibility study in accelerated pavement testing facility. In Proceedings of the 6th International Conference on Maintenance and Rehabilitation of Pavement and Technological Control (MAIREPAV), Torino, Italy, 8–10 July 2009.
34. Doré, G.; Duplain, G.; Pierre, P. Monitoring mechanical response of in service pavements using retrofitted fibre optic sensors. In Proceedings of the Advanced Characterization of Pavement and Soil Engineering, Athens, Greece, 20–22 June 2007; Loizos, A., Scarpas, T., Al-Qadi, I.L., Eds.; CRC Press Taylor & Francis Group: London, UK, 2007; pp. 883–891, ISBN 978-0-415-44882-6.
35. Grellet, D.; Dore, G.; Bilodeau, J.-P. Comparative study on the impact of wide base tires and dual tires on the strains occurring within flexible pavements asphalt concrete surface course. *Can. J. Civ. Eng.* **2012**, *39*, 526–535. [CrossRef]
36. OpSens Solutions White-Light Polarization Interferometry Technology. Available online: https://opsens-solutions.com/wp-content/uploads/sites/4/2015/04/IMP0002-WLPI-REV2.5.pdf (accessed on 17 June 2019).
37. Loizos, A.; Papavasiliou, V.; Plati, C. Investigating in situ stress-dependent behavior of foamed asphalt-treated pavement materials. *Road Mater. Pavement Des.* **2012**, *13*, 678–690. [CrossRef]

38. Loizos, A.; Plati, C.; Papavasiliou, V. Fiber optic sensors for assessing strains in cold in-place recycled pavements. *Int. J. Pavement Eng.* **2013**, *14*, 125–133. [CrossRef]
39. Loizos, A.; Papavasiliou, V.; Plati, C. Effectiveness of FWD to simulate traffic loading in recycled pavements. *J. Perform. Constr. Facil.* **2016**, *30*, 04014193. [CrossRef]
40. Chapeleau, X.; Blanc, J.; Hornych, P.; Gautier, J.-L.; Carroget, J. Use of distributed fiber optic sensors to detect damage in a pavement. In Proceedings of the 7th European Workshop on Structural Health Monitoring, Nantes, France, 8–11 July 2014; pp. 1847–1854.
41. Grellet, D.; Doré, G.; Bilodeau, J.-P. Effect of tire type on strains occurring in asphalt concrete layers. In Proceedings of the 11th International Conference on Asphalt Pavements, Nagoya, Japan, 1–6 August 2010; p. 10.
42. Grellet, D.; Doré, G.; Kerzreho, J.-P.; Piau, J.-M.; Chabot, A.; Hornych, P. Experimental and Theoretical Investigation of three-dimensional strain occurring near the surface in asphalt concrete layers. In *7th RILEM International Conference on Cracking in Pavements*; Scarpas, A., Kringos, N., Al-Qadi, I.A.L., Eds.; Springer: Dordrecht, The Netherlands, 2012; pp. 1017–1027. [CrossRef]
43. Grellet, D.; Doré, G.; Bilodeau, J.-P.; Gauliard, T. Wide-base single-tire and dual-tire assemblies: Comparison based on experimental pavement response and predicted damage. *Transp. Res. Rec.* **2013**, *2369*, 47–56. [CrossRef]
44. Grellet, D.; Doré, G.; Chupin, O.; Piau, J.-M. Experimental evidence of the viscoelastic behavior of interfaces in bituminous pavements—An explanation to top-down cracking? In *8th RILEM International Conference on Mechanisms of Cracking and Debonding in Pavements*; Springer: Dordrecht, The Netherlands, 2016; pp. 575–580.
45. Garg, N.; Bilodeau, J.-P.; Doré, G. Experimental study of asphalt concrete strain distribution in flexible pavements at the national airport pavement test facility. In Proceedings of the 2014 FAA Worldwide Airport Technology Transfer Conference, Galloway, NJ, USA, 5–7 August 2014.
46. Zhou, Z.; Liu, W.; Huang, Y.; Wang, H.; He, J.; Huang, M.; Ou, J. Optical fiber Bragg grating sensor assembly for 3D strain monitoring and its case study in highway pavement. *Mech. Syst. Signal Process.* **2012**, *28*, 36–49. [CrossRef]
47. Tan, Y.; Wang, H.; Ma, S.; Xu, H. Quality control of asphalt pavement compaction using fiber Bragg grating sensing technology. *Constr. Build. Mater.* **2014**, *54*, 53–59.
48. Van der Vegt, J. Over de Haalbaarheid van Fiber Bragg Grating Sensoren in Het Asfaltverwerkingsproces. Bachelor's Thesis, UTwente, Enschede, The Netherlands, 2016. (In Dutch). Available online: https://essay.utwente.nl/71035/1/Vegt-Jurian.pdf (accessed on 12 May 2019).
49. Van den bergh, W.; Jacobs, G.; Kara De Maeijer, P.; Vuye, C.; Arimilli, S.; Couscheir, K.; Lauriks, L.; Baetens, R.; Severins, I.; Margaritis, A.; et al. Demonstrating innovative technologies for the Flemish asphalt sector in the CyPaTs project. In *Proceedings of the 3rd World Multidisciplinary Civil Engineering, Architecture, Urban Planning Symposium, Prague, Czech Republic, 18–22 June 2018*; IOP Publishing: Bristol, UK, 2019; Volume 471, p. 022031. [CrossRef]
50. FBGS, Strain Measurement Wire SMW-01, Technical Datasheet. 2015. Available online: http://www.fbgs.com/products/strain-sensors/smw-01/ (accessed on 17 June 2019).
51. Prismusz, P.; Peterfalvi, J.; Marko, G.; Toth, C. Effect of pavement stiffness on the shape of deflection bowl. *Acta Silv. Lign. Hung.* **2015**, *11*, 39–54. [CrossRef]
52. Duong, N.S.; Blanc, J.; Hornych, P. Analysis of the behavior of pavement layers interfaces from in situ measurements. In *Bearing Capacity of Roads, Railways and Airfields*; Taylor & Francis Group: London, UK, 2017; pp. 1503–1510, ISBN 978-1-138-29595-7.

infrastructures

Article

Sensitivity of the Flow Number to Mix Factors of Hot-Mix Asphalt

Md Rashadul Islam *, Sylvester A. Kalevela and Shelby K. Nesselhauf

Department of Engineering Technology, Colorado State University, Pueblo, CO 81001, USA; sylvester.kalevela@csupueblo.edu (S.A.K.); sk.nesselhauf@pack.csupueblo.edu (S.K.N.)
* Correspondence: md.islam@csupueblo.edu; Tel.: +1-719-549-2612

Received: 25 April 2019; Accepted: 4 June 2019; Published: 7 June 2019

Abstract: In the design of pavement infrastructure, the flow number is used to determine the suitability of a hot-mix asphalt mixture (HMA) to resist permanent deformation when used in flexible pavement. This study investigates the sensitivity of the flow numbers to the mix factors of eleven categories of HMAs used in flexible pavements. A total of 105 specimens were studied for these eleven categories of HMAs. For each category of asphalt mixture, the variations in flow number for different contractors, binder types, effective binder contents, air voids, voids in mineral aggregates, voids filled with asphalt, and asphalt contents were assessed statistically. The results show that the flow numbers for different types of HMA used in Colorado vary from 47 to 2272. The same mix may have statistically different flow numbers, regardless of the contractor. The flow number increases with increasing effective binder content, air voids, voids in mineral aggregates, voids filled with asphalt, and asphalt content in the study range of these parameters.

Keywords: hot-mix asphalt; flow number; effective binder content; air voids; voids in mineral aggregates; voids filled with asphalt; asphalt content

---

1. Introduction

The flow number (N) is an empirical way of characterizing a hot-mix asphalt (HMA) mixture's rutting potential. To determine the flow number, a cyclic load in haversine form is applied on a cylindrical specimen axially as shown in Figure 1. The duration of the load pulse is 0.1 s, followed by a rest period of 0.9 s. The permanent axial deformation measured at the end of the rest period is monitored during repeated loading, and the strain is calculated by dividing by the initial gauge length. The test may be conducted with or without confining pressure. However, if confining pressure is used, it is kept constant while the flow number is tested. If confining pressure is used, it remains constant during the test.

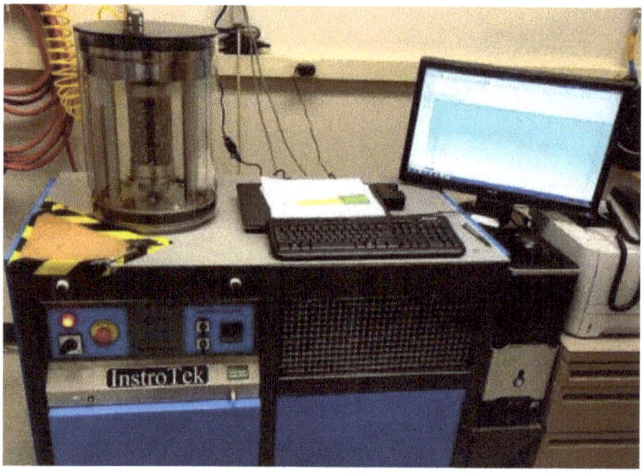

Figure 1. Flow number test setup.

In the flow number test, the permanent strain at each cycle is measured while constant deviator stress is applied at each load cycle on the test sample. Permanent deformation of asphalt pavements has three stages [1]: (i) primary or initial consolidation; (ii) secondary; and (iii) tertiary or shear deformation.

Figure 2 shows the three stages of permanent deformation. The first stage of deformation is due to the initial filling of voids, particle rearrangement, etc. The second region is the actual deformation of the aggregates, asphalt film, etc. The tertiary region is the zone where drastic shear failure of the mix occurs. The *N*-value is the number of load cycles at which tertiary flow begins, i.e., where permanent deformation occurs non-linearly. Tertiary flow can be differentiated from secondary flow by a marked departure from the linear relationship between cumulative strain and number of cycles in the secondary zone, as shown in Figure 2. It is assumed that in tertiary flow, the specimen's volume remains constant. The *N*-value can be correlated with rutting potential. The higher the flow number, the better the mix is in terms of its rutting resistance.

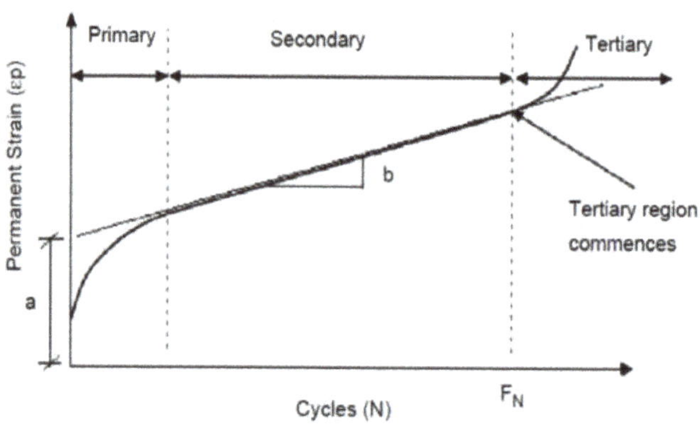

Figure 2. Relationship between permanent strain and number of load cycles [1].

Mathematically, the *N*-value can be determined using the Francken model.

The Francken model is currently built into the Asphalt Mixture Performance Tester (AMPT) software (Federal Highway Administration, Washington DC, USA). At the beginning, the entire permanent strain curve is fitted using nonlinear least squares optimization, as shown in Equation (1).

The flow number is then determined from the second derivative of the best fitted curve. The flow number is the number of cycles where the second derivative, Equation (2), changes from negative to positive.

$$\varepsilon_p = A\left(n^B\right) + C\left(e^{Dn} - 1\right), \quad (1)$$

where ε_p is the permanent strain (%); n is the number of cycles; and A, B, C, and D are fitting parameters.

$$\frac{d^2\varepsilon_p}{dn^2} = AB(B-1)n^{B-2} + CD^2e^{Dn}, \quad (2)$$

where $\frac{d^2\varepsilon_p}{dn^2}$ is the second derivative of permanent strain with respect to the number of loading cycles.

The final evaluation is an evaluation of the rutting resistance of the mixture using the flow number test defined by the American Association of State Highway and Transportation Officials (AASHTO) TP 79 [2] using the AMPT. The test is conducted at the "high" pavement temperature calculated by the long-term pavement performance (LTPP) Bind 3.1 software program (Federal Highway Administration, Washington DC, USA) for a specific project location. An unconfined flow number test with a repeated deviatoric stress of 600 kPa (87 psi) and a contact deviatoric stress of 30 kPa (4.4 psi) was used in this study. The test was conducted on specimens that were short-term conditioned for two hours at the compaction temperature to simulate the binder absorption and stiffening that occurs during construction. The flow number criteria for HMA as a function of the traffic level are summarized in Table 1.

Table 1. Flow number criteria for hot-mix asphalt (HMA) [2].

Traffic Level, Million Equivalent Single Axle Load (ESAL, 80 kN (18 kips))	Flow Number
Less than 3.0	NA
3.0 to less than 10	50
10 to less than 30	190
More than 30	740

The effects of different mix factors on the flow number have also been studied by different researchers [3–6]. Kaloush [3] determined that the flow number increases with the viscosity of binder and decreases with the test temperature, effective binder content, and air voids. Kvasnak et al. [4] determined that the flow number increases with gyrations and the viscosity of the binder and decreases with voids in mineral aggregates (VMA) when using Wisconsin dense graded mixtures. Both researchers determined that aggregate gradation also affects the flow number. Christensen [5] applied various statistical techniques to relate the flow number with the applied stress level and observed that the flow number decreases with increasing applied deviator stress. Rodezno et al. [6] determined that the flow number increases with the viscosity of the binder; however, it decreases with the test temperature and air voids and is affected by aggregate gradation. The current study was not intended to investigate the viscosity of the binder or the test temperature. Other factors such as the VMA, voids filled with asphalt (VFA), effective binder content, contractors, testing time, mix gradation, and binder types were investigated. There are some other researches on flow number. For example, Irfan et al. [7] evaluated the flow numbers for static and dynamic creep tests and then correlated it with the rutting. Ogundipe [8] studied the flow numbers of lime-modified asphalt concrete. Irfan et al. [9] investigated the flow numbers of fiber-added stone mastic asphalt concrete mixtures. Leiva-Villacorta et al. [10] evaluated the flow numbers for High-modulus asphalt concrete (HMAC) mixtures for use as base course. Tripathi [11] studied the economic benefits of fiber-reinforced asphalt mixtures by flow number including some other tests. Ziari and Divandari [12] developed a flow number prediction model using artificial neural network. Siswanto et al. [13] investigated the flow numbers of Asphalt Concrete Using Crumb Rubber Modified of Motorcycle Tire Waste. All these studies investigated the flow number under different conditions. However, none of the studies investigated the sensitivity of different mix factors such as such as VMA, void-filled with asphalt (VFA), effective binder content, contractors, testing time, mix gradation, and binder types on the flow number of asphalt concrete.

Air void (V_a) is the total volume of the small pockets of air between the coated aggregate particles expressed as a percentage of the bulk volume of the compacted mixture. The volume of the void space among aggregate particles of a mixture that includes the air voids and the effective asphalt content is known as VMA. The portion of the voids in the mineral aggregate that contain asphalt binder is known as VFA. The total asphalt binder content of the mix less the portion of asphalt binder that is lost by absorption into the aggregate is called the effective asphalt content (V_{be}). This portion of binder is coated on the aggregate surface and takes part in binding aggregates. The total asphalt binder used in a mix is called the asphalt content (AC).

The flow number (N) test procedure recommended in the National Cooperative Highway Research Program (NCHRP) project 9-19 is a simple performance test for rutting evaluation of asphalt mixtures. The test showed good correlation with the rutting performance of mixtures at WesTrack, MnROAD, and the Federal Highway Administration's (FHWA's) accelerated loading facility. Subsequent NCHRP studies allowed the development of a provisional standard. AASHTO TP 79 [2] includes test parameters for stress, temperature, specimen conditioning, and minimum flow number criteria that were established for HMA and for warm-mix asphalt (WMA) based on the traffic level.

The current study used the testing conditions and criteria for N testing described in AASHTO TP 79 [2] for unconfined tests. The recommended test temperature, determined by LTPP Bind Version 3.1 software, was the average design high pavement temperature at 50% reliability for cities in Colorado. Tests were conducted at a temperature of 55 °C with an average deviator stress of 600 kPa (87 psi) and a minimum (contact) axial stress of 30 kPa (4.4 psi). For conditioning, samples were kept in a conditioning chamber at the testing temperature for 12 h prior to testing.

To confirm again, the objectives of this study were to study the effects of mix factors such as VMA, VFA, effective binder content (V_{be}), contractors, testing time, mix gradation, and binder types on the flow number of asphalt concrete.

2. Materials

The eleven types of mixtures studied are listed in Table 2 along with their basic information such as nominal maximum aggregate size (NMAS), performance grade (PG) binder type, and number of gyrations used while designing the mixes. Superpave performance grading is reported using two numbers: the first being the average seven-day maximum pavement temperature (°C) and the second being the minimum pavement design temperature likely to be experienced (°C). Thus, a PG 64-22 is intended for use where the average seven-day maximum pavement temperature is 64°C and the expected minimum pavement temperature is −22°C. The letter "S" denotes an NMAS of 0.75 in. (19 mm). The letters "SX" denote an NMAS of 0.5 in. (12.5 mm). SMA is the abbreviated form of the stone mix asphalt mixture. The numbers in parentheses are the numbers of gyrations used in the mix design. All the mixes were designed following the Superpave requirements for all parameters. Every group of mixes had identical aggregate gradations.

Table 2. A list of the eleven mixtures used in this study

Mix ID	NMAS, in. (mm)	Binder	Number of Gyrations	Number of Specimens
S(100) PG 64-22	0.75 (19)	PG 64-22	100	6
S(100) PG 76-28	0.75 (19)	PG 76-28	100	5
SMA PG 76-28	0.50 (12.5)	PG 76-28	100	12
SX(75) PG 58-28	0.50 (12.5)	PG 58-28	75	8
SX(75) PG 58-34	0.50 (12.5)	PG 58-34	75	4
SX(75) PG 64-22	0.50 (12.5)	PG 64-22	75	8
SX(75) PG 64-28	0.50 (12.5)	PG 64-28	75	4
SX(100) PG 58-28	0.50 (12.5)	PG 58-28	100	2
SX(100) PG 64-22	0.50 (12.5)	PG 64-22	100	15
SX(100) PG 64-28	0.50 (12.5)	PG 64-28	100	10
SX(100) PG 76-28	0.50 (12.5)	PG 76-28	100	31

S = NMAS of 0.75 in. (19 mm); SX = NMAS of 0.5 in. (12.5 mm); SMA = stone mix asphalt mixture.

3. Effects on the Flow Number

3.1. *Same Mix by the Same Contractor*

To investigate the variation in the flow number within the work of a single contractor for the same mix, the following mixes were selected randomly. The information regarding the paving contractor, binder supplier, and aggregate pits is kept confidential. The mixes were manufactured in 2014.

The N-values vary from 120 to 531 with an average value of 261 and standard deviation of 125, as shown in Figure 3. The values shown are for each individual specimen. To determine whether these data are statistically significant or not, a one-sample t-test was conducted. The t-test requires the data to be normally distributed. The t-test showed the 95% Confidence Interval (CI) boundaries to be 150 and 372 with a mean value of 261. This means that all the mixes, except for 19655 P21 14 and 19655 P87 14, are statistically the same. Therefore, a conclusion can be made that the same mix may have statistically different flow numbers for the same contractor. Note that Colorado Department of Transportation (CDOT) uses a 10-digit format to express the mix identity, such as 19655 P21 14. The first five numbers denote the project and the last five digits denote the site and specimen number.

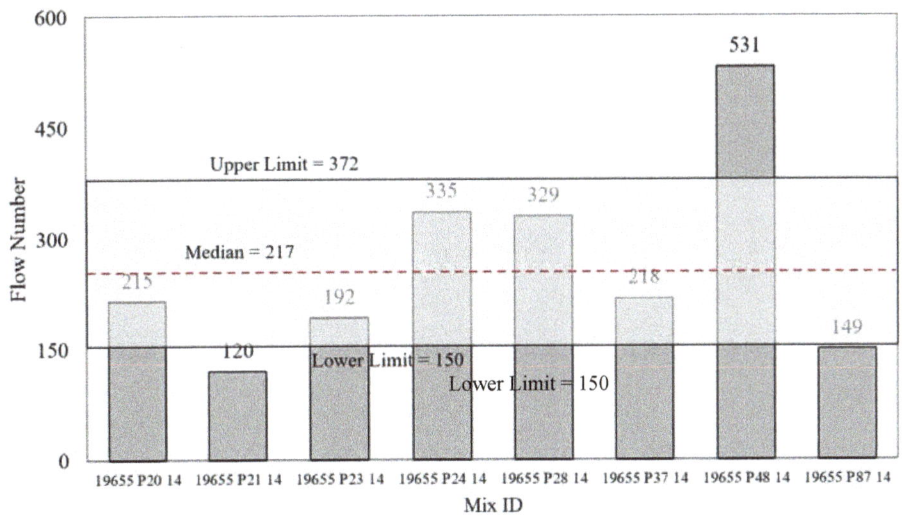

Figure 3. Flow numbers of eight specimens of SX(100) PG 64-28 mix.

3.2. *Same Mix by Different Contractors*

To investigate the differences in flow number for the same mix prepared by different contractors, SX(100) PG 76-28 mix was selected. The average flow numbers from four contractors, 19128, 18842, 19458, and 19677, are presented in Figure 4.

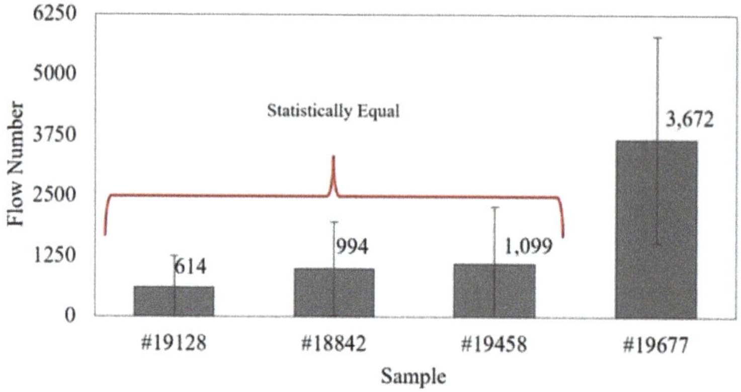

Figure 4. Flow numbers of a mix by different contractors.

The pairwise comparison test result shows that the mixes by 19128, 18842, and 19458 are statistically the same (Table 3). Therefore, a conclusion can be made that the same mix may have statistically different flow numbers for different contractors. This is due to variations in the aggregate structures, shapes, orientation, smoothness, etc.

Table 3. Pairwise comparisons using *t*-tests to determine statistical difference.

	19128	**18842**	**19458**
18842	Equal	-	-
19458	Equal	Equal	-
19677	Different	Different	Different

3.3. *Groupwise Comparison*

The flow number variation for each group of mix is described in this section. An effort was made to examine whether all specimens' flow numbers were statistically equal or not. A 95% Confidence Interval (CI) was used to indicate reliability. Next a statistical regression analysis was conducted to find the effects of different mix factors.

3.3.1. S(100) PG 64-22

The flow numbers for the S(100) PG 64-22 mix varied from 110 to 252, with an average number of 155, median of 130, and standard deviation of 59. As per AASHTO [2], a mix is good for 3 to 10 million equivalent single axle loads (ESALs) if it has a flow number greater than 50. The current mix had an average flow number of 103; however, about one-third of the test results indicated flow numbers less than 50. Despite this, the S(100) PG 64-22 mix would be considered good for pavement designs with traffic between 3 and 10 million ESALs. The *t*-test showed the 95% CI boundaries to be 47 and 262. Four mixes, 18695 P9 15, 18695 P13 14, 18695 P25 14, and 18465 P9 14, were found to be statistically the same, although the paving contractors, binder supplier, and production dates differed. The 18465 P9 14 mix, for example, had a much lower VFA compared to the others. Thus, definitive conclusions cannot be made from this mix.

3.3.2. S(100) PG 76-28

The flow numbers for the S(100) PG 76-28 mix varied from 626 to 2065 with an average number of 1223, median of 1101, and standard deviation of 528. As per AASHTO [2], a mix is good for more than 30 million ESALs if it has a flow number greater than 740. The current mix had an average flow number of 1223. All the test results showed flow numbers greater than 740. Therefore, this mix is considered good for traffic greater than 30 million ESALs. The *t*-test showed the 95% CI boundaries to be 253 and 2193. All the values were within the 95% CI boundaries; thus, they are statistically equal. The generic information shows that the mix factors of all mixes were very close to each other. Comparing the above two mixtures, S(100) PG 64-22 and S(100) PG 76-28, both have the same aggregate size; the only difference is the binder type. With an increase in binder grade, the flow number increased. The mix parameters show that all four mixes (17800 P17 14, 17800 P26 14, 17800 P38 14, and 17800 P53 14) had similar properties, and their flow numbers are statistically equal.

3.3.3. SMA PG 76-28

The flow numbers for the SMA PG 76-28 mix varied from 426 to 4311, with an average number of 2272, median of 2219, and a standard deviation of 1182. All the test results, with the exception of one outlier judged by visual inspection, showed the flow numbers to be greater than 740. Therefore, this mix is considered to be good for traffic greater than 30 million ESALs. The *t*-test showed the 95% CI boundaries to be 1487 and 3057. Comparing the two mixtures S(100) PG 76-28 and SMA PG 76-28, both have the same binder, but they have different aggregate sizes. An increase in aggregate size increased the flow number as observed from these two mixtures. The coarser aggregate shows better resistance to deformation due to its aggregate-to-aggregate interlocking. The mix parameters dictatethat mixes with similar properties (such as V_{be}, V_a, VMA, VFA, AC) can have statistically different flow numbers.

3.3.4. SX(75) PG 58-28

The flow numbers for the SX(75) PG 58-28 mix varied from 29 to 220, with an average number of 91, median of 81, and standard deviation of 55. All the test results, apart from one outlier, showed a flow number greater than 50. The current mix had an average flow number of 91; therefore, per AASHTO, this mix is considered good for traffic of 3 to 10 million ESALs. The *t*-test showed the 95% CI boundaries to be 42 and 140. All mixes except 19489 P51 14 and 19879 P113 14 did not have statistically similar flow numbers. One mix (19489 P51 14) had a unique aggregate source (Chambers), and another mix (19879 P113 14) had a unique contractor and aggregate source (Ralston, Firestone). All the other properties were similar to those of the mixes whose flow numbers were statistically the same. Therefore, contractor or aggregate source may be a factor in the variation of the flow number.

3.3.5. SX(75) PG 58-34

The flow numbers for the SX(75) PG 58-34 mix varied from 19 to 75, with an average flow number of 47 and standard deviation of 28. As per AASHTO, a mix is considered good for traffic of less than 3 million ESALs if it has a flow number less than 50. The current mix had an average flow number of 47; therefore, this mix is good for traffic of less than 3 million ESALs. However, we had data for only two mixes from the current mix. Therefore, no sensitivity analysis was conducted on this mix.

3.3.6. SX(75) PG 64-22

The flow numbers for the SX(75) PG 64-22 mix varied from 19 to 123, with an average number of 59, median of 49, and standard deviation of 34. As per AASHTO, a mix is considered good for traffic between 3 and 10 million ESALs if it has a flow number greater than 50. The current mix had an average flow number of 59; therefore, this mix is considered good for traffic between 3 and 10 million ESALs. The *t*-test showed the 95% CI boundaries to be 28 and 90. The mix parameters show that mixes with similar properties have statistically different flow numbers. For example, mixes with the prefix

19935 have the same binder supplier, aggregate source, region, and volumetric properties, though they have statistically different flow numbers.

3.3.7. SX(75) PG 64-28

The flow numbers for the SX(75) PG 64-28 mix varied from 32 to 311, with an average number of 106, median of 41, and standard deviation of 118. As per AASHTO, a mix is considered good for traffic between 3 and 10 million ESALs if it has a flow number greater than 50. The current mix had an average flow number of 99; therefore, this mix is considered good for traffic between 3 and 10 million ESALs. The same observations were noted for binders SX(75) PG 58-28 and SX(75) PG 58-34.

3.3.8. SX(100) PG 58-28

Only a single sample with a flow number of 128 was tested for this mix. No statistical test or sensitivity analysis was conducted on this mix due to insufficient data.

3.3.9. SX(100) PG 64-22

The flow numbers for the SX(75) PG 64-22 mix varied from 23 to 388, with an average number of 112, median of 97, and standard deviation of 92. As per AASHTO, a mix is considered good for traffic between 3 and 10 million ESALs if it has a flow number greater than 50. The current mix had an average flow number of 112; therefore, this mix is considered good for traffic between 3 and 10 million ESALs. Nonetheless, a *t*-test was conducted, and the results showed the 95% CI boundaries to be 59 and 164. The mix parameters show that mixes with different properties had statistically similar flow numbers. On the other hand, mixes by the same contractor with the same aggregate source had statistically different flow numbers.

3.3.10. SX(100) PG 64-28

The flow numbers for the SX(100) PG 64-28 mix varied from 77 to 531, with an average of 241, median of 215, and standard deviation of 131. As per AASHTO, a mix is considered good for traffic between 10 and 30 million ESALs if it has a flow number greater than 190. The current mix had an average flow number of 240; therefore, this mix is considered good for traffic between 10 and 30 million ESALs. The *t*-test showed the 95% CI boundaries to be 134 and 347. Three mixes were not statistically the same; however, two mixes had similar properties to the statistically similar mixes. Therefore, flow numbers can be statistically different for the same mix by the same contractor.

3.3.11. SX(100) PG 76-28

The flow numbers for the SX(100) PG 76-28 mix varied from 82 to 6343, with an average number of 1578, median of 810, and a standard deviation of 1837. As per AASHTO, a mix is considered good for traffic greater than 30 million ESALs if it has a flow number greater than 740. Although the average flow number was 1482, nearly half of the samples had a flow number less than 740. Therefore, it is very difficult to conclude whether this mix is considered good for traffic greater than 30 million ESALs. Comparing this result with those for the previous binders, the flow number increased with an increase in the high-temperature grade of the binder. Similar observations were noted for the SX(75) mix. The *t*-test showed the 95% CI boundaries to be 893 and 2262. Out of 33 specimens, only 7 specimens were within the 95% CI boundaries.

3.4. *Analysis Summary*

The flow numbers for each group and their variations, 95% CI boundaries, etc., are presented in Table 4 and Figure 5. They show that SX(75) PG 58-34 had the lowest flow number and SMA PG 76-28 had the highest flow number. Comparing the average flow numbers with the above-listed values, the following may be concluded:

- Only two types of mixtures, SX(100) PG 76-28 and SMA PG 76-28, had flow numbers greater than 740. Thus, only these mixtures are considered good for traffic greater than 30 million ESALs.
- S(100) PG 76-28 had an average flow number of more than 190; thus, it is considered good for traffic between 10 and 30 million ESALs.
- SX(100) PG 64-22, SX(100) PG 64-28, and SX(100) PG 58-28 are considered good for traffic between 3 and 10 million ESALs.
- The other five mixtures—S(100) PG 64-22, SX(75) PG 58-28, SX(75) PG 58-34, SX(75) PG 64-22, and SX(75) PG 64-28—had flow numbers less than 50; thus, they are considered good for traffic of less than 3 million ESALs.
- Comparing SX(100) PG 64-28 and SX(100) PG 76-28, the flow number of HMA increases with an increase in the high-temperature grade of the binder.
- Variable results were observed as to whether the flow number increases or decreases with an increase in the low-temperature grade of the binder. For example, when comparing SX(75) PG 64-22 and SX(75) PG 64-28, the flow number increases with an increase in the low-temperature grade of the binder; however, when comparing SX(75) PG 58-34 and SX(75) PG 58-28, the flow number decreases with an increase in the low-temperature grade of the binder.
- An SX mix has 0.5 in. (12.5 mm) nominal aggregate size, and an S mix has 0.75 in. (19 mm) nominal aggregate size. SX mixes have larger flow numbers, i.e., smaller aggregate size produces a larger flow number, from the comparisons of the flow numbers of SX(100) PG 64-22 with S(100) PG 64-22, and SX(100) PG 76-28 with that of S(100) PG 76-28. However, the differences between these pairs are not statistically significant.
- The (75) and (100) refer to the number of gyrations during design. Greater number of gyrations produce greater flow numbers, as shown from the comparisons of SX(75) PG 58-28 with SX(100) PG 58-28, SX(75) PG 64-22 with SX(100) PG 64-22, and SX(75) PG 64-28 with SX(100) PG 64-28. However, the differences between these pairs were not statistically significant.

Table 4. Groupwise average flow numbers with 95% boundaries.

Mixes	Lowest Value	Highest Value	95% CI Lower Limit	95% CI Upper Limit	Average
S(100) PG 64-22	110	252	47	262	155
S(100) PG 76-28	626	2065	253	2193	1223
SMA PG 76-28	426	4311	1487	3057	2272
SX(75) PG 58-28	29	220	42	140	91
SX(75) PG 58-34	19	75	0	403	47
SX(75) PG 64-22	19	123	28	90	112
SX(75) PG 64-28	32	311	0	323	106
SX(100) PG 58-28	-	-	-	-	128
SX(100) PG 64-22	23	388	59	164	112
SX(100) PG 64-28	77	531	134	347	241
SX(100) PG 76-28	82	6343	893	2263	1578

The sensitivity analysis summary presented in Table 5 shows that the effects of V_{be}, V_a, VMA, VFA, and AC on the flow number are inconsistent. For example, six mixes show that the flow number increases with V_{be}, two mixes show the opposite, and one mix shows it is insensitive to V_{be}. This inconsistency is true for V_a, VMA, VFA, and AC as well. The reason behind this may be the effects of the paving contractor, manufacture date, and/or aggregate source. Using most scores, the flow number increases with an increase in V_{be}, V_a, VMA, VFA, and AC for the range studied in this study. The study by Kaloush [3] showed that the flow number decreases with an increase in air voids, which is contradictory to the results of the current study. This is due to the study range of air voids. The current study only investigated air void proportions between 3% and 6%.

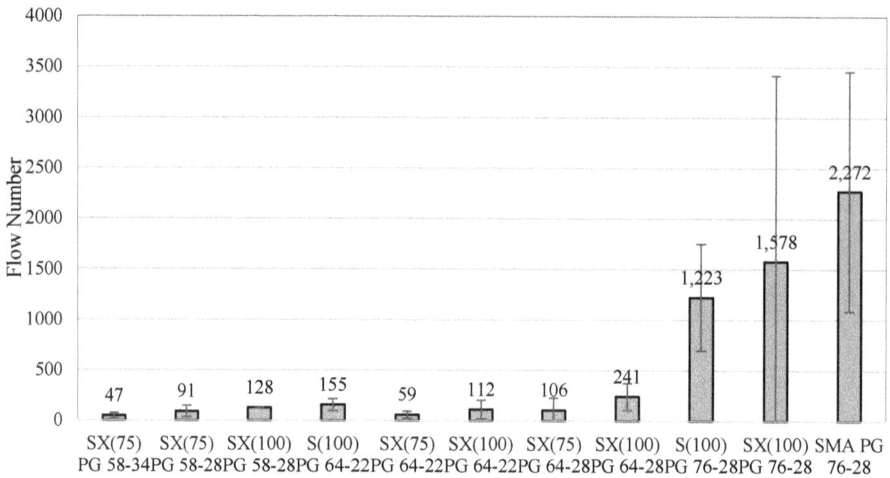

Figure 5. Groupwise average flow numbers.

Table 5. Summary of the effect of mix factors on the flow number of HMA.

	V_{be} (%)	V_a (%)	VMA (%)	VFA (%)	AC (%)
S(100) PG 64-22	Decreases	Increases	Increases	Increases	Decreases
S(100) PG 76-28	-	Decreases	Increases	Decreases	-
SMA PG 76-28	Decreases	Increases	Increases	Increases	Decreases
SX(75) PG 58-28	Increases	Decreases	Decreases	Increases	Increases
SX(75) PG 58-34	NA	NA	NA	NA	NA
SX(75) PG 64-22	Increases	Decreases	Increases	Decreases	Increases
SX(75) PG 64-28	Increases	Decreases	-	Increases	-
SX(100) PG 58-28	NA	NA	NA	NA	NA
SX(100) PG 64-22	Increases	Increases	Decreases	Decreases	Increases
SX(100) PG 64-28	Increases	Increases	Decreases	Decreases	Increases
SX(100) PG 76-28	Increases	Increases	Decreases	Increases	Increases
Summary	6 Increases 2 Decreases 1 Insensitive	5 Increases 4 Decreases 2 N/A	5 Increases 4 Decreases 1 Insensitive	5 Increases 4 Decreases 2 N/A	5 Increases 2 Decreases 2 Insensitive

4. Conclusions

This study evaluates the effects of mix factors such as VMA, void-filled with asphalt, effective binder content, etc. on the flow number of asphalt concrete. Laboratory testing was performed, and test results were analyzed using the statistical tools. The following conclusions can be made from this study:

1. The same mix may have statistically different flow numbers, and this is independent of the contractor.
2. The flow number increased with increasing Vbe, Va, VMA, VFA, and AC for the range studied in this study.
3. Only two types of mixtures, SX(100) PG 76-28 and SMA PG 76-28, had flow numbers greater than 740. Thus, only these mixtures are considered good for traffic greater than 30 million ESALs.
4. S(100) PG 76-28 had an average flow number of more than 190; thus, it is considered good for traffic between 10 and 30 million ESALs.
5. SX(100) PG 64-22, SX(100) PG 64-28, and SX(100) PG 58-28 are considered good for traffic between 3 and 10 million ESALs.

6. The other five mixtures—S(100) PG 64-22, SX(75) PG 58-28, SX(75) PG 58-34, SX(75) PG 64-22, and SX(75) PG 64-28—had flow numbers less than 50; thus, they are considered good for traffic of less than 3 million ESALs.

Our recommendation for future research is that a flow number predictive model should be developed to determine the flow number of a new mix with more laboratory testing on a pre-planned test matrix.

Author Contributions: M.R.I. is the primary investigator of this research article. He is the lead researcher with collecting the research ideas, pursuing funding, execution, delivery and publication. S.A.K. supervised all aspects of this research including editing and proofreading. S.K.N. helped in data collection, analysis, and helped the team finally publish it.

Funding: This research is funded by the Colorado Department of Transportation (CDOT), Grant No. CDOT 417.01.

Acknowledgments: The Colorado State University—Pueblo (CSU-Pueblo) research team appreciates the research funding by the Colorado Department of Transportation (CDOT). It would like to express its sincere gratitude and appreciation to Jay Goldbaum, Michael Stanford, Aziz Khan, Melody Perkins, Keith Uren, Vincent Battista, Skip Outcalt, Bill Schiebel, and Roberto E. DeDios from the CDOT.

Conflicts of Interest: The authors declare no conflict of interest. The funders had no role in the design of the study; in the collection, analyses, or interpretation of data; in the writing of the manuscript, or in the decision to publish the results.

References

1. Biligiri, K.; Way, G. Predicted E* dynamic moduli of the Arizona mixes using asphalt binders placed over a 25-year period. *Constr. Build. Mater.* **2014**, *54*, 520–532. [CrossRef]
2. AASHTO TP 79. *Standard Method of Test for Determining the Dynamic Modulus and Flow Number for Asphalt Mixtures Using the Asphalt Mixture Performance Tester (AMPT)*; American Association of State Highway and Transportation Officials (AASHTO): Washington, DC, USA, 2015; pp. 1–21.
3. Kaloush, K.E. Simple Performance Test for Permanent Deformation of Asphalt Mixtures. Ph.D. Thesis, Department of Civil and Environmental Engineering, Arizona State University, Tempe, AZ, USA, 2001.
4. Kvasnak, A.; Robinette, C.J.; Williams, R.C. Statistical development of a flow number predictive equation for the Mechanistic-Empirical Pavement Design Guide. In Proceedings of the TRB Annual Meeting, Washington, DC, USA, 21 January 2007; [CD-ROM]. TRB: Washington, DC, USA, 2007.
5. Christensen, D.W.; Bonaquist, R.; Jack, D.P. *Evaluation of Triaxial Strength as a Simple Test for Asphalt Concrete Rut Resistance*; Final Report to the Pennsylvania Department of Transportation, Report No. FHWA-PA-2000-010+97-04 (19); The Pennsylvania Transportation Institute: University Park, PA, USA, August 2000.
6. Rodezno, M.C.; Kaloush, K.K.; Corrigan, M.R. Development of a Flow Number Predictive Model. In *Transportation Research Board [CD-ROM]*; TRB: Washington, DC, USA, 2010.
7. Irfan, M.; Ali, Y.; Iqbal, S.; Ahmed, S.; Hafeez, I. Rutting Evaluation of Asphalt Mixtures Using Static. Dynamic, and Repeated Creep Load Tests. *Arab. J. Sci. Eng.* **2018**, *43*, 5143–5155. [CrossRef]
8. Ogundipe, O.M. Marshall Stability and Flow of Lime-modified Asphalt Concrete. *Trans. Res. Proc.* **2016**, *14*, 685–693. [CrossRef]
9. Irfan, M.; Ali, Y.; Ahmed, S.; Iqbal, S.; Wang, H. Rutting and Fatigue Properties of Cellulose Fiber-Added Stone Mastic Asphalt Concrete Mixtures. *Advan. Mat. Sci. Eng.* **2019**, *2019*, 5604197. [CrossRef]
10. Leiva-Villacorta, F.; Taylor, A.; Willis, R. *High-Modulus Asphalt Concrete (HMAC) Uixtures for Use as Base Course*; NCAT Report 17-04; 2017; National Center for Asphalt Technology, Auburn University: Auburn, AL, USA, June 2017.
11. Tripathi, A. Mechanistic analysis and economic benefits of fiber-reinforced asphalt mixtures. Master's Thesis, Department of Civil Engineering, The University of Texas, Tyler, TX, USA, 2018.
12. Ziari, H.; Divandari, H. Presenting asphalt mixtures flow number prediction model using gyratory curves. *Int. J. Civ. Eng.* **2013**, *11*, 125–133.

13. Siswanto, H.; Supriyanto, B.; Pranoto, P.R.C.; Hakim, A. Marshall Properties of Asphalt Concrete Using Crumb Rubber Modified of Motorcycle Tire Waste. In Proceedings of the Green Construction and Engineering Education (GCEE) Conference, East Java, Indonesia, 8–9 August 2017.

infrastructures

MDPI

Article

Determination of Layers Responsible for Rutting Failure in a Pavement Structure

Nathan Chilukwa [1,*] and Richard Lungu [2]

[1] Department of Civil Engineering and Construction, Copperbelt University, Kitwe 21692, Zambia
[2] Graduate of Civil Engineering, Copperbelt University, Kitwe 21692, Zambia; richardlungu03@gmail.com
* Correspondence: chilukwa2008@yahoo.com or nathan.chilukwa@cbu.ac.zm

Received: 11 February 2019; Accepted: 22 May 2019; Published: 26 May 2019

Abstract: Rutting is one of the most common distresses in asphalt pavements in Zambia. The problem is particularly prevalent at intersections, bus stops, railway crossings, police checkpoints, climbing lanes and other heavily loaded sections, where there is deceleration, slow moving or static loading. The most widely used methods to identify the source of rutting among flexible pavement layers are destructive methods; field trenching and coring methods. The Transverse Profile Analysis method (TPAM), which is a non-destructive method, was suggested by White et al. in 2002 as an alternative method, to avoid the expensive and destructive nature of the traditional methods. In this method, data from the transverse profile of the rutted section is used to deduce the layer of the pavement structure responsible for rutting failure. This study used the TPAM to determine the layers of pavement responsible for rutting on sections of the Chibuluma and Kitwe-Chingola Roads in Zambia. The method was first validated using the trenching method on the Kitwe-Ndola Road. Results from the TPAM showed good comparability with those from the trenching method. It was established that most of the rutting emanated from the surfacing layer. This is consistent with recent research indicating that most rutting occurs in the upper part of the asphalt surfacing. It was also established that the TPAM was a simpler, faster and less costly method of determining the source of rutting failure compared to the traditional methods.

Keywords: transverse profile analysis; rutting; asphalt; trenching; coring

1. Introduction

Zambia is a key transit country in the north–south corridor as it sits between and borders eight countries in the region, viz. Tanzania, the Democratic Republic of Congo, Botswana, Namibia, Zimbabwe, Mozambique, Angola and Malawi. With a virtually non-functional rail system, various mining products and equipment are transported in and out of the mines in the Copperbelt Province to South Africa, Tanzania and other countries in the region mostly by road.

Zambia has a total gazetted road network of 67,671 km, 40,454 km of which comprises the Core Road Network (CRN). The CRN is defined as the bare minimum network that Zambia requires to be maintained continuously and on a sustainable basis to realize its social and economic potential. The CRN infrastructure in Zambia consists of a sparsely interconnected network of Trunk (T), Main (M), District (D), Primary Feeder (PF) and Urban (U) roads. Table 1 shows the breakdown of the CRN in Zambia [1].

The traffic growth rate on major highways, in particular the truck traffic, has been increasing day by day in Zambia due to its geostrategic location and international trade corridor, subsequently leading to the premature failure of roads [2]. A recent study on the condition of the roads showed that as of 2014, less than 25% of the CRN was paved, and close to half of the paved road network was in a fair to poor condition [1].

Table 1. Zambia's Core Road Network (CRN) (2014).

Road Category	Core Road Network (km)
Trunk (T)	3116
Main (M)	3701
District (D)	13,707
Urban (U)	5597
Primary Feeder (PF)	14,333
Total	40,454

One of the most prominent defects, particularly on major highways, is rutting failure [2]. This type of failure is prominent in sections where traffic is forced to stop or move at a slow pace, thereby increasing the loading on the road pavement. Thus, the failure is load-related and it eventually escalates to include other defects such as cracking, potholing, etc.

Rutting is defined as a longitudinal surface depression occurring in the wheel paths of roadways due to repeated traffic loading. It accumulates incrementally with small permanent deformations from each load application over the pavement's service life. It is often followed in later stages by an upheaval along the sides of the rut. The decreased thickness in the rutted portions may accelerate fatigue cracking and eventual loss of the surfacing [3]. Rut depth (RD) is one of the most commonly used index variables for quantifying pavement surface rutting. This index has been traditionally measured manually, using a gauge with either a straightedge or a wire. The method is considered "a reliable and low budget option" [4].

Rutting failure not only reduces the lifespan of the road but is also a serious safety issue for road users. When the vehicle moves along the rutted portion of the pavement, steering becomes difficult and it reduces driving comfort. If rainwater pools in the rutted wheel path, it can result in hydroplaning and spray that reduces visibility. To this extent, rutting has become such a serious problem of modern-day roads that countries such as the United States have taken it as one of the design criteria for asphalt pavements [5].

Most Zambian roads are constructed of flexible pavements. Flexible pavements are roads constructed of several layers of natural or treated granular material covered with one or more waterproof bituminous surface layers (e.g., asphalt). They are so named because the total pavement structure deflects or flexes under loading. The objective with the design of a flexible pavement is to avoid the excessive flexing of any layer, the failure of which will result in the over-stressing of the layer, which ultimately will cause the pavement to fail [6].

The layer(s) in which rutting occurs is influenced by the loading magnitude and the relative strengths of the pavement layers. Stresses within the layer of a pavement structure are determined by applied load, individual and combined layer thickness and layer material properties [7].

The Mechanistic-Empirical Design Guide (MEPDG) has defined three distinct stages for the permanent deformation behavior of asphalt pavements under a given set of material, load and environmental conditions. This is shown in Figure 1 [5]. The primary stage has a high initial level of rutting, with a decreasing rate of plastic deformation, predominantly associated with volumetric change. The secondary stage has a small rate of rutting, exhibiting a constant rate of change of rutting that is also associated with volumetric changes; however, shear deformations increase at an increasing rate. The tertiary stage has a high level of rutting predominantly associated with plastic (shear) deformations under no volume change conditions.

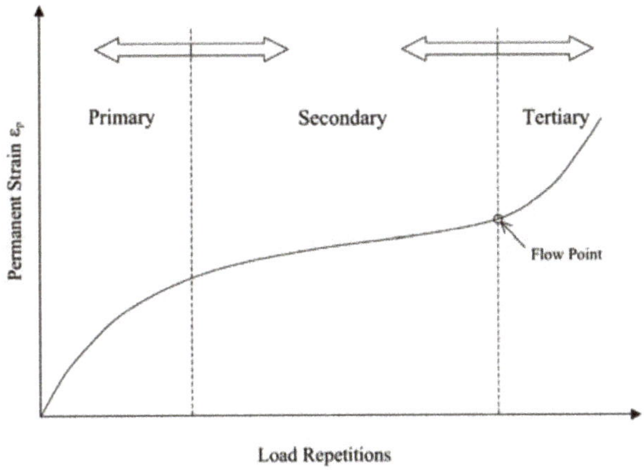

Figure 1. Typical repeated load permanent deformation behavior of pavement materials.

Generally, there are two causes of rutting in asphalt pavements—the accumulation of permanent deformation in the asphalt-surfacing layer and the permanent deformation of subgrade or underlying layers. In the past, subgrade deformation was considered the primary cause of rutting and many pavement design methods applied limiting criteria on vertical strain at the subgrade level. However, recent research has indicated that most of the rutting occurs in the upper part of the asphalt surfacing layer. Nonetheless, these two causes of rutting can act in combination, i.e., the rutting could be the sum of permanent deformation in all the layers [2,8,9].

The determination of which layer in the pavement structure is responsible for rutting failure is the first step toward arriving at a remedial measure.

The destructive methods of trenching and coring have traditionally been used for rutting failure investigations. Field trenching is conducted to expose the layers of the pavement so that they can be studied. The thickness of the layers, i.e., asphalt, the base and the subbase, are measured and plotted to show the deflections in the layers. Alternatively, core samples can be obtained at a constant spacing on a lane to determine the layer thicknesses of the pavement layers and the results can be plotted (similar to trenching) to determine the contribution of each layer to rutting due to changes in thickness and deflections of the pavement layers. The Dynamic Cone Penetrometer (DCP) Test has also been used to determine the strength of the pavement layers by calculating the California Bearing Ratio (CBR) of the pavement layers. The knowledge of the strengths of the existing pavement layers at a rutted section of the road can be used to deduce whether the ruts have been caused by weak structural layers, the subgrade or whether it is restricted to the surface. These methods are, however, time consuming and costly and may lead to weaknesses in the pavement structure if not repaired properly.

TPAM is one of the non-destructive techniques developed to determine the source of rutting in a pavement structure. The method was suggested by White et al. [7], who conducted extensive computer analyses using Finite Element Methods (FEM) to simulate rutting failures in Hot-Mix-Asphalt (HMA) surface mixtures, base courses and subgrades. Transverse surface profile characteristics indicative of failure within specific structural layers were then determined in the form of simple distortion parameters, and specific criteria were developed for these distortion parameters. The criteria were applied to an analysis of full-scale accelerated pavement test (APT) data, confirming that the relative contributions of the layers to rutting in an HMA pavement could be determined from an analysis of its transverse surface profile. This followed works such as [10], wherein the authors hypothesized that the area under the transverse surface profile could be used to predict the source of rutting from within the pavement structure. The authors of [11] also suggested that quantifying transverse profile measurements presented a potential method of determining the cause of rutting in the absence of traffic, structural and environmental data.

Another study [12] found that TPAM is one of the most accurate and precise methods of establishing the cause of rutting failure. The study used trenching, coring and Falling Weight Deflectometer (FWD) tests to validate the method. The authors of [13] also used the TPAM to determine the source of rutting failure on a 300 m section of the N5 National Highway in Pakistan. The study determined that most rutting was mainly due to the shear failure of the HMA layer. It was also observed that an increase in rut depth resulted in an increase in negative area and a reduction in positive area of the profile. The results were validated using the trenching method.

The TPAM was also used to determine the source of rutting on the Belbis-Zagazig Road in Egypt. A study determined that 60% of the rutting resulted from the HMA, 30% from the base layer and 10% from the subgrade. The authors concluded that the TPAM proved to be a good diagnostic tool for determining the source of most of the rutting failure [14].

The goal of this publication is to highlight the results of a study undertaken to establish the layer(s) responsible for rutting failure using the TPAM on some selected sections of road in the Copperbelt Province of Zambia.

The traditional methods of coring and trenching are the most commonly used methods of determining the source of rutting failure. However, these two methods are destructive, in that the subsurface layers must be exposed by excavating a trench or coring and then restored after rut investigations. This may lead to weaknesses at the joints, resulting in the shortening of the lifespan of the pavement if not repaired properly. Furthermore, these destructive methods are time-consuming, costly and inconveniencing to road users.

TPAM has been suggested as a simple yet effective way of determining the rutting failure source within the pavement structure through theoretical and analytical analyses of surface profiles, with little to none of the challenges associated with trenching and/or coring.

2. Materials and Methods

A typical transverse profile is shown in Figure 2. The method of analysis with the TPAM is explained below.

The first and last points of the profile line are joined to establish a reference line (indicated by the dotted line). The portion of the area above the line is considered as positive area (A_p) and the portion below the reference line is considered as negative area (A_n). The maximum rut depth (D) is computed by firstly drawing a line joining the high points of the profile known as the wire line; the maximum rut depth is then equal to the line with the maximum length drawn between the profile line and the wire line in a direction perpendicular to the wire line.

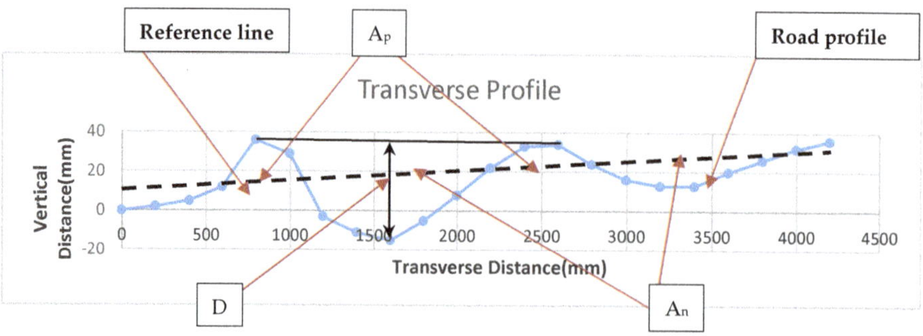

Figure 2. Plot of transverse profile from the edge of a shoulder to the center line of a lane. A_p = area above dotted line; A_n = area below dotted line; D = maximum rut depth.

The positive and negative areas are calculated by plotting the profile to scale in AutoCAD or other software.

The distortion parameters, i.e., the total area, termed 'distortion' and the ratio of the positive or negative area to the total area, termed 'ratio of distortion' are computed by the following equations [7].

$$A = A_p + A_n \tag{1}$$

$$R = A_p/A_n \tag{2}$$

where

A = total area (mm^2);
R = ratio of area;
A_p = positive area (mm^2);
A_n = negative area (mm^2).

In predicting the layer responsible for rutting distress, it is also necessary to compute some critical coefficients by the following equations:

$$C_1 = (-858.21)\ D + 667.58 \tag{3}$$

$$C_2 = (-1509)\ D - 287.78 \tag{4}$$

$$C_3 = (-2120)\ D - 407.95 \tag{5}$$

where

C_1 = theoretical average total area for HMA failure (mm^2);
C_2 = theoretical average total area for base/subbase failure (mm^2);
C_3 = theoretical average total area for subgrade failure (mm^2);
D = maximum rut depth (mm).

These equations are called equations of trend lines for HMA, base and subgrade failure modes in failure criteria.

Based on the characteristics of a given surface profile and the criteria described above, the mode of failure can be inferred [7] as follows:

1. Failure has occurred in the HMA layer if both the following conditions are satisfied:

$$R > 0.05 \tag{6}$$

$$A > (C_1 + C_2)/2 \tag{7}$$

2. Failure has occurred in the base/subbase layer if both the following conditions are satisfied:

$$R < 0.05 \tag{8}$$

$$A > (C_2 + C_3)/2 \tag{9}$$

3. Subgrade failure has occurred if no failure can be determined from the previous comparisons.

The method of analysis is summarized in the flow chart shown in Figure 3.

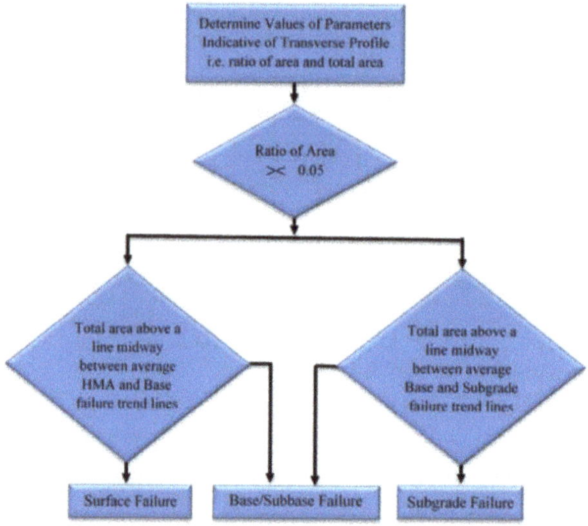

Figure 3. Illustration of the procedure for failed layer prediction [15].

A visual condition survey was undertaken to identify and select road pavements exhibiting rutting failure that were thus suitable for investigation. Three roads were selected, namely Kitwe-Ndola Road, Kitwe-Chingola Road and Chibuluma Road, all within the Copperbelt Province of Zambia.

2.1. Kitwe-Ndola Road

On the Kitwe-Ndola Road, the TPAM and trenching were conducted on a selected rutted section. The results of the profile analysis were compared to the trenching method for purposes of validation.

The pavement transverse profile was analyzed by measuring rut depths on a heavily rutted section of the road. This was done using the straight edge and gauge method. Four measurements were taken every 15 m on the rutted section. Profile measurements were taken at intervals of 200 mm from the shoulder edge to the center line of a lane. Four trenches were excavated at each point where a profile analysis was conducted. The excavation was done from the edge of the shoulder to the center line of a lane. The thickness of the asphalt and underlying layers were measured at intervals of 200 mm equal to the spacing used for rut depth measurements.

Figures 4–7 show the straight edge and gauge measurements and trench measurements on the Kitwe-Ndola Road.

Figure 4. Trench measurements of Pit 1 (Kitwe-Ndola).

Figure 5. Trench measurements of Pit 2 (Kitwe-Ndola).

Figure 6. Profile measurements on Kitwe-Ndola Road.

Figure 7. Trench measurements of Pit 3 (Kitwe-Ndola).

2.2. Chibuluma Road and Kitwe-Chingola Road

After the validation, assessments were also conducted on the Chibuluma and Kitwe-Chingola Roads. The TPAM was undertaken on an approximately 800 m section of the Chibuluma Road and a 135 m section of the Kitwe-Chingola Road to predict the layer(s) responsible for rutting failure.

The two roads were chosen due to the high levels of traffic and the high degree of rutting on the selected sections (see Figures 8 and 9). Chibuluma Road is used by heavy traffic to bypass the central business district of Kitwe, while Kitwe-Chingola Road is a key route connecting the two towns and goes further to the Kasumbalesa border and the mining region of the northwestern province. The two roads are interconnected somewhat, in that the heavy traffic that passes through Chibuluma Road eventually uses Kitwe-Chingola Road to connect to the Kasumbalesa border or the northwestern province. A recent study found that the traffic levels on the two roads averaged over 35 million equivalent standard axles.

Figure 8. Profile measurements (Chibuluma).

Figure 9. Extent of rutting (Kitwe-Chingola).

3. Results

3.1. Validation of Kitwe-Ndola Road

Results of the measurements after profile analysis and field trenching at Chainages; 60 + 365, 60 + 427, 60 + 437 and 60 + 487 of Kitwe-Ndola Road are shown in Figures 10–17.

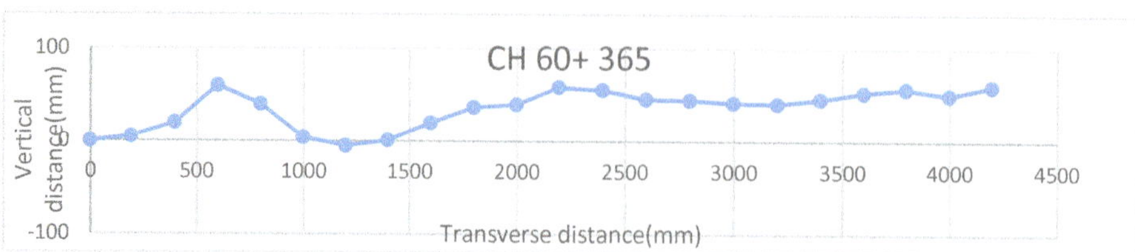

Figure 10. Transverse profile of 60 + 365 km.

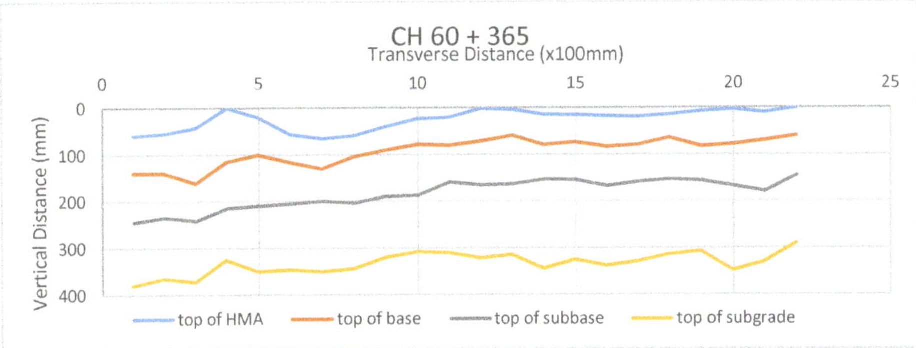

Figure 11. Profile of trench measurements of 60 + 365 km.

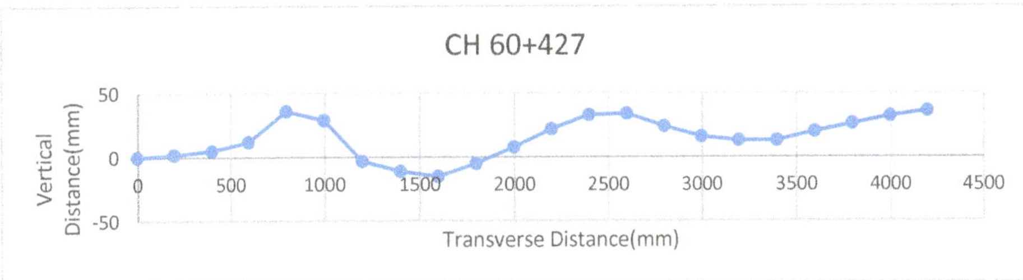

Figure 12. Transverse profile of 60 + 427 km.

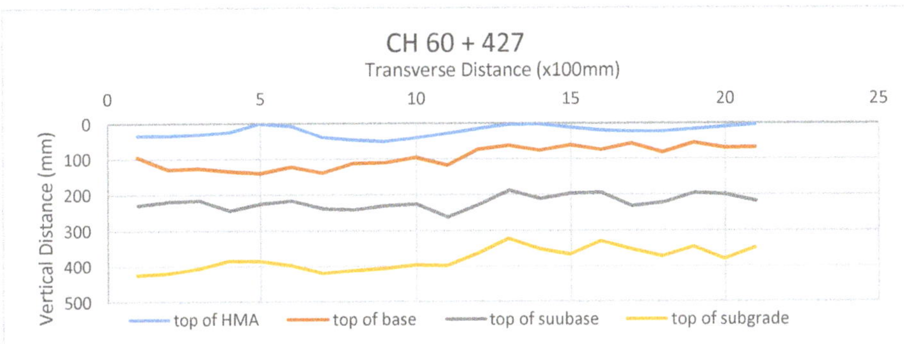

Figure 13. Profile of trench measurements of 60 + 427 km.

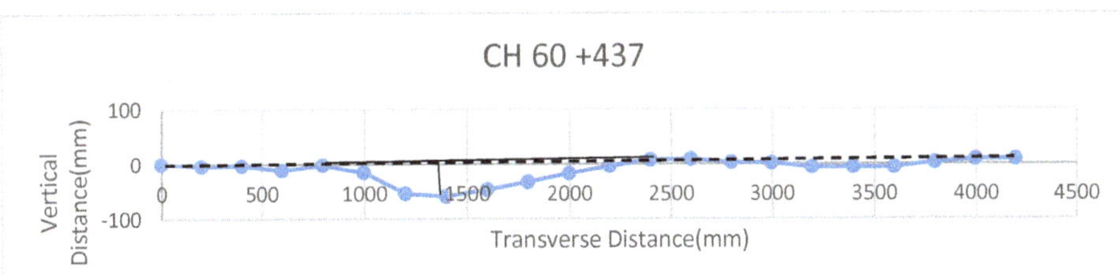

Figure 14. Transverse profile of 60 + 437 km.

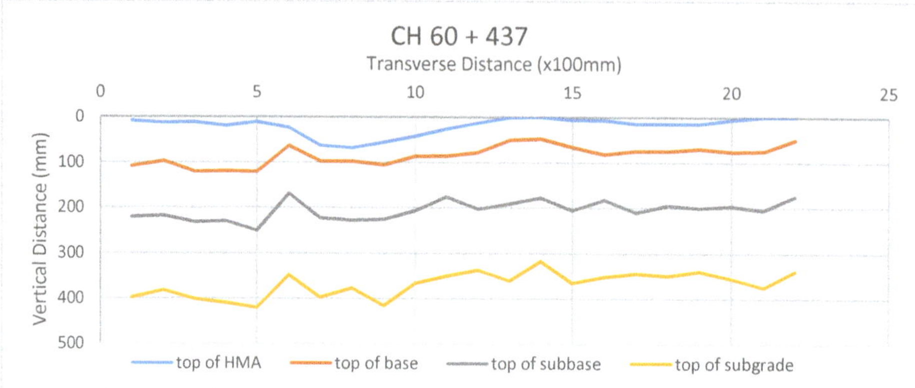

Figure 15. Profile of trench measurements of 60 + 437 km.

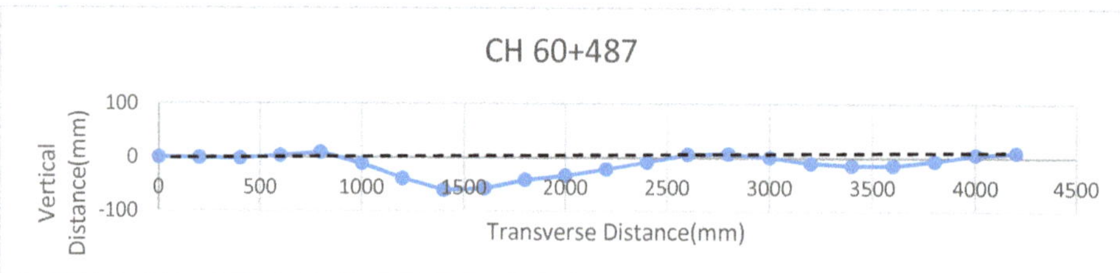

Figure 16. Transverse profile of 60 + 487 km.

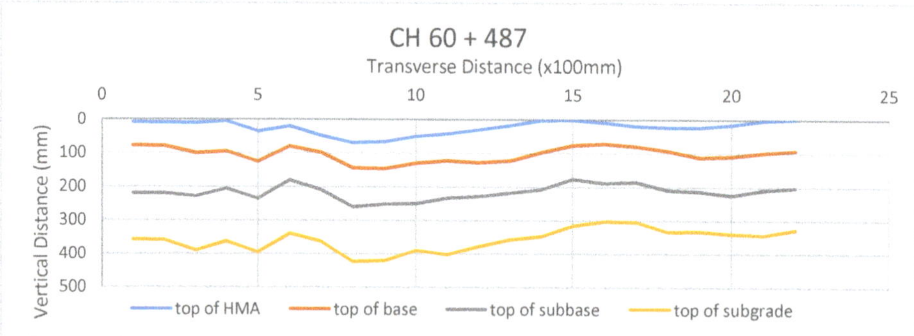

Figure 17. Profile of trench measurements of 60 + 487 km.

From the profiles it is evident that the rutting at 60 + 365 km and 60 + 427 km resulted from within the asphalt surfacing. This is evidenced by the shoving and upheaval of the asphalt, which is clearly visible in Figures 10 and 12. It can also be seen from Figures 11 and 13 that the profile of the top of the asphalt surfacing is different from that of the top of the base, subbase and subgrade.

The profiles at 60 + 437 km and 60 + 487 km, on the other hand, show that the underlying layers are responsible for the rutting failure. The asphalt profile is almost equivalent to the underlying layers and there is little to no shoving and/or upheaval of the asphalt.

The TPAM was undertaken to verify these observations. Distortion parameters where determined and calculated as outlined in Section 2. For example, for profile Chainage (CH) 60 + 365;

$A_p = 35{,}963\ mm^2$,
$A_n = -12{,}563\ mm^2$,
$D = 64\ mm$.

Total area (A):

$$A = A_p + A_n = 35{,}963 + (-12{,}563) = 23{,}400\ mm^2.$$

Ratio of Area (R):

$$R = |A_p/A_n| = 2.8626.$$

Calculate critical coefficients by Equations (3)–(5).

$C_1 = (-858.21)*64 + 667.58 = -54{,}257.9\ mm^2$,
$C_2 = (-1509)*64 - 287.78 = -96{,}863.8\ mm^2$,
$C_3 = (-2120)*64 - 407.95 = -13{,}6094\ mm^2$.

First check is applied by using Equations (6) and (7);

$R = 2.8626 > 0.05$ ✓**checked**

for $A = 23{,}400\ mm^2$.

and $(C_1 + C_2)/2 = (-54{,}257.9 - 96{,}863.8)/2 = -75{,}560.85\ mm^2$,

thus $A > (C_1 + C_2)/2$ ✓**checked**

Both criteria were met, indicating that the failure occurred in the asphalt surfacing.

The procedure was repeated for all the profiles on the road (Table 2). It was established that at 60 + 365 km and 60 + 427 km the rutting was in the asphalt while at 60 + 437 km and 60 + 487 km the rutting emanated from the underlying base/subbase. This is consistent with the observed profile shape.

Table 2. Kitwe-Ndola Road profile analysis.

Failed Layer Prediction from Transverse Profile Analysis									
Road Name/Chainage	**A (mm^2)**	**A_p (mm^2)**	**A_n (mm^2)**	**R**	**C_1 (mm^2)**	**C_2 (mm^2)**	**C_3 (mm^2)**	**D (mm)**	**Failed Layer**
Kitwe-Ndola Road									
CH 60 + 365	23,400	35,963	12,563	2.863	−54,257.9	−96,863.8	−136,094	64	HMA
CH 60 + 427	−13,800	16,075	29,875	0.538	−41,384.7	−74,228.8	−104,293	49	HMA
CH 60 + 437	−61,785	882	62,667	0.0143	−47,392.2	−84,791.8	−119,134	56	BASE
CH 60 + 487	−78,995	1370	80,365	0.0173	−56,832.5	−101,391	−142,455	67	BASE

The results of the profile analysis were compared to measurements from the trenching. The results showed consistency between the two methods (Table 3).

Table 3. Comparison of trench measurements vs. profile analysis.

KITWE-NDOLA ROAD						
Chainage	**A (mm^2)**	**R**	**D (mm)**	**Failed Layer (Transverse Profile Analysis)**	**Failed Layer (Trenching Method)**	**Agreement**
CH 60 + 365	23,400	2.863	64	HMA	HMA	YES
CH 60 + 427	−13,800	0.538	49	HMA	HMA	YES
CH 60 + 437	−61,785	0.0143	56	BASE/SUBBASE	BASE/SUBBASE	YES
CH 60 + 487	−78,995	0.0173	67	BASE/SUBBASE	BASE/SUBBASE	YES

3.2. *Chibuluma and Kitwe-Chingola Roads*

After the validation, the profile analysis was conducted on sections of the Chibuluma and Kitwe-Chingola Roads. The results are shown in Tables 4 and 5, respectively.

Table 4. Chibuluma Road profile analysis.

Failed Layer Prediction from Transverse Profile Analysis									
Road Name/Chainage	A (mm^2)	A_p (mm^2)	A_n (mm^2)	R	C_1 (mm^2)	C_2 (mm^2)	C_3 (mm^2)	D (mm)	Failed Layer
Chibuluma Road									
CH 0 + 880	−37,100	8275	45,375	0.1824	−43,959.3	−78,746.8	−110,653	52	HMA
CH 0 + 895	17,200	36,255	19,055	1.9027	−51,683.2	−92,327.8	−129,734	61	HMA
CH 0 + 910	8775	23,950	15,175	1.5783	−41,384.7	−74,219.8	−104,293	49	HMA
CH 0 + 925	34,499	38,344	3845	9.9724	−41,384.7	−74,219.8	−104,293	49	HMA
CH 0 + 940	35,300	62,213	26,913	2.3116	−57,690.7	−10,2891	−144,575	68	HMA
CH 1 + 655	−4900	35,503	40,403	0.8787	−48,250.4	−86,291.8	−121,254	57	HMA
CH 1 + 670	−4600	28,568	33,168	0.8613	−38,810.1	−69,692.8	−97,932.6	46	HMA
CH 1 + 685	25,600	46,101	20,501	2.2487	−39,668.3	−71,201.8	−100,053	47	HMA
CH 1 + 700	24,676	40,656	15,980	2.5442	−45,675.8	−81,764.8	−114,893	54	HMA
CH 1 + 715	23,300	40,653	17,353	2.3427	−43,101.1	−77,237.8	−108,533	51	HMA

Table 5. Kitwe-Chingola Road profile analysis.

Failed Layer Prediction from Transverse Profile Analysis									
Road Name/Chainage	A (mm^2)	A_p (mm^2)	A_n (mm^2)	R	C_1 (mm^2)	C_2 (mm^2)	C_3 (mm^2)	D (mm)	Failed Layer
Kitwe-Chingola Road									
CH 2 + 220	−2363	56,200	58,563	0.9597	−129,780	−229,647	−322,663	152	HMA
CH 2 + 235	−3400	54,559	88,559	0.6161	−137,504	−243,228	-3,417,044	161	HMA
CH 2 + 250	−47,001	23,339	70,340	0.3318	−99,743	−176,832	−248,460	117	HMA
CH 2 + 265	−21,400	33,698	55,098	0.6116	−92,019.1	−163,251	−229,379	108	HMA
CH 2 + 280	29,400	34,706	5306	6.5409	−35,806.4	−64,411.3	-90,512.2	42.5	HMA
CH 2 + 295	13,600	16,468	2868	5.7421	−15,638.4	−28,949.8	−40,689.9	19	HMA
CH 2 + 310	32,000	41,851	9851	4.2484	−28,511.6	−51,584.8	−72,491.4	34	HMA
CH 2 + 325	−7800	19,743	27,543	0.7168	−41,384.7	−74,219.8	−104,293	49	HMA
CH 2 + 340	7400	23,487	16,087	1.46	−27,635.4	−50,075.8	−70,371.3	33	HMA
CH 2 + 355	1454	15,970	14,516	1.1002	−32,802.6	−59,129.8	−83,091.9	39	HMA

4. Discussion

Results indicated that in all cases, the ratio of area (R) was greater than 0.05 (the limit between HMA and base/subbase failure). This is an indication of the profiles having significant positive areas. It was also determined that the total area (A) was significantly greater than the combination of the critical coefficients ((C_1+ C_2)/2), as shown in Tables 4 and 5 for Chibuluma and Kitwe-Chingola Roads, respectively. This meets both criteria for HMA failure and thus indicates that the asphalt was mainly responsible for the rutting failure. These results are consistent with the observed shape of the rutted road sections which were relatively deeply depressed and accompanied by upheavals along the sides of the wheelpath. The results are also consistent with recent research suggesting that rutting mainly occurs in the asphalt (surfacing) layer.

These results are an indication of inadequacy in the performance of the asphalt. Zambia, like most developing countries, is still using empirical pavement design methods whose focus is mostly on protecting the subgrade, thus leaving rutting to be assessed during the mix design of the asphalt. The results therefore call for particular attention to be paid to the asphalt mix design and construction in order to avoid/minimize this problem. Some of the measures that should be considered, as suggested in the literature as well, include:

1. Mix design method: The Marshall mix design method used in empirical designs has proven to be unsuitable for present day traffic, as evidenced by the steady increase in rutting problems [2]. A study by Verhaeghe et al. [16] suggested that the Marshall mix design method can be used for lower traffic or non-rut potential situations, but for high traffic or rut potential situations, gyratory-based design methods should be used, possibly coupled with performance tests.

2. Bitumen binder: A higher bitumen content is needed for improved fatigue life and durability of the asphalt mix, but it tends to enhance the rutting and shoving problems. Therefore, the mix needs to be maximized for fatigue and permanent deformation through a compromise. It has also been established that bitumen binder with a high viscosity at 60 °C will be more resistant to horizontal thrust as far as plastic flow in a mix is concerned, as compared to a low viscosity bitumen binder [17].
3. Gradations: An increase in the −75 microns fraction of the mix will tend to stiffen (increase the viscosity) the binder. However, finer gradations or over sanded mixes are more susceptible to permanent deformation than coarser gradations. Therefore, a compromise needs to be reached on the quantity of fine particles in the mix. It has also been demonstrated that mixtures utilizing angular manufactured sand and fine aggregates are more resistant to permanent deformation than mixes produced with rounded or sub-rounded natural sand. Larger-size aggregates (such as 19.5 mm) mixed in the wearing course also tend to be more resistant to rutting [17,18].
4. Air void content: Mixtures with low voids in the mineral aggregate (VMA) and higher bitumen contents tend to have very low air void contents after densification by traffic. Low air voids result in a loss of mixture stability and rutting due to the buildup of pore pressure in the asphalt mixture under traffic loading, thus leading to loss of strength and flow [19].
5. Improved bond between pavement layers: The authors of [20] conducted FEM simulations to determine the effect of bonding between pavement layers and rutting. It was concluded that a good bond between the pavement layers was essential in minimizing rutting failure in road pavements.
6. Construction: One of the main causes of accelerated rut development, especially when the asphalt mat is still fresh, is insufficient compaction at the time of construction, which not only would result in higher levels of consolidation under traffic, but also could render the mix more susceptible to shear deformation in the early life of the HMA layer. Hence, during construction every attempt should be made to meet the density specifications of HMA. For continuously graded mixes, minimum density specifications are usually set at 93% of the maximum theoretical relative density (MTRD) [16].

5. Conclusions

Rutting failure is a common problem on most Zambian roads. The determination of which of the pavement layers is responsible for the failure is the first step in remediating the problem. The objective of this study was to determine the source of rutting failure in some selected sections of road using the TPAM.

It was established on both Chibuluma Road and Kitwe-Chingola Road that rutting failure mostly emanated from the surfacing layers. This is consistent with the observed shape of the rutted sections. The sections of the Kitwe-Ndola Road analyzed indicated a half split with two sections indicating HMA failure and the other two indicating base/subbase failure.

It was also established that the TPAM is an effective way of determining the pavement layer(s) responsible for rutting failure. The method showed very good comparability with the trench method. This indicates that the inferences made in the TPAM criteria are consistent with the traditional methods of rutting failure determination and can therefore perform the same function.

The TPAM is simple and less time consuming than traditional methods. A task which would have taken at least a couple of days was completed within a matter of a few hours. There was also less disturbance to traffic and, most importantly, the proposed method eliminated the need for expensive destructive testing. This tool will therefore be invaluable to road agencies, particularly in developing countries.

Author Contributions: Conceptualization, N.C.; Investigation, N.C. and R.L.; Project administration, N.C.; Supervision, N.C.; Writing—original draft, R.L.; Writing—review & editing, N.C.

Funding: This research received no external funding.

Conflicts of Interest: The authors declare no conflict of interest.

References

1. Road Development Agency. *Road Condition Report*; Road Development Agency: Lusaka, Zambia, 2014.
2. Mwanza, A.D. Modelling of Rutting Prediction for Flexible Pavements in Zambia. Ph.D. Thesis, Chang'an University, Xi'an, China, 2008.
3. Schwartz, C.W.; Carvalho, L.R.; Von Quintus, H.L.; Mallela, J.; Bonaquist, R. *Calibration of Rutting Models for HMA Structural and Mixture Design—Appendix K: Advanced Materials Characterization and Modeling*; Transportation Research Board: Washington, DC, USA, 2012. [CrossRef]
4. Serigos, P.A.; Murphy, M.; Prozzi, J.A. *Evaluation of Rut-Depth Accuracy and Precision Using Different Automated Systems for Texas Conditions*; TRB: Washington, DC, USA, 2013; pp. 13–117. Available online: https://pdfs.semanticscholar.org/336f/faf6a5d3342795f99cf586b5670403d5bd41.pdf (accessed on 13 January 2019).
5. American Association of State Highway and Transportation Officials. *Mechanistic-Empirical Pavement Design Guide: A Manual of Practice, Interim edition*; American Association of State Highway and Transportation Officials: Washington, DC, USA, July 2008; p. 212. Available online: https://fenix.tecnico.ulisboa.pt/downloadFile/563568428712666/AASHTO08.pdf (accessed on 27 December 2018).
6. Sethi, K.C.; Sangita; Tiwari, D.; Baral, D. Investigation on the Causes of Failure in Flexible Pavements. In Proceedings of the International Conference on Engineering Science and Management, Goa, India, 17–18 December 2016; p. 264.
7. White, T.D.; Haddock, J.E.; Hand, A.J.T.; Fang, H. *Contributions of Pavement Structural Layers to Rutting of Hot Mix Asphalt Pavements*; (NCHRP report No. 468); Transport Res Board, National Academy Press: Washington, DC, USA, 2002; Available online: http://onlinepubs.trb.org/onlinepubs/nchrp/nchrp_rpt_468-a.pdf (accessed on 27 December 2018).
8. Ogundipe, O.M. Marshall Stability and Flow of Lime-Modified Asphalt Concrete. *Transp. Res. Arena* **2016**, *14*, 685–693. [CrossRef]
9. Garba, R. Permanent Deformation Properties of Asphalt Concrete Mixtures. Ph.D. Thesis, Department of Road and Railway Engineering, Norwegian University of Science and Technology, Trondheim, Norway. 28 June–2 July 2002.
10. Simpson, A.L. Characterization of Transverse Profile. *Transp. Res. Rec.* **1999**, *1655*, 185–191. [CrossRef]
11. Lenngren, C.A. Some Approaches in Treating Automatically Collected Data on Rutting. In *Transportation Research Record: Journal of the Transportation Research Board*; No. 11 96; Transportation Research Board of the National Academies: Washington, DC, USA, 1988; pp. 20–26. ISBN 0309047714.
12. Villiers, C.; Roque, R.; Dietrich, B. Interpretation of Transverse Profiles to Determine the Source of Rutting within an Asphalt Pavement System. *Transp. Res. Rec.* **2005**, *1905*, 73–81. [CrossRef]
13. Hussan, S.; Kamal, M.A.; Khan, M.B.; Irfan, M.; Hafeez, I. Determining the Contribution of Different Structural Layers of Asphalt Pavement System to Rutting Using Transverse Profile Analysis. *Am. J. Civ. Eng. Archit.* **2013**, *1*, 174–180. [CrossRef]
14. Solyman, M.; Salama, H. Field Investigation of Flexible Pavement Rutting Damage Using the Transverse Surface Profile. *J. Am. Sci.* **2012**. Available online: http://www.jofamericanscience.org/journals/am-sci/am0808/007_9834am0808_44_50.pdf (accessed on 27 December 2018).
15. Haddock, J.E.; Hand, A.J.T.; Fang, H.; White, T.D. Determining Layer Contributions to Rutting by Surface Profile Analysis. *J. Transp. Eng.* **2005**, *131*, 131–139. [CrossRef]
16. Verhaeghe, B.M.J.A.; Myburgh, P.A.; Denneman, E. Asphalt Rutting and its prevention. In Proceedings of the 9th Conference on Asphalt Pavements in Southern Africa, Gaborone, Botswana, 2–5 September 2007; Available online: http://hdl.handle.net/10204/1241 (accessed on 18 April 2019).
17. Mahan, H.M. Behavior of Permanent Deformation in Asphalt Concrete Pavements under Temperature Variation, Al-Qadisiya. *J. Eng. Sci.* **2013**, *6*, 62–73.
18. Kandhal, P.S.; Mallick, R.B.; Brown, E.R. *Hot Mix Asphalt for Intersection in Hot Climates*; Presented at the NCAT Report 98-06; National Center for Asphalt Technology, Auburn University: Auburn, AL, USA, 1998.
19. Lushinga, N.; Jiang, X. Effect of Horizontal Shear Load on Pavement Performance. In Proceedings of the 2nd International Conference on Geological and Civil Engineering, Singapore, 13–14 March 2015.

20. Aluga, M. Finite Element Modelling of Rutting Failure of Flexible Pavements. Master's Thesis, Science in Engineering, Copperbelt University, Kitwe, Zambia, 2016.

infrastructures

Article

Characterization of Recovered Bitumen from Coarse and Fine Reclaimed Asphalt Pavement Particles

Saeed S. Saliani [1,*], Alan Carter [1], Hassan Baaj [2] and Peter Mikhailenko [2]

1 Department of Construction Engineering, École de Technologie Supérieure, Montreal, QC H3C 1K3, Canada; Alan.Carter@etsmtl.ca
2 Department of Civil and Environmental Engineering, University of Waterloo, Waterloo, ON N2L 3G1, Canada; hassan.baaj@uwaterloo.ca (H.B.); Peter.Mikhailenko@empa.ch (P.M.)
* Correspondence: saliani.saeed@gmail.com

Received: 9 April 2019; Accepted: 8 May 2019; Published: 11 May 2019

Abstract: In the current era of road construction, it is common to add a small amount of reclaimed asphalt pavement (RAP) in asphalt mixes without significantly changing properties such as stiffness and low-temperature cracking resistance. Not only can these mixes be better for the environment, but they can also improve certain properties like rutting resistance. However, there is no clear understanding of how RAP gradation and bitumen properties impact the mixture properties. In this study, a single RAP source was separated into coarse and fine particles and added into a hot mix asphalt (HMA). Fourier transform infrared (FTIR) spectrometry was used to evaluate the chemical properties of the bitumen, while environmental scanning electron microscopy (ESEM) image analysis was used to visualize the differences of the virgin and RAP bitumen at a microscopic level. The observed results indicated that the recovered bitumen from coarse RAP did not have the same characteristics as the fine RAP bitumen, and the interaction of RAP bitumen with virgin bitumen significantly depended on RAP particle size. The amount of active RAP bitumen in coarse RAP particles was higher than in fine RAP particles.

Keywords: hot mix asphalt; recycled asphalt; RAP gradation; ignition test; FTIR; ESEM

1. Introduction

Nowadays, sustainability and environmental matter are the main issues for most research studies on construction materials. One of the solutions is to use reclaimed asphalt pavement (RAP) as the main recycled material used in hot mix asphalts. Some researchers [1–4] have studied the use of RAP in road construction and concluded that because of the large environmental potential and cost benefits, it is important to maximize RAP content in the mix design. According to Al-Qadi et al. [5], the performance of pavements using properly prepared RAP was found to be satisfactory in terms of fatigue, rutting, thermal resistance, and durability.

The required thickness for pavement design depends on the pavement design method and material characteristics [6]. Saliani et al. [7] concluded that the surface area of mix components, like virgin aggregates, fiber, RAP, etc., impact bitumen absorption. Mix design factors, like mix duration and mix temperature, influence the rheological properties of mixes with RAP particles. Because of this, it is important to precisely characterize RAP according to particle size. Previous studies have shown that active bitumen content in coarse RAP particles is more than that for fine RAP particles [8].

It is desirable to have a higher recycle material content in hot mix asphalt (HMA), but incorporating large quantities can make the mix stiffer and more brittle. Consequently, these mixes are less workable, harder to compact in the field, and prone to crack [9–13]. Various strategies have been developed in order to increase RAP content such as incorporating a warm mix asphalt additive, adding rejuvenator agent, or a combination of mix designs [14]. In addition, RAP size is one of the critical factors that impact the level of blending between RAP and virgin particles. The degree of blending has not been quantified clearly. McDaniel and Anderson assumed full blending happened during mixing [15]. Some investigations were carried out to evaluate the level of interaction between RAP and virgin asphalt [16–18] and enhance the understanding of the interaction degree, but they could not quantify it. In all above studies, RAP size by itself was not studied precisely.

On the other hand, film thickness of bitumen that surrounds RAP aggregate particles can significantly impact HMA performance. Bressi et al. [19] proposed an initial approach to show different aging levels in RAP binder film thickness, and they proposed a methodology to detect the existence of a cluster phenomenon in RAP binder. Aging level and cluster phenomena can also impact the level of interaction between RAP and virgin materials.

Gardiner [20] concluded that the complex modulus was not solely controlled by the stiffness of the binder, whereas several other factors, including the gradation and angularity of the aggregate, played a main role in stiffness.

The term available or effective RAP binder refers to the binder that is released from RAP, becomes fluid, and blends with a virgin binder under typical mixing temperatures [21]. The stiffer the RAP binder, the less it will blend with the virgin binder. The binder stiffens with time mostly because of its reaction with oxygen (i.e., oxidization).

Oxygen availability depends on the pavement structure and varies from outer layers to pavement sublayers. Oxygen diffusion depth is considerable [22]. Also, asphalt film thickness affects oxygen diffusion. Sufficient methodology has been developed to estimate the diffusion depth [23]. Basically, using RAP in new mixes is a good idea for sustainable development, but information is still needed to understand the interaction between different components in a mix with RAP to optimize the properties of those recycled mixes. In order to have more RAP in HMA, as a greener alternative, blending degree and aging rate need to be well-understood. The objective of the present research is to study bitumen characteristics of fine and coarse RAP particles from the same RAP source and to verify the interaction between RAP bitumen and virgin bitumen according to RAP particle size.

2. Materials and Methods

In order to reach the goal of this study, a single RAP source was selected and split in two sizes, fine RAP (FR) and coarse RAP (CR), before extracting the bitumen (with solvent). The two RAPs were then used to make mixes with virgin aggregate and virgin binder before extracting the mixed bitumen. The following steps were performed:

- Performance grading (PG) for CR and FR bitumen;
- Performance grading (PG) for binder blended with CR and FR (coarse RAP mix (CRM) and fine RAP mix (FRM));
- Rheology behavior for CR, FR, CRM, FRM, and control mix;
- A Fourier transform infrared spectroscopy (FTIR)–ATR (attenuated total reflection) analysis was done on all bitumen and bitumen mixes;
- Environmental scanning electron microscopy (ESEM) image analysis was used to visualize bitumen differences on a micro scale.

In this study, a base course 20 mm mix was made with 4.5% total bitumen. The virgin bitumen used was a PG 64-28 from Bitumar (Montreal, QC, Canada). Also, RAP was supplied by Bauval (Montreal, QC, Canada).

To have a clear and better understanding of the differences between fine and coarse RAP, a single source of RAP from a specific area was chosen. In this way, the potential errors according to the RAP source were reduced. Two mixes were designed and tested. These mixes contained 35% fine RAP or 54% coarse RAP. Those mixes were referred to as fine RAP (FR) mix and coarse RAP (CR) mix in this paper. Initially, it was assumed that all RAP bitumen participated in the mix. The bitumen content of fine RAP was measured with the ignition oven at 6.67%, and it was 4.33% for the coarse RAP (Figure 1).

The RAP content for each type of RAP was selected to achieve the same bitumen content of 4.5% for all mixes.

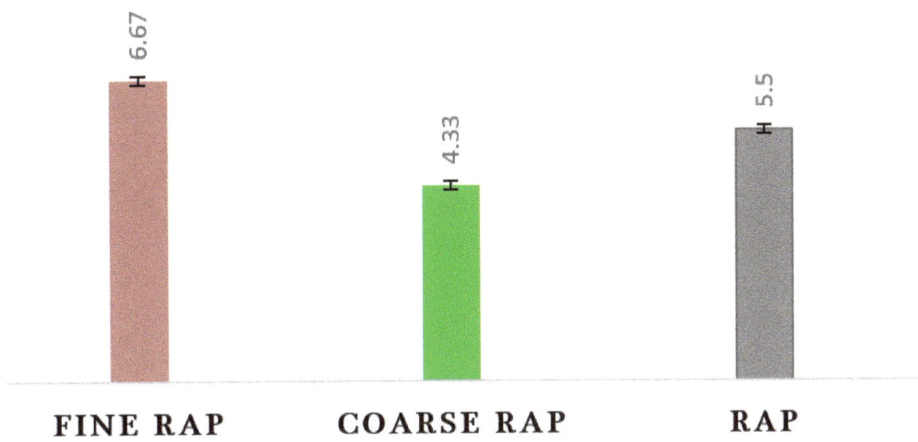

Figure 1. Bitumen content of fine reclaimed asphalt pavement (FR), coarse reclaimed asphalt pavement (CR), and reclaimed asphalt pavement (RAP) from the ignition test.

Once the mixes were done, the bitumen was extracted and recuperated according to ASTM D2172 standards. Bitumen was also extracted from the two RAP sources. The five resulting bitumen were:

- Virgin bitumen PG64-28;
- Recovered bitumen from coarse RAP particles (CR);
- Recovered bitumen from fine RAP particles (FR);
- Recovered bitumen from a mixture of virgin bitumen with CR (CRM); and
- Recovered bitumen from a mixture of virgin bitumen with FR (FRM).

The rheology of the bitumen was tested in terms of dynamic shear modulus, and their chemical characteristics were tested with infrared spectrometry. The difference between recovered coarse RAP bitumen and recovered fine RAP bitumen was also visualized by electron microscopy.

2.1. Bitumen Characterization

Extracted bitumens were classified by their performance grade (PG). Bitumen classification in North America is based on performance-grade (PG H–L), which is selected according to the temperature. The H value corresponds to the maximum pavement temperature, at which the bitumen has enough cohesion to minimize plastic deformation of the asphalt mix, and the L value corresponds to the minimum air temperature, where the bitumen has enough elasticity to keep asphalt mix flexible at low temperatures to reduce cracking potential [24].

2.2. Rheological Behavior

A dynamic shear rheometer (DSR) was used to determine the complex shear modulus (G*). All five bitumens were tested. AASHTO-T-315 [25] was adopted for the analysis. This test method covered the determination of the dynamic shear modulus and phase angle of asphalt binder when tested in dynamic (oscillatory) shear using parallel plate test geometry. It was applicable to asphalt binders that have dynamic shear modulus values in the range from 100 Pa to 10 MPa. This range of modulus was typically obtained between 6 °C and 88 °C at an angular frequency of 10 rad/s. This test method was intended for determining the linear viscoelastic properties of asphalt binders, as required for specification testing, and was not intended as a comprehensive procedure for full characterization of the viscoelastic properties of asphalt binder.

The shear modulus responses (G*) of the bitumen measured in the laboratory were first modelled with the 2S2P1D model, a generalization of the Huet–Sayegh model [26]. Referring to the 2S2P1D model, the values of the DBN (Di Benedetto-Neifar) bodies were fixed in the Linier Visco-Elastic (LVE) domain. An alternative general model "2S2P1D" (generalization of the Huet–Sayegh model), valid for both the bituminous binders and mixes and based on a simple combination of physical elements (spring, dashpot, and parabolic elements), was proposed.

The introduced 2S2P1D model had seven constants (Figure 2). At a given temperature T, the seven constants of the 2S2P1D model can be determined. G* is calculated following Olard and Di Benedetto [26]:

$$E^*_{(i\omega\tau)} = E_{00} + (E_0 - E_{00})/\left(1 + \delta\,(i\omega\tau)^{(-k)} + (i\omega\tau)^{(-h)} - (i\omega\beta\tau)^{(-1)}\right), \tag{1}$$

where:

- h and k are exponents such as $0 < k < h < 1$;
- E_{00} is the static modulus obtained when $\omega\tau \rightarrow 0$ (at low frequencies and high temperatures) with $\omega = 2\pi$ frequency;
- E_0 is the glassy modulus when $\omega\tau \rightarrow \infty$ (at high frequencies and low temperatures), MPa
- τ is the characteristic time, which this value varies only with temperature, accounting for the time–temperature superposition principle (TTSP); and
- η is the Newtonian viscosity, $\eta = (E_0 - E_{00})\beta\tau$.

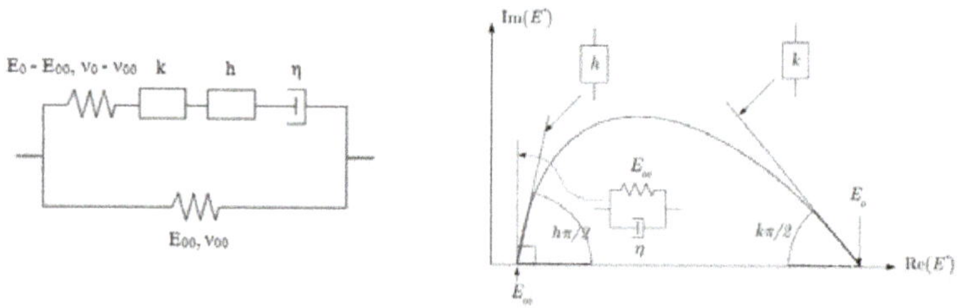

Figure 2. 2S2P1D schematic model.

When $\omega\tau \rightarrow 0$, the $E^*_{(i\omega t)} \rightarrow E_{00} + i\omega\,(E_0 - E_{00})\beta\tau \times \beta$.

2.3. Fourier Transform Infrared Spectroscopy–Attenuated Total Reflection (FTIR–ATR) Spectrometry

The FTIR-ATR analysis was performed with a Bruker Tensor 27 and a diamond ATR crystal. The analysis was performed 16 times per specimen with a 4 cm^{-1} resolution and a band range of 4000–600 cm^{-1}. Around 0.5 g of bitumen was placed with a spatula (at ambient temperature) on the crystal. The crystal was cleaned with a bitumen remover after each test followed by ethanol to remove traces of the bitumen remover.

Oxidation of hydrocarbons was associated, notably, with the increase of carbonyl C=O (around 1700 cm^{-1}) and sulfoxide S=O (around 1030 cm^{-1}) bonds. The carbonyl index (%) and sulfoxide index (%) are defined in Table 1 for C=O and S=O, respectively, with a higher value indicating relatively more oxidation. The peaks around 1460 cm^{-1} and 1376 cm^{-1} were for aliphatic C-CH3 groups, which served as baselines for the analysis, as they change relatively little during aging [27].

Table 1. Fourier transform infrared (FTIR) bands with bitumen aging [27].

Chemical Group	Bond	Approximate Wavenumber (cm^{-1})	Change with Aging	Intensity	Expression
Sulfoxide	S=O	1030	Increases in short-term	Weak	$\frac{A_{1030}}{A_{1460}+A_{1376}}$
Carbonyl	C=O	1700	Increases in long-term	Weak to medium	$\frac{A_{1700}}{A_{1460}+A_{1376}}$
Aliphatics (Asymmetric)	C-CH_3	1460, 1376	Relatively constant, small decrease	Medium	$\frac{A_{1700}}{A_{1460}+A_{1376}}$

2.4. *Environmental Scanning Electron Microscopy (ESEM) Analysis*

Environmental scanning electron microscopy (ESEM) observations were conducted in accordance with the settings developed previously for bitumen by Mikhailenko et al. [28]. Bitumen specimens were observed at room temperature immediately after being removed from the cooler with a FEI Quanta 250 FEG ESEM.

3. Results and Discussion

3.1. *Bitumen Performance Grade (PG)*

Rheological characterizations of all extracted and recovered bitumen from FR, CR, FRM, CRM, and control mix were performed following AASHTO T 315. Figure 3 presents the rheological properties of each asphalt bitumen using a dynamic shear rheometer (DSR).

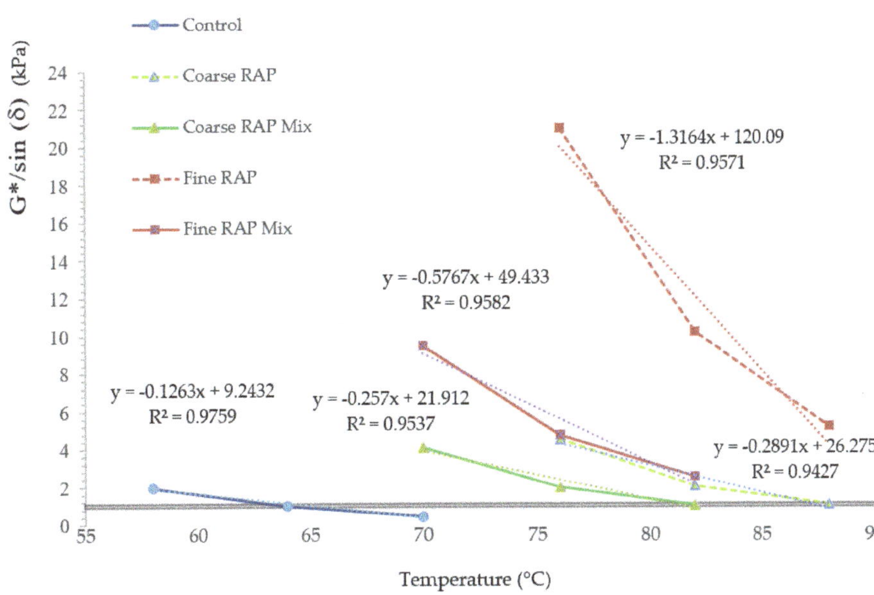

Figure 3. Performance grade of recovered bitumen by dynamic shear rheometer (DSR) at 1.6 Hz.

It was important to note that the recovered bitumen from the fine RAP was much stiffer than the recovered bitumen from the coarse RAP, even for the same RAP source that was simply split into two different groups. For example, at 76 °C and 1.6 Hz, the fine RAP bitumen was 4.6 times stiffer than the coarse RAP bitumen. Basically, the slope of each trend showed the sensitivity of that bitumen to temperature change. The control mix had the lowest sensitivity to temperature changes. Recovered bitumen from CR and CRM showed almost the same sensitivities, but recovered bitumen from FRM had less sensitivity than FR. Both CRM and FRM had the same virgin bitumen quantity (2.2%) and RAP bitumen content (2.3%). It was seen that FRM was stiffer than CRM at 76 °C (4.8 and 2 kPa, respectively), but the CRM bitumen was less sensitive than FRM to temperature change (0.25 and 0.57, respectively). In terms of the coarse and fine RAP before and after mixing with virgin bitumen, there was a huge difference from FR to FRM. This may be due to the amount of active bitumen in FR. As it was shown in a previous work [8], more RAP bitumen was transferred from coarse RAP than fine RAP particles because fine RAP particles were covered by a clump of mastic that did not tend to participate as active bitumen in the mixture. Despite the fact that fine RAP particles had higher aged bitumen content, participation by volume was less for the bitumen surrounding the coarse RAP particles in HMA.

One method that has been extensively used to evaluate the PG of virgin bitumen, in adding to HMA containing a high amount of RAP, is the blending chart [29]. In blending, it is assumed that there is a linear relationship between the amount of RAP bitumen and the rigidity (or DSR results) at a given temperature. Figure 4 shows a blending chart for the tested mix. To get results for all mixes at the same temperature, linear best fit curves were drawn from the results shown in Figure 3 (all with R2 > 0.95), and results were calculated for missing temperatures. At 70 °C, G*/sin(d) value of control mix (0.4 KPa) was referred to as 0% coarse rap content, and G*/sin(d) value of CR (6.038 KPa) was referred to as 100% coarse rap content. With these two values, the linear relationship between coarse RAP bitumen content and rigidity was plotted in Figure 4. The linear best fit curve for fine rap content was drawn at 70 °C and 1.6 Hz as well as in Figure 4.

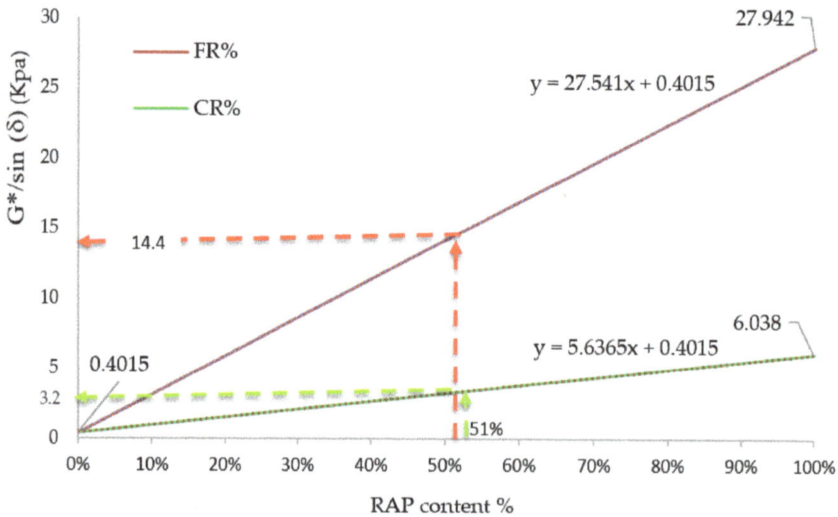

Figure 4. Blending charts for the fine and coarse RAP mixes at 70 °C and 1.6 Hz.

As shown in Figure 4, if we compare mixes with 51% FR or CR RAP, we have G*/sin (d) values of 14.4 and 3.2 KPa respectively. It showed that a mix with CR binder had a huge difference in shear modulus from the same mix with FR. It might be because the bitumen aging rate around finer particles is faster than the binder surrounding coarse particles, but more studies are needed to precisely verify this hypothesis. It may also be because fine particles have more surface area than coarse particles. Clearly, 51% CR had the same impact as 11% FR in mixes. Thus, it was important to clarify properly characterized RAP bitumen for a given particle size before mixing with virgin materials.

3.2. Shear Modulus (G*)

Figure 5 shows the 2S2P1D Cole–Cole plot of the shear complex modulus at different temperatures and frequencies for all recovered bitumen from CR, CRM, FR, FRM, and control mix. 2S2P1D was adopted to model the shear modulus of recovered bitumen. From Figure 5, it seemed that the virgin bitumen (PG64-28) was more viscous than the other bitumen.

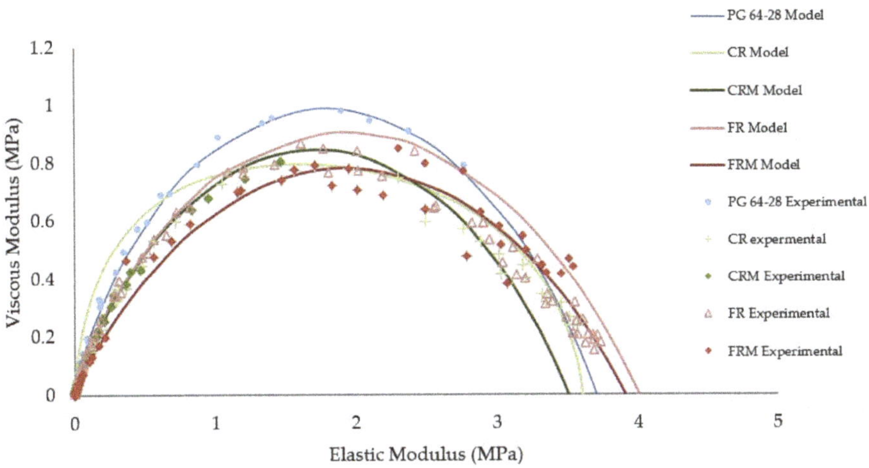

Figure 5. Cole–Cole shear modulus of recovered bitumen (2S2P1D model).

Table 2 shows the seven constant values for all four mixes. As mentioned earlier, these seven constants represented the bitumen characteristics. In general, the same binder was supposed to have equal values for the constants. Table 2 shows that the constants were not the same, so it could be concluded that recovered binder from FR was not the same as CR. Virgin bitumen was expected to have higher viscous components. Aged binder lost some parts of its viscous components and became brittle, so it mixed with aged bitumen and had a lower viscosity than virgin bitumen.

Table 2. The 2S2P1D model constants for all recovered binder.

	G_0 (Pa)	G_∞ (Pa)	k	h	δ	τ_E (s)	β
virgin	2	3,700,000	0.59	0.99	3.90	0.00015	5000
CR	100	5,000,000	0.38	0.86	4.90	0.00080	5000
FR	500	4,000,000	0.51	0.89	3.90	0.06000	5000
Coarse RAP Mix (CRM)	2	3,500,000	0.56	0.98	7.00	0.00250	5000
Fine RAP Mix (FRM)	10	3,900,000	0.48	0.90	10.00	0.00700	5000

Figure 6 shows the master curve of shear modulus (Pa). The G* values of FR and CR were higher than the other mixes. Asphalt production, lifetime, and presence of air impacted the aging progress. A single source of RAP was used for CR and FR. Then, since all these factors were the same for CR and FR, the recovered bitumen was supposed to have the same master curve. Figure 6 shows that, except for the mentioned factors, particle size may change the rheology of bitumen. This was because FR and CR mixes had higher aged bitumen content than CRM and FRM at intermediate and lower frequencies. FR was stiffer than CR, which meant that the finer RAP particles were covered by a stiffer bitumen. This could be explained as bitumen that surrounded fine particles aged faster than when they surrounded coarse particles; a higher surface area of the bitumen was exposed to oxidative aging. After mixing the coarse and fine RAP with virgin materials, master curves from recovered bitumen were plotted. Here, we again saw that FRM was stiffer than CRM at intermediate and lower frequencies.

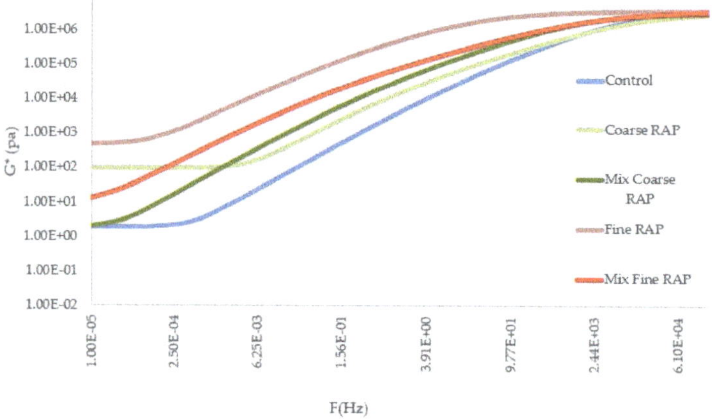

Figure 6. G* master curve at 10 °C of recovered binder (2S2P1D model).

The Reduced G* (RG*) index was looked at to clarify the interaction of RAP bitumen with virgin bitumen according to RAP particles. G* was determined for CR and FR as well as for CRM and FRM. According to CRM (FRM) mix design, CR (or FR) was added to the 2.2% virgin bitumen. The difference between CR (or FR) and CRM (FRM) showed the impact of virgin bitumen on total bitumen stiffness. (RG*) is calculated by Equation (2) as a percentage:

$$RG^* = ((G_i^* - G_{iM}^*))/(G_i^*), \quad (2)$$

where RG^* is the percentage of reduced G*, G^*_i is the shear modulus of pure RAP, and G^*_{iM} is the shear moduli of RAP and virgin bitumen mixes. The difference of G^*_i and G^*_{iM} showed how well the bitumen mixed together. As the aged bitumen was stiffer than virgin bitumen, eventually G^*_i was supposed to be higher than G^*_{iM}. Figure 7 indicates the rheology transition before and after mixing RAP with virgin materials. Both CR and FR mixed with 2.2% virgin binder. Most of the influence of aged binder was observed at low frequency. The difference between aged bitumen and mixture bitumen changed from 0% to 90%. These differences meant rigidity was changed. FRM was 90% softer than FR; it increased rapidly and was constant over a wide frequency range. The impact of CR clearly differed from FR. The different trends in CR and FR could translate to the impact of each type of RAP on virgin bitumen. CR and CRM had the same rigidities at high frequency and gradually increased at low frequency. CR and FR bitumen recovered from a single source of recycled asphalt were supposed to have the same characteristics when blended with a specific amount of virgin bitumen, but the interactions were different. Aging rates were not the same between fine gradation and coarse gradation.

3.3. FTIR–ATR Spectrometry

The results for the FTIR–ATR analysis are presented in Figures 8–10. Asphalt aging was characterized by oxidization indices shown in Table 1. The C=O and S=O indices both increased significantly with RAP and blended bitumen relative to the virgin bitumen. There was a small increase in the RAP bitumen compared to the blended bitumen in the indices. The aging indices were higher for the fine RAP, likely because of its higher surface area, compared to the course RAP, which allowed for a greater degree of aging. The sulfoxide indices for all bitumen were higher, as was typical for asphalt bitumen [30].

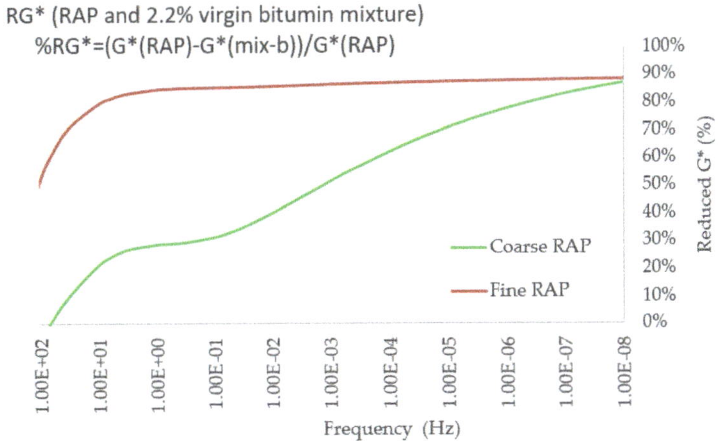

Figure 7. Reduced G* for CR and FR at 10 °C.

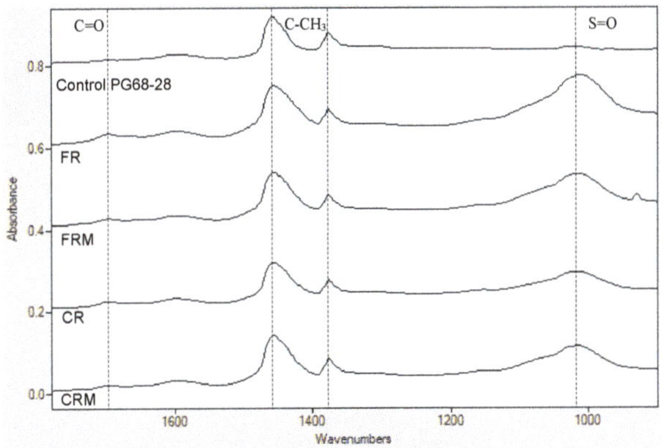

Figure 8. Compilation of FTIR spectra featuring C=O and S=O oxidation bands.

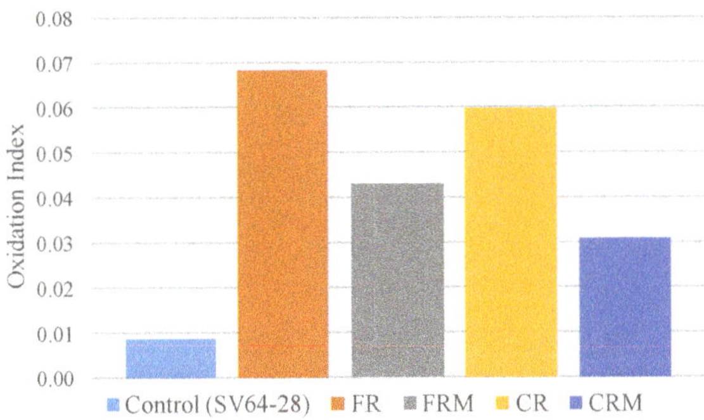

Figure 9. FTIR carbonyl oxidation indices.

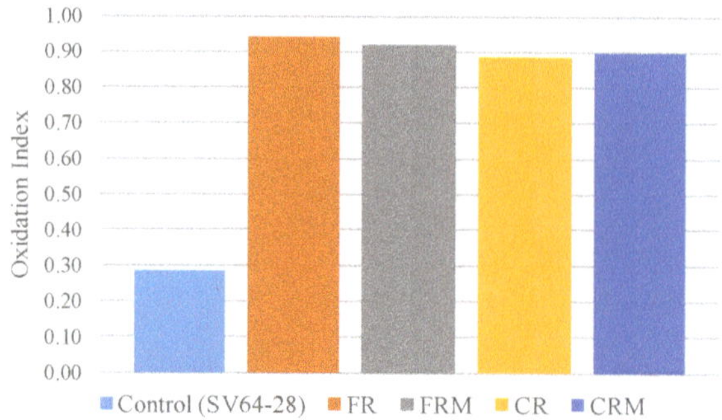

Figure 10. FTIR sulfoxide oxidation indices.

3.4. Environmental Scanning Electron Microscopy (ESEM) Analysis

The images from the ESEM analysis are shown in Figure 11, before electron beam irradiation and after stabilization of the sample. The bitumen before irradiation showed a 'bee' type structure, similar to the one found in AFM observations [31,32], which was denser in the RAP bitumen compared to the virgin and blended bitumen.

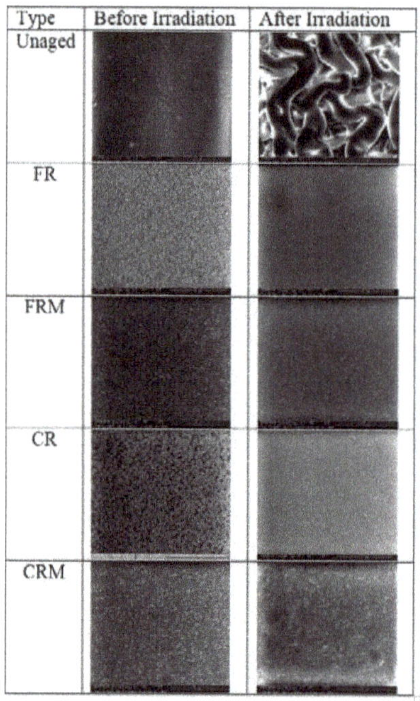

Figure 11. Environmental scanning electron microscopy (ESEM) Images of Samples at 1000× before Electron Beam Irradiation and After Image Stabilized.

After the sample was exposed to the electron beam, and the accumulation of energy irradiated the sample, the asphalt bitumen tended to reveal a 'fibril' microstructure as lighter components of the bitumen dispersed [33], which was the case for the virgin bitumen. However, both the blended and RAP bitumen responded much less to beam irradiation. This was likely because the microstructure of the bitumen stiffened from aging, and the molecular mobility was reduced. This made revealing the microstructure by ESEM irradiation a slower and more difficult process with the settings that were used.

4. Conclusions

It is common to use green alternatives in asphalt mix designs such as recycled asphalt, crumb rubber, etc. Several studies showed the advantages of RAP in hot mix asphalt. Still, there is no clear vision of the impact RAP size has on HMA behavior. In this study, a single Quebec source of RAP was separated into coarse and fine particles and mixed in with bitumen pavement mixture. Properties of the asphalt mixes were evaluated by the complex modulus test and DSR. Also, the ignition test (ASTM D6307) was used to quantify bitumen content in the RAP, FTIR spectrometry was used for chemical properties of the bitumen, while ESEM image analysis was used to visualize the differences of the virgin and RAP bitumen at a microscopic level.

FRM was designed with 35% fine RAP particles, including 2.3% aged bitumen and 2.2% virgin bitumen. CRM was designed with 54% coarse RAP particles, including 2.3% aged bitumen and 2.2% virgin bitumen. Recovered bitumen (4.3%) was measured for CR particles, and bitumen content was 6.6% for FR particles. Designed mixes were compared with the control mix, and the following results were achieved:

- FR-recovered bitumen is more sensitive to temperature changes than virgin and CR-recovered bitumen.
- FR-recovered bitumen is stiffer than CR-recovered bitumen. Also, FRM-recovered bitumen is stiffer than CRM-recovered bitumen. This shows that the bitumen grade of coarse particles is not the same as fine particles in RAP.
- Reduced G* (RG*) was looked at to clarify the interaction of RAP bitumen with virgin bitumen according to RAP particles. The CR reduced factor was higher than FR; CR bitumen and CRM bitumen were more similar than FR and FRM. This factor might be applied in mixing designs to predict the stiffness of HMA with the inclusion of RAP, which is the case for future studies.
- Aging rates in fine particles are faster than coarse particles because fine RAP has a higher surface area.
- Results for the FTIR–ATR analysis showed aging indices are higher for the fine RAP, likely because of its higher surface area compared to the course, which allowed for a greater degree of aging.
- Images from the ESEM analysis shows both the blended and RAP bitumen respond much less to beam irradiation. This is likely because the bitumen microstructure stiffened from aging.

More work is needed to better understand the impact of RAP gradation. The results shown here definitely give new information about different RAP bitumen contained in different sizes, but new tests with several other RAP size are needed to confirm the results presented here.

Author Contributions: S.S.S. conceived of the presented idea. A.C. and H.B. developed the theory and verified the analytical methods and supervised the project. S.S.S. wrote the original draft; and P.M. wrote, reviewed, and edited the manuscript. All authors discussed the results and contributed to the final manuscript.

Funding: This research was funded by The Pavements and Bituminous Materials Laboratory (LCMB).

Acknowledgments: This work was supported by The Pavements and Bituminous Materials Laboratory (LCMB) and the Centre for Pavement and Transportation Technology (CPATT). The authors would like to thank the companies in Quebec that provided us with the materials for the project.

Conflicts of Interest: The authors declare no conflict of interest.

References

1. McDaniel, R.S.; Shah, A.; Huber, G.A.; Copeland, A. Effects of reclaimed asphalt pavement content and virgin binder grade on properties of plant produced mixtures. *Road Mater. Pavement Des.* **2012**, *13* (Suppl. 1), 161–182. [CrossRef]

2. Mogawer, W.; Austerman, A.; Mohammad, L.; Kutay, M.E. Evaluation of high RAP-WMA asphalt rubber mixtures. *Road Mater. Pavement Des.* **2013**, *14* (Suppl. 2), 129–147. [CrossRef]
3. Sias Daniel, J.; Gibson, N.; Tarbox, S.; Copeland, A.; Andriescu, A. Effect of long-term ageing on RAP mixtures: Laboratory evaluation of plant-produced mixtures. *Road Mater. Pavement Des.* **2013**, *14* (Suppl. 2), 173–192. [CrossRef]
4. Visintine, B.; Khosla, N.P.; Tayebali, A. Effects of higher percentage of recycled asphalt pavement on pavement performance. *Road Mater. Pavement Des.* **2013**, *14*, 432–437. [CrossRef]
5. Al-Qadi, I.L.; Carpenter, S.H.; Roberts, G.; Ozer, H.; Aurangzeb, Q.; Elseifi, M. *Determination of Usable Residual Asphalt Binder in RAP*; Illinois Center for Transportation (ICT): Rantoul, IL, USA, 2009.
6. Perraton, D.; Baaj, H.; Carter, A. Comparison of Some Pavement Design Methods from a Fatigue Point of View: Effect of Fatigue Properties of Asphalt Materials. *Road Mater. Pavement Des.* **2010**, *11*, 833–861. [CrossRef]
7. Saliani, S.S.; Carter, A.; Baaj, H.; Badeli, S. Investigation of the tensile strength of hot mix asphalt incorporating pulp aramid fiber. In Proceedings of the Sixth-Second Annual Conference of the Canadian Technical Asphalt Association (CTAA), Halifax, NS, Canada, 12–15 November 2017.
8. Saliani, S.S.; Carter, A.; Baaj, H. Investigation of the impact of rap gradation on the effective binder content in hot mix asphalt. In Proceedings of the CSCE Annual Conference, London, ON, Canada, 1–6 June 2016.
9. Kim, S.; Byron, T.; Sholar, G.A.; Kim, J. *Evaluation of Use of High Percentage of Reclaimed Asphalt Pavement (RAP) for Superpave MixturesI*; Report No. FL/DOT/SMO/07-507; Florida Department of Transportation: Tallahassee, FL, USA, 2007; Volume 103.
10. Mogawer, W.; Bennert, T.; Daniel, J.; Bonaquist, R.; Austerman, A.J.; Booshehrian, A. Performance characteristics of plant produced high RAP mixtures. *Road Mater. Pavement* **2012**, *13* (Suppl. 2), 183–208. [CrossRef]
11. Munoz, J.S.C.; Kaseer, F.; Arambula, E.; Martin, A.E. Use of the resilient modulus test to characterize asphalt mixtures with recycled materials and recycling agents. *Transp. Res. Rec. J. Transp. Res. Board* **2015**, *2506*, 45–53. [CrossRef]
12. Kaseer, F.; Yin, F.; Arámbula-Mercado, E.; Martin, A.E. Stiffness characterization of asphalt mixtures ith high recycled materials contents and recycling agents. *Transp. Res. Rec. J. Transp. Res. Board* **2017**, *2633*, 58–68. [CrossRef]
13. Kaseer, F.; Cucalon, L.G.; Arámbula-Mercado, E.; Martin, A.E.; Epps, J. Practical tools for optimizing recycled materials content and recycling agent dosage for improved short-and long-term performance of rejuvenated binder blends and mixtures. *J. Assoc. Asph. Paving Technol.* **2018**, *33–34*, 87.
14. Kaseer, F.; Martin, A.E.; Arámbula-Mercado, E. Use of recycling agents in asphalt mixtures with high recycled materials contents in the United States: A literature review. *Constr. Build. Mater.* **2019**, *211*, 974–987. [CrossRef]
15. McDaniel, R.S.; Soleymani, H.; Anderson, R.M.; Turner, P.; Peterson, R. *Recommended Use of Reclaimed Asphalt Pavement in the SuperPave™ Mixture Design Method*; NCHRP Final Report (9–12), TRB; National Research Council: Washington, DC, USA, 2000.
16. Bennert, T.; Dongre, R. *A Back Calculation Method to Determine "Effective" Asphalt Binder Properties of RAP Mixtures*; Transportation Research Board: Washington, DC, USA, 2010.
17. Bonaquist, R. New Approach for the Design of High RAP HMA. Presented at the 2005 Northeast Asphalt User's Producer's Group Meeting, Burlington, NJ, USA, 19–20 October 2005.
18. Rowe, G. Phase angle determination and interrelationships within bituminous materials. In Proceedings of the 7th International RILEM Symposium on Advanced Testing and Characterization of Bituminous Materials, Rhodes, Greece, 27–29 May 2009.
19. Bressi, S.; Cavalli, M.C.; Partl, M.N.; Tebaldi, G.; Dumont, A.G.; Poulikakos, L.D. Particle clustering phenomena in hot asphalt mixtures with high content of reclaimed asphalt pavements. *Constr. Build. Mater.* **2015**, *100*, 207–217. [CrossRef]
20. Stroup-Gardiner, M.; Wagner, C. Use of reclaimed asphalt pavement in Superpave hot-mix asphalt applications. *Transp. Res. Rec.* **1999**, *1681*, 1–9. [CrossRef]
21. Kaseer, F.; Arámbula-Mercado, E.; Martin, A.E. *A Method to Quantify Reclaimed Asphalt Pavement Binder Availability (Effective RAP Binder) in Recycled Asphalt Mixes*; Transportation Research Record: Washington, DC, USA, 2019; p. 0361198118821366.

22. Jin, X.; Cui, Y.; Glover, C.J. Modeling Asphalt Oxidation in Pavement with Field Validation. *Pet. Sci. Technol.* **2013**, *31*, 1398–1405. [CrossRef]
23. Rose, A.A.; Arambula, E.; Howell, T.; Glover, C.J. An X-Ray CT Validated Laboratory Measurement Method for Air Voids Distribution over Depth in Asphalt Pavement: A Step toward Simplified Oxidation Modeling. *Pet. Sci. Technol.* **2014**, *32*, 3020–3028. [CrossRef]
24. American Association of State Highway and Transportation Officials-PP6. *Practice for Grading or Verifying the Performance Grade of an Asphalt Binder*; AASHTO Provisional Standards; AASHTO: Washington, DC, USA, 1994.
25. American Association of State Highway and Transportation Officials-T-315. *Standard Method of Test for Determining the Rheological Properties of Asphalt Binder Using a Dynamic Shear Rheometer (DSR)*; AASHTO Provisional Standards; AASHTO: Washington, DC, USA, 2012.
26. Olard, F.; Di Benedetto, H. General "2S2P1D" model and relation between the linear viscoelastic behaviours of bituminous binders and mixes. *Road Mater. Pavement Des.* **2003**, *4*, 185–224.
27. Mikhailenko, P.; Bertron, A.; Ringot, E. *Methods for Analyzing the Chemical Mechanisms of Bitumen Aging and Rejuvenation with FTIR Spectrometry*; RILEM Publications: Ancona, Italy, 2015. [CrossRef]
28. Mikhailenko, P.; Kadhim, H.; Baaj, H.; Tighe, S. Observation of asphalt binder microstructure with ESEM. *J. Microsc.* **2017**, *267*, 347–355. [CrossRef] [PubMed]
29. McDaniel, R.; Soleymani, H. *Recommended Use of Reclaimed Asphalt Pavement in the Superpave Mix Design Method*; NCHRP Web Document; National Cooperative Highway Research Program: Washington, DC, USA, 2000; Available online: http://onlinepubs.trb.org/Onlinepubs/nchrp/nchrp_w30-a.pdf (accessed on 5 May 2017).
30. Hofko, B.; Porot, L.; Falchetto Cannone, A.; Poulikakos, L.; Huber, L.; Lu, X.; Grothe, H. *FTIR* spectral analysis of bituminous binders: Reproducibility and impact of ageing temperature. *Mater. Struct.* **2018**, *51*, 45. [CrossRef]
31. Das, P.K.; Baaj, H.; Tighe, S.; Kringos, N. Atomic force microscopy to investigate asphalt binders: A state-of-the-art review. *Road Mater. Pavement Des.* **2016**, *17*, 693–718. [CrossRef]
32. Oenen, H.; Besamusca, J.; Fischer, H.R.; Poulikakos, L.D.; Planche, J.-P.; Das, P.K.; Chailleux, E. Laboratory investigation of bitumen based on round robin DSC and AFM tests. *Mater. Struct.* **2013**, *47*, 1205–1220. [CrossRef]
33. Mikhailenko, P.; Kou, C.; Baaj, H.; Poulikakos, L.; Cannone-Falchetto, A.; Besamusca, J.; Hofko, B. Comparison of ESEM and physical properties of virgin and laboratory aged asphalt binders. *Fuel* **2019**, *235*, 627–638. [CrossRef]

infrastructures

MDPI

Article

Peat Fibers and Finely Ground Peat Powder for Application in Asphalt

Patricia Kara De Maeijer [1,*], Hilde Soenen [2], Wim Van den bergh [1], Johan Blom [1], Geert Jacobs [1] and Jan Stoop [1]

1 EMIB Research Group, Faculty of Applied Engineering, University of Antwerp, 2020 Antwerp, Belgium; wim.vandenbergh@uantwerpen.be (W.V.d.b.); johan.blom@uantwerpen.be (J.B.); geert.jacobs@uantwerpen.be (G.J.); jan.stoop@uantwerpen.be (J.S.)

2 Nynas NV, Bitumen Research, 2020 Antwerp, Belgium; hilde.soenen@nynas.com

* Correspondence: patricija.karademaeijer@uantwerpen.be; Tel.: +32-3-265-8851

Received: 20 November 2018; Accepted: 28 December 2018; Published: 4 January 2019

Abstract: In this study, the feasibility of a natural peat fiber and finely ground peat powder as a modifier for bitumen was investigated. Initially, the as-received peat material was characterized in detail: the material was ground to various degrees, separated into fiber and powder fractions, and the gradation of the powder fraction as well as the size of the fibers were determined. A possible solubility in bitumen, the moisture content, and the density of both fractions were evaluated, and a limited chemical characterization of the fibers was conducted. Secondly, the rheological behavior of the powder and the fibers when blended with bitumen was evaluated. Additionally, a limited asphalt study was conducted. The rheological data showed the stiffening effects of the powder fraction and the presence of a fiber network, which were obvious as a plateau modulus towards lower frequencies. The fiber network was strain-dependent and showed elastic effects. This was further confirmed by the multiple stress creep recovery (MSCRT) tests. These tests also indicated that the fibers should improve the rutting resistance, although it was not possible to confirm this in asphalt rutting tests. Asphalt drainage tests demonstrated that adding dry peat, whether this is ground or not, is effective in reducing the binder drainage. However, the data also revealed that the amount of added peat fibers and powder should be limited to avoid difficulties in the compaction of these asphalt mixes.

Keywords: peat; asphalt; rheology; drainage; wet process; rutting

1. Introduction

Peat is an accumulation of partially decayed vegetation or organic matter formed in wetlands: *fens* with typical plants, such as bushes and trees, which are fed by ground water rich in nutrients; and *bogs* with typical plants, such as mosses, cotton grass, and heather, which are fed by rain water poor in nutrients [1]. The most-used material is Sphagnum moss peat, which is the main material building up in bogs in the Northern hemisphere. In some countries, peat is regarded as a slow-renewable material, although the rate of extraction and usage of peat far exceeds the rate of reforming. In Finland in 2016, bogland usage was 9.39 million ha with peat usage of 3 million m^3 and 3% of the annual energy production was provided by peat [2]. Peat, apart from usage as an energy provider, has agricultural applications, such as increasing the water-holding capacity of sands, and industrial applications, such as an oil absorbent or as an efficient filtration medium for mine waste streams, municipal storm drainage, and septic systems [3].

The major distresses that occur in asphalt pavement are related to crack formation, permanent deformation, and water damage. Moreover, the properties of asphalt change with time, mainly due to ageing effects occurring in the binder phase. Additives, such as polymers, crumb rubber, waxes, and surface-active components, have been used to prevent distresses and improve the durability of the pavement. Among the additives, fibers have also been used. Fibers, in particular cellulose fibers, are added to avoid binder drainage during transportation from the asphalt plant to the construction site, typically in open mixes that contain a high binder percentage. Fibers have also been added for other reasons, such as increasing the viscosity, and, related to this, the rutting resistance. Glass fibers, for example, have a potential to improve fatigue life and deformation characteristics by increasing the rutting resistance [4]. The application of natural fibers, such as banana [5], bamboo [6], cellulose [7], coconut [8,9], hemp [10], jute [11], kapok [12], peat [13,14] and sisal [15], has so far been used for improving the drainage, water sensitivity, and stability, and increasing the tensile strength, of the asphalt pavement. Typically, the optimum added fiber content into the asphalt mix is 0.3–0.5% by weight of the asphalt mix [16]. In terms of workability, mixes with fibers showed a slight increase in the optimum binder content compared to the control mix. This is comparable to the addition of very fine aggregates. The proper quantity of bitumen to coat the fibers is dependent on the absorption and the surface area of the fibers. Therefore, this content is affected not only by the fiber concentration but also by the fiber type [17]. In addition, the degree of homogeneity or dispersion of the fibers within the mix is also important and determines the strength of the resulting mixes [4]. If the fibers are longer, typically more than 40 mm, a so called "balling" problem may occur, i.e., some of the fibers may lump together, and other fibers may not blend well with bitumen. Short fibers may not provide any reinforcement effect and can serve just as a filler in the mix [17]. The inclusion of fibers during the mixing process as a stabilizing agent has several advantages, including the possibility of using an increased binder content, creating an increased film thickness around the aggregate, an increased mix stability, and interlocking between the fibers and the aggregates, which improves the strength and reduces the possibility of drain down during transport and paving. Peat has already been applied in asphalt as a stabilizing additive for peat-based asphalt–concrete mixes, providing high performance to the road surface at a low cost [13].

Peat has perhaps a unique trait. Wettability, or the hydrophilic property of the peat, is observed as long as the peat contains a minimum moisture level (depending on the peat type, this is around 70%). Below this level, the hydrophilic character of peat weakens, and it becomes hydrophobic, meaning it will expel water [18]. For the application of fibers in bitumen or asphalt, hydrophobic characteristics are preferred.

This study will discuss the results of an investigation on peat itself and on the effects of adding peat powder and peat fibers into bituminous binders and the asphalt mix, where peat is seen as an example of a natural additive, such as cellulose and many other fiber types.

2. Materials and Methods

2.1. Materials

The Finnish company *VAPO Fibers* provided different sizes of peat material: medium (PM), long (PL), and extra-long (PEL) peat fibers with approximate lengths of <8 mm, <16 mm, and >16 mm, respectively. Extra-long peat fibers (PEL) with a moisture content of 20 wt% and 55 wt% were chosen for the present study, and denoted as PEL20 and PEL55, respectively. Peat was dried for at least 2 h in an oven at 110 °C before applying it in the experiments. Peat was ground in a blender "Philips ProBlend 6", which resulted in a fine powder mixed with a fiber fraction (10–30 mm). Ground peat was subsequently sieved to separate the fiber from the powder fraction (see Figure 1a,b).

Three different powders: PEL20, PEL20(3), and fine FPEL20, with a grinding time of 1, 3, and 10 minutes, respectively, were obtained (see Figure 2a). The granulometric composition of these three powders in comparison to Duras II filler is shown in Figure 2b. It can be seen in Figure 3a that powders after grinding for 3 min still may contain very fine fiber fractions (below 1 mm). Grinding for a longer time excluded this possibility. During the sieving of peat and the separation of fibers and powder, it was observed that in the as-received peat the fiber content is around 3%, and the powder content around 97%.

(a) (b)

Figure 1. The preparation of peat material for testing: (**a**) grinding and (**b**) sieving.

(a) (b)

Figure 2. (**a**) The obtained peat powders and (**b**) the granulometric composition of the peat powders and filler.

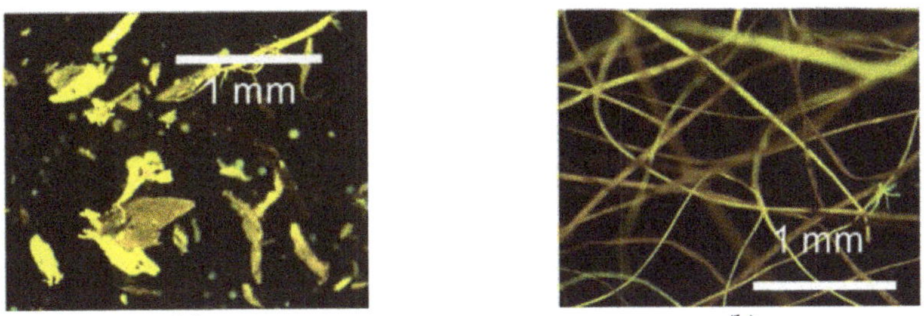

(a) (b)

Figure 3. An example of fluorescence images of PEL20: (**a**) powder and (**b**) fibers (scale 50×).

The base bitumen was a standard 70/100 unmodified binder, with a penetration of 70 dmm and a softening point of 47.2 °C.

2.2. Test Methods

The density of peat fibers and powders was defined according to the standard NBN EN 15326+A1 [19]. In this paper, toluene was used as the test liquid.

Fourier Transform Infrared Spectroscopy (FT-IR) combined with attenuated total reflection (ATR) was used. The instrument was a Nicolet IS 1, with a diamond cell (smart-orbit).

Fluorescence microscopy was performed with a Carl Zeiss Axioskop 40Fl microscope equipped with a digital camera DeltaPix DP200. In fluorescence mode, a high-pressure mercury arc lamp HBO50, which transmits intense light with a wavelength between 450 and 490 nm, was used. Microscopy images of peat powder FPEL20 and fibers are shown in Figure 3a,b. These images give an indication of the variation in the particle and fiber shape and size.

Scanning Electron Microscopy (SEM) images were scanned using a Coxem EM 30 P. A high vacuum and Tungsten element with an accelerating voltage of 20 KV and a magnification of 236 times were used. In Figure 4, a SEM image of the fiber fraction is shown, indicating the thickness and thickness variability within this fraction, and a minor porosity effect on the fibers.

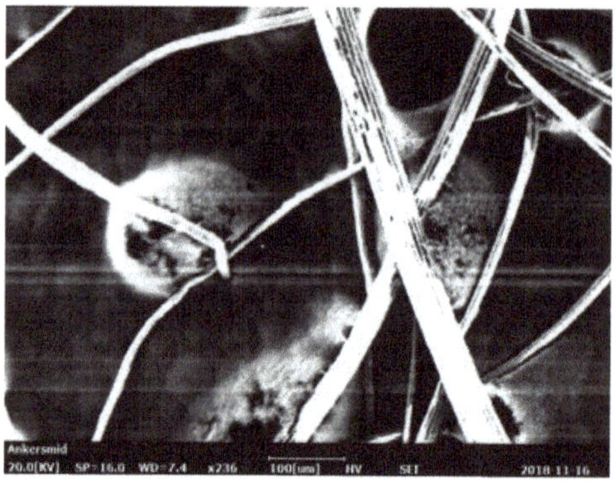

Figure 4. An SEM image of peat fibers (scale 236×) with a fiber thickness of 50 μm (the appearance of bubbles in the background is the result of evaporation of the glue on the heated surface).

The rheological properties were determined by an Anton Paar MCR 500 rheometer. Measurements were conducted at 50 °C using the 25 mm plate geometry to relate with the rutting test. The base binder was investigated at a gap setting of 1 mm, while samples modified with peat were investigated with a gap setting of 1.5 mm. Stress sweeps, frequency sweeps, and repeated creep measurements were performed. For some samples, frequency sweeps were also recorded at 40 °C, 60 °C, and 70 °C to observe the effect of temperature. The specimens were prepared in silicon moulds and afterwards transferred to the rheometer plate. Special precautions to obtain repeatable results were taken, which will be discussed in the results section.

The binder drainage test was performed in accordance with NBN EN 12697-18 [20] using drainage baskets constructed from 3.15 mm perforated stainless-steel sheets, in accordance with ISO 3310-2, on the side and base, to form 100-mm cubes with feet at each corner of the base. The asphalt mixes consisted of 1100 g batches of (loose) stone mastic asphalt (SMA) mix with the following composition: 70.7% crushed porphyry aggregates with the maximum size of 10 mm, 20.4% crushed porphyry sand, 8.90% Duras II filler, and 6.9% bitumen of standard penetration grade 70/100.

The rutting resistance of asphalt mixes was evaluated using the wheel tracking test in accordance with standard NBN EN 12697-22 [21]. The same mix composition as for the drainage tests was used for all of the measurements. Six slabs (see Figure 5a) with dimensions 18 × 50 × 5 cm were produced. The asphalt mixing temperature was 150 °C.

Wheel tracking tests were performed with the LCP rut tester at 50 °C. The slabs were conditioned at 50 °C for a period of 12 h prior to the testing. The rut depth in the slabs was measured manually, using a specifically designed setup (see Figure 5b) at 15 predetermined locations. Rut depths were measured after 1 000, 3 000, 5 000, 10 000, 20 000, and 30 000 load cycles.

The volumetric properties of the asphalt slabs were determined according to NBN EN 12697 [22,23].

(**a**) (**b**)

Figure 5. The rutting test: (**a**) the prepared slabs for testing and (**b**) the rut depth measuring procedure.

3. Results and Discussion

Before using the peat material in bitumen or asphalt, several simple tests were carried out to investigate whether the peat is blending or interacting with, or possibly even modifying, the bitumen, and which properties it may influence. To have an indication of the solubility of the peat or peat fractions in bitumen, the solubility in several organic solvents, such as toluene ($\rho = 870$ kg/m^3), acetone ($\rho = 781$ kg/m^3), and heptane ($\rho = 684$ kg/m^3), was investigated (see Figure 6). It was concluded that there is no solubility in any of these solvents, and therefore no solubility in bitumen is expected. However, for the fine peat powder FPEL20, it was observed that the solution was slightly colored in toluene and acetone (see Figure 6a, the photograph was taken after 24 h). That indicates that a minor solubility of this powder FPEL20 may occur in bitumen. The solubility tests also demonstrate that the density of the peat fibers and powder is between 1 and 0.87 kg/m^3. A more exact determination, according to NBN EN 15326+A1, showed that the average density for the powders varied from 929 kg/m^3 to 989 kg/m^3, and for the fibers it was 989 kg/m^3.

(**a**)

(**b**)

Figure 6. The solubility test of peat (**a**) powder and (**b**) fibers.

Natural materials often contain moisture, and when blending such materials in asphalt it is important to remove this moisture or at least to know the percentage of moisture. According to the literature, three types of water could be present in peat: chemically bound, capillary, and free water. The chemically bound water is usually the most difficult to remove. Regarding the moisture content of peat, when heating an initial mix of peat fibers and powder, a quick decrease in weight was observed, probably due to the free water fraction, which could be very easily removed. This stage is followed by a much slower weight decrease, which is probably related to capillary water, which is more difficult to remove. It can be noted that, after drying, the peat mix weight is again increasing, indicating that the material reabsorbs moisture again from the environment. Therefore, peat powder and fibers were kept in a dry atmosphere after drying and before applying them in bitumen or asphalt tests (see Figure 7).

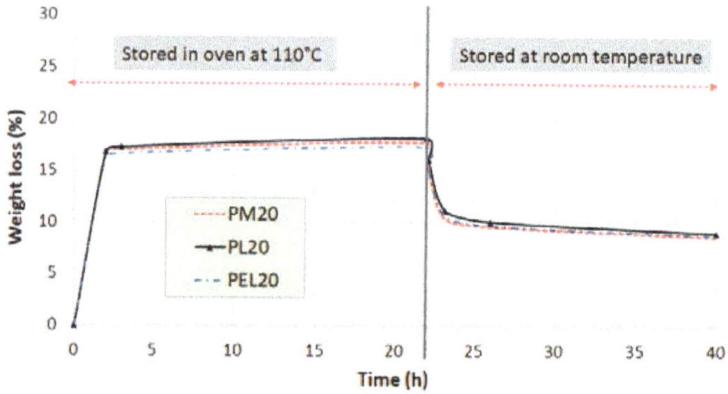

Figure 7. Weight losses of peat during drying (weight loss compared to the as-received sample) [24].

In FT-IR spectroscopy, water can be determined by the broad hydrogen bonding region (3500–3000 cm^{-1}); this was used to investigate the water content differences between PEL20 and PEL55. The spectra of PEL20 and PEL55 are shown in Figure 8. The data confirm differences in moisture content between both samples. For each moisture content, three repeat tests were performed, and they showed little variation. Besides water, the spectra also indicate that peat consists of saturated organic groups, alcoholic groups, a very small amount of C=O groups, and probably also inorganic material, indicated by the large signal at 1000 cm^{-1}. Moreover, there were no differences in the spectra of the powder and the fiber fraction, indicating that both fractions are chemically identical. Similarities to the spectrum of cellulose were also observed, indicating that the findings for the peat material may be very similar to what is seen for cellulose fibers.

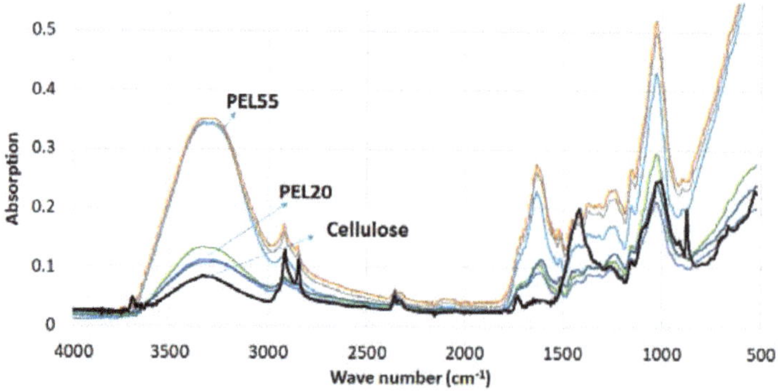

Figure 8. FT-IR spectra of medium peat fibers containing two water contents [24].

Peat bitumen blends were investigated by rheological test methods. In the initial tests, the peat bitumen samples were prepared in 100 mL cans, and from these cans, specimens for the rheological tests were taken. However, the repeatability of the so-obtained data was not very high, showing a difference of about 30% between tests on material from the same can, but different specimens. This was only the case for the fiber-modified samples. For the bitumen modified with powder, the repeatability was on the same level as for unmodified binders. Therefore, several actions were taken to improve the repeatability of the fiber-modified samples. Test samples were prepared directly in a silicon mold with the same dimensions as the specimen needed for the rheometer, in a sample size of 1 g. In this way, it was assured that the added amounts were also present in the test sample. Additionally, the fibers were cut to about 5 mm in length.

In rheological tests, it is common practice to test inside the linear viscoelastic (LVE) range of the material. Within this region, weak structures and networks will stay intact and particles, if present in the sample, will not be deformed. As an example, stress sweeps at 50 °C and 0.1 Hz are plotted in Figure 9. The evolution of the complex modulus and the phase angle with strain are shown for various bitumen (base binder is denoted as REF), fiber and powder combinations. As these tests were conducted on heterogeneous materials, the samples were observed very carefully to assure that the sample radius and shape were still intact after each test. The obtained data indicate a nice LVE range for the unmodified and the powder-modified binders. However, the fiber-modified binders are extremely strain-sensitive even at low strain levels, and the modulus and elasticity decrease quickly with increasing strain, indicating damage or rearrangements in the sample structure.

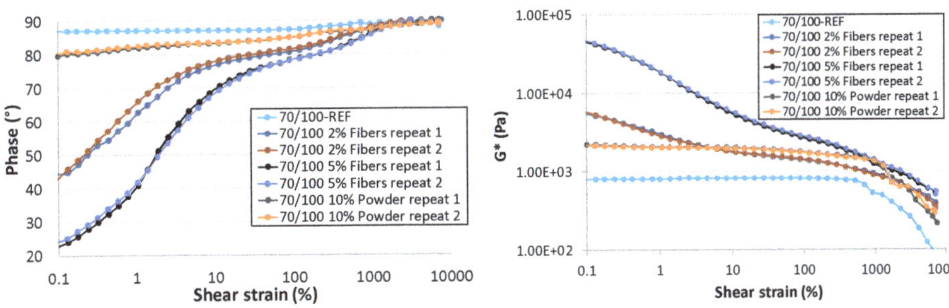

(a) (b)

Figure 9. Stress sweeps taken at 50 °C and 0.1 Hz: (**a**) complex modulus and (**b**) phase angle.

For one sample, the effect of what happens after a high strain was investigated further. In this test, the changes after applying a strain sweep were followed with time by applying an oscillation at low strain (0.1%). The strain sweeps as well as the "recovery" period at low strain are shown in Figure 10a,b. It is obvious that, after the strain sweep, which in this case went up to 200%, the fiber-modified sample can recover the modulus and phase angle almost back to its starting value. In these tests, the sample was visually checked and remained unchanged, so edge effects can be excluded.

Frequency sweeps at 50 °C were also performed using a strain of 1% for all the samples. The results are shown in Figure 11a,b. When adding powder to bitumen, there is mainly a stiffening effect. However, when adding fibers, there is an effect on the elasticity, especially at low frequencies. All measurements were made in duplicate, resulting in very similar results.

The formation of a plateau modulus at low frequencies in the fiber-modified samples was further investigated. For the reference binder and the 2% fiber-modified sample, the frequency sweep at 50 °C was extended, based on horizontal shift factors, by frequency sweeps at 40 °C, 60 °C, and 70 °C. This is shown in Figure 12. These data show that the storage and loss modulus are moving to a plateau modulus if the frequency decreases or the temperature increases, in the investigated temperature range.

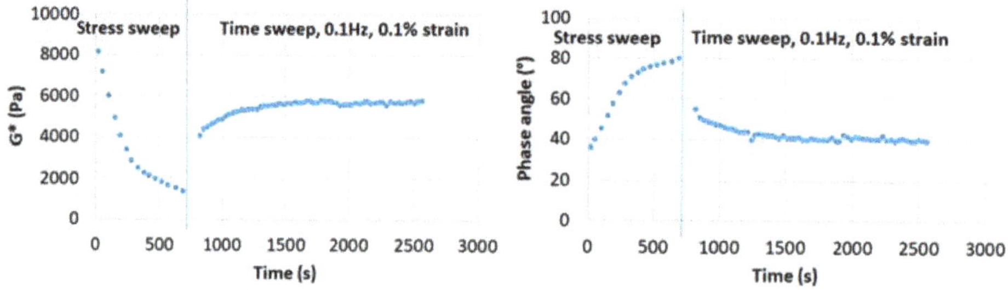

(a) (b)

Figure 10. The recovery after stress sweeps on a 2% fiber sample (50 °C, 0.1 Hz): (**a**) complex modulus and (**b**) phase angle.

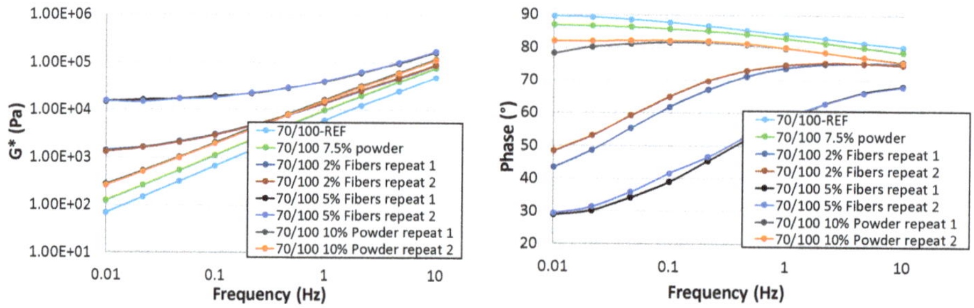

(a) (b)

Figure 11. Frequency sweeps at 50 °C, 1% strain: (**a**) complex modulus and (**b**) phase angle.

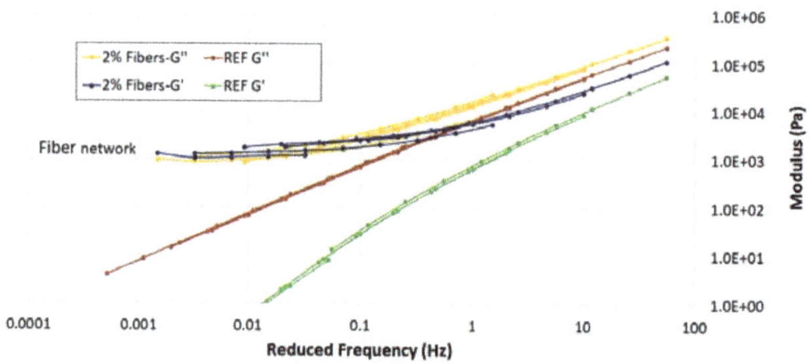

Figure 12. The master curves at 50 °C for the reference binder and the 2% fiber sample.

Finally, repeated creep tests were performed at 50 °C, according to the multiple stress creep and recovery test (MSCRT) specification [25]. This test is developed as a performance-based binder test to predict asphalt rutting. The test results are plotted in Figure 13, while the calculated parameters are shown in Table 1. These data indicate that, as the modification level with fibers increases, the rutting resistance could improve. Even adding a small percentage of fibers (0.1 wt% fibers, see Figure 13), already has an effect. As the fiber percentages increase, the percentage recovery increases and the non-recoverable compliance, as well as the end strain level, decreases. In the MSCRT tests, the recovery part and the non-recoverable part are dependent on the stress level. This corresponds to the behavior observed in the stress sweeps.

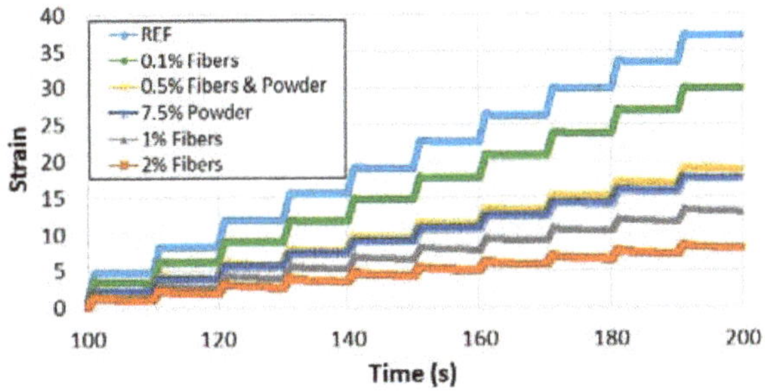

Figure 13. The multiple stress creep and recovery (MSCRT) test results, 50 °C (only repeated creep tests at 3200 Pa are shown).

Table 1. The parameters derived from the MSCRT test for various bitumen–peat combinations at 50 °C.

Parameter	REF	2% Fibers	1% Fibers	0.5% Fibers and 0.5% Powder	0.1% Fibers	7.5% Powder
Average recovery 100 Pa (%)	6.0	56.7	53.3	36.8	22.2	13.5
Average recovery 3200 Pa (%)	2.8	40.5	26.9	16.8	5.8	6.2
Average non-recoverable creep compliance 100 Pa (1/Pa)	1.1	0.1	0.2	0.4	0.7	0.5
Average non-recoverable creep compliance 3200 Pa (1/Pa)	1.1	0.2	0.4	0.6	0.9	0.5
End strain	37.1	8.0	13.1	18.7	29.8	17.6

When interpreting the rheological data, it is important to consider that these samples consist of very different phases. The fibers, and the powder, are embedded in an almost unmodified bitumen phase, there are fiber–fiber contacts, and the fibers are long (long enough to bridge the gap between the plates). The effects of adding powder or fibers to a base binder are quite different: powders stiffen the binder, while the fibers also induce an elastic effect. In the fiber-added blends, the elastic effects are due to the fiber skeleton. This skeleton becomes obvious, under test conditions of frequency and temperature, where the binder stiffness is in the same range or below that of the skeleton. For example, for the blend modified with 2% fibers, the modulus level of the fiber network is around 1 kPa, and for 5% fibers it is around 10 kPa. In the frequency sweeps, there is a frequency range where the binder stiffness is in the same range as the one of the fiber network; at 50 °C, this frequency is around 0.1 Hz for 2% fibers and 1 Hz for 5% fibers.

When tested at higher frequencies, the binder stiffness determines the behavior; if the frequency is reduced, the modulus of the binder drops below the one of the fiber network. In that case, the fiber network is determining the rheological test result, and, as this network is stiff, it is rather independent of frequency or loading time which results in an elasticity effect. As the network points formed by the fibers are just loose contact points, they can start to move under high strain, explaining the high strain dependency that was observed. Surprising is the recovery seen in the creep-recovery tests and after the stress sweep. This indicates that although the network points in the fiber skeleton are loose they still tend to move back to their initial position after deformation. This could be related to the stiff character of these fibers.

It is known that fibers are used to prevent drainage. The optimal amount of fibers (%) per asphalt mix composition was defined in [14] to keep an asphalt batch mass homogeneous by manual mixing. A variation in fibers of 0.3–1.5% was applied, and, according to the obtained results, it was decided to keep 0.5% of peat fibers per mix. According to the drainage test results, the amount of drained material for the reference mix was 0.33% (see Figure 14b), and no drainage was observed for mixes modified with PEL20. The data show that peat fibers (ground/non-ground) are effective in reducing or even preventing drainage.

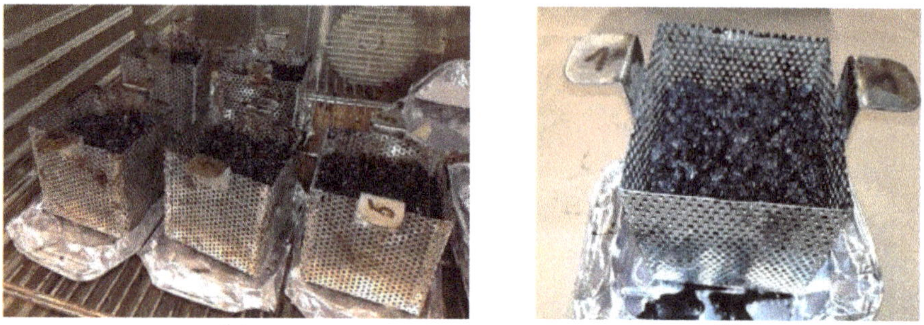

(**a**) (**b**)

Figure 14. The drainage test: (**a**) asphalt mixes with peat and (**b**) the REF asphalt mix [24].

During the drainage test, it was observed that a direct addition of peat fibers into the asphalt mix, and mixing it manually, does not really result in a homogeneous spread of the fibers in the asphalt mix. To obtain a more homogeneous mixing, two approaches were compared: (1) a dry process: direct mixing of peat fibers with hot aggregates, before or after adding bitumen and filler in the mixer (see Figure 15a); and (2) a wet process: blending the peat fibers with a Janke & Kunkel RW-20 variable-speed stirrer at speed of 300 rpm in hot bitumen at the temperature of 150 °C with specially designed steel blades before adding it to the hot aggregates mix (see Figure 15b). To reach a homogeneous distribution of peat fibers in the asphalt mix, mixing should be performed only by the second approach; with the first approach, this homogeneity cannot be reached (see Figure 16). This is remarkable, since, considering the type and length of fibers, it has been reported in the literature that dry mixing was supposed to be sufficient [4,9,15,26] or it was not mentioned at all how fibers were mixed in the asphalt mixes [5–8,11,16].

(**a**) (**b**)

Figure 15. Mixing of peat fibers in the asphalt mix: (**a**) the first approach and (**b**) the second approach.

It was mentioned by Kumar et al. [15] that natural fibers tend to mix thoroughly with dry heated aggregates. During the present experimental study, it was observed that natural fibers might tend to mix thoroughly with dry heated aggregates, depending on the length of the fibers (no more than 10 mm in length); however, as soon as the binder is added into the mix, the homogeneous fiber distribution disappears, which can be clearly observed in Figure 16 on the slab surface. Button and Lytton already in 1987 stated that the proper quantity of bitumen for a consistent coating of all particles is different not only for different concentrations but also for different types of fibers. This is likely due to the variations in surface area of the different fiber types. In addition, mix design procedures showed that the incorporation of fibers in an asphalt mix will increase the resulting air void content when the compaction effort remains constant [17].

Figure 16. An example of an asphalt slab produced by the dry approach (mixing peat with aggregates in the asphalt mix).

In this study, several slabs were produced: (1) two reference slabs (denoted as REF), (2) two slabs with 0.5% peat fibers and 0.5 % peat powder (FPEL20), and (3) two slabs with 1% peat fibers in the asphalt mix. In these tests, the mix design was not changed since the fiber and powder percentages were small. The rut depth results are shown in Figure 17. It can be seen that there is no significant difference among the results. The addition of peat fibers seems to reduce the rut depth, as was predicted from rheological tests, but the effect is in fact too small to be considered.

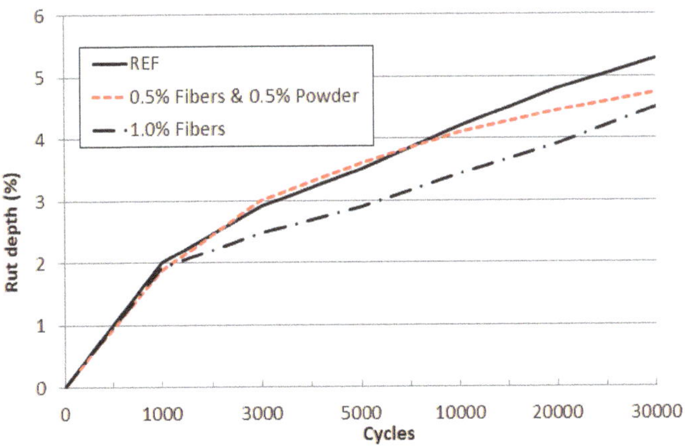

Figure 17. Examples of wheel tracking tests (50 °C).

Finally, the air voids (AV) for the various slabs was determined. The results are shown in Table 2. It is very clear that adding peat fibers or peat powder or very fine peat powder leads to a larger void content and shows that compaction becomes more difficult. When adding this material to asphalt, the mix design needs to be re-optimized.

Table 2. Air void properties of the slabs.

Mix	AV, %
REF	5.15
0.5% Fibers and 0.5% Powder	6.53
1.0% Fibers	8.80

4. Conclusions

In this study, the applicability of a natural peat fiber, as a modifier for bitumen, was investigated. It was noticed that the peat fiber material did not dissolve in bitumen, not even partly. Only for the very fine powder fraction some color change was observed in toluene and in acetone. Therefore, the peat products, whether this is the powder or the fiber, cannot be regarded as a bitumen extender or modifier; they should be treated as solid additives.

Rheological tests were carried out, and despite the heterogenous nature of the investigated blends of bitumen, powder, and fiber, the tests showed repeatable and consistent results, indicating some trends: the powder part mainly acts as a filler and the fiber part introduces an increment of elasticity. The data also show that the elastic action of the fibers is weak and quickly destroyed at higher stresses. As both the fiber part and the powder part show similar FT-IR spectra, the differences in the rheological behavior are related to the size difference of these components. A fiber skeleton is observed in the rheological tests when the binder has a stiffness that is similar or lower as compared to this network. Under test conditions where the binder is stiffer, a binder phase is observed, while, as soon as the binder stiffness drops or is in the same range as the fiber network, this network will influence and determine the rheological behavior. MSCRT tests were also conducted and indicated recovery effects, again mainly when adding fibers. This would suggest that especially the fibers could improve the rutting resistance; however, it was not possible to confirm this in asphalt rutting tests.

The asphalt drainage tests have shown that adding dry peat, whether this is ground or not, is effective in reducing the binder drainage, and, consequently, from this point of view would allow for thicker binder films around the aggregate. However, if the mix design is not adapted, the amount of added peat fibers and powder should be limited to keep the mixes workable. Compaction problems became obvious and resulted in high void contents.

When mixing peat fibers and powder with asphalt, it was observed that, to reach a homogeneous distribution of peat fibers in the asphalt mix, mixing should be performed only by the wet method; i.e., by blending the peat fibers in hot bitumen before adding them to the hot aggregates in the mixer. Direct mixing of peat fibers with hot aggregates and filler before adding the bitumen does not guarantee an even distribution of the fibers in the asphalt mix.

As a continuation, it could be worthwhile to investigate how powder and fibers influence the asphalt performance, when a thicker binder film is used, and when the mix design is re-optimized. In this way, modified asphalt mixes with sufficient workability and compactability could be compared. Also, possible beneficial effects of using peat fibers on the cracking resistance and investigation of the distribution of peat fibers in the asphalt mix with X-ray computed tomography (X-ray CT), which have not been investigated in this study, could be interesting to provide more information.

Author Contributions: Conceptualization, P.K.D.M. and H.S.; Data curation, H.S. and P.K.D.M.; Investigation, H.S., P.K.D.M., J.S., and G.J.; Resources, H.S. and W.V.d.b.; Writing—original draft preparation and editing, P.K.D.M. and H.S.; Writing—review, W.V.d.b. and J.B.

Funding: This research was funded by NYNAS.

Acknowledgments: The authors would like to thank VAPO (https://www.vapo.com/en) for providing peat material; the UAntwerp Mechanics Workshop for preparing perforated steel boxes for the drainage test, the rut depth measuring setup for rutting test, and steel blades for the blending of bitumen with peat; and Ankersmid M&C Bvba (www.ankersmid.eu) for providing the SEM.

Conflicts of Interest: The authors declare no conflict of interest.

References

1. VAPO. *Peat Production*; VAPO: Jyväskylä, Finland, 2016.
2. Energy in Finland. *Communication and Information Services Statistics Finland*. 2016. Available online: http://www.stat.fi (accessed on 1 November 2018).
3. U. S. Geological Survey. Mineral Commodity Summaries. 2017. Available online: https://minerals.usgs.gov/minerals/pubs/mcs/2017/mcs2017.pdf (accessed on 1 November 2018).
4. Mahrez, A.; Karim, M.R.; Katman, H.Y. Fatigue and deformation properties of glass fiber reinforced bituminous mixes. *J. East Asia Soc. Trans. Stud.* **2005**, *6*, 997–1007. [CrossRef]
5. Bindu, C.S.; Beena, K.S. Influence of additives on the drain down characteristics of stone matrix asphalt mixtures. *Int. J. Res. Eng. Technol.* **2014**, *3*, 83–88.
6. Sheng, Y.; Zhang, B.; Yan, Y.; Li, H.; Chen, Z.; Chen, H. Laboratory investigation on the use of bamboo fiber in asphalt mixtures for enhanced performance. *Arab J. Sci. Eng.* **2018**, 1–10. [CrossRef]
7. Tamburini, C.; Ziyani, L.; Dony, A.; Rohart, C.; Toraldo, E. Promotion of bitumen-impregnated cellulose fibers from lightweight roofing tiles in stone mastic asphalt. In *RILEM 252-CMB: RILEM 252-CMB-Symposium on Chemo-Mechanical Characterization of Bituminous Materials*; Poulikakos, L.D., Falchetto, A.C., Wistuba, M.P., Hofko, B., Porot, L., di Benedetto, H., Eds.; Springer: Cham, Switzerland, 2019; Volume 20, pp. 306–3011. ISBN 978-3-030-00475-0. [CrossRef]
8. Awanti, S.S.; Habbal, A.; Hiremath, P.N.; Tadibidi, S.; Hallale, S.N. Characterization of stone matrix asphalt with cellulose and coconut fiber. In *Advance in Civil Engineering and Building Materials*; Chang, S., al Bahar, S.K., Zhao, J., Eds.; Taylor & Francis Group: London, UK, 2012; pp. 639–642. ISBN 9780203388075.
9. Ting, T.L.; Jaya, R.P.; Hassan, N.A.; Yaacob, H.; Jayanti, D.S. A review of utilization of coconut shell and coconut fiber in road construction. *Jurnal Teknologi (Sci. Eng.)*, **2015**, *76*, 121–125. [CrossRef]
10. Delgado, H.; Arnaud, L. Investigation of the fatigue properties of asphalt mixtures reinforced with natural fibers. In Proceedings of the International Conference Bituminous Mixtures and Pavements, Thessaloniki, Greece, 1–3 June 2011; ISBN 978-9-609-992206.
11. Kumar, P.; Sikdar, P.K.; Bose, S.; Chandra, S. Use of jute fiber in stone matrix asphalt. *Road Mater. Pavement* **2004**, *5*, 239–249. [CrossRef]
12. Dong, T.; Xu, G.; Wang, F. Adsorption and adhesiveness of kapok fiber to different oils. *J. Hazard. Mater.* **2015**, *296*, 101–111. [CrossRef]
13. Kudrjashov, A.P.; Kudrjashov, I.V.; Kudrjashov, P.A.; Germashev, V.G.; Jadykina, V.V. Stabilising Additive for Peat-Based Asphalt-Concrete Mixture (Versions) and Method of Producing Structure-Forming Agent Thereof. Patent RU 2479524 C2, 20 April 2013.
14. Soenen, H.; Kara De Maeijer, P.; Blom, J.; Van den bergh, W. Peat as an example of a natural fiber in bitumen. In *RILEM 252-CMB: RILEM 252-CMB-Symposium on Chemo-Mechanical Characterization of Bituminous Materials*; Poulikakos, L.D., Falchetto, A.C., Wistuba, M.P., Hofko, B., Porot, L., di Benedetto, H., Eds.; Springer: Cham, Switzerland, 2019; Volume 20, pp. 300–305, (first online on 13 September 2018); ISBN 978-3-030-00475-0. [CrossRef]
15. Kumar, N.R.; Sunitha, V. Experimental investigation of stone mastic asphalt with sisal fiber. *Int. J. Eng. Res. Technol. (IJERT)* **2016**, *5*, 546–550.
16. Oda, S.; Fernandes, J.L.; Ildefonso, J.S. Analysis of use of natural fibers and asphalt rubber binder in discontinuous asphalt mixtures. *Constr. Build. Mater.* **2012**, *26*, 13–20. [CrossRef]
17. Button, J.W.; Lytton, R.L. Evaluation of Fabrics, Fibers and Grids in Overlays. In Proceedings of the Sixth International Conference on the Structural Design of Asphalt Pavements, Ann Arbor, MI, USA, 13–17 July 1987; pp. 925–934. Available online: https://static.tti.tamu.edu/tti.tamu.edu/documents/TTI-1987-ID19631.pdf (accessed on 2 November 2018).
18. Perdana, L.R.; Ratnasari, N.G.; Ramadhan, M.L.; Palamba, P.; Nasruddin, N.; Nugroho, Y.S. Hydrophilic and hydrophobic characteristics of dry peat. *IOP Conf. Ser. Earth Environ. Sci.* **2018**, *105*, 012083. [CrossRef]
19. *NBN EN 15326+A1 Bitumen and Bituminous Binders—Measurements of Density and Specific Gravity—Capillary-Stoppered Pycnometer Method*; Bureau for Standardisation: Brussels, Belgium, 2009.

20. *NBN EN 12697-18, Bituminous Mixtures—Test Methods for Hot Mix Asphalt—Part 18: Binder Drainage*; Bureau for Standardisation: Brussels, Belgium, 2017.
21. *NBN EN 12697-22, Bituminous Mixtures—Test Methods for Hot Mix Asphalt—Part 22: Wheel Tracking*; Bureau for Standardisation: Brussels, Belgium, 2007.
22. *NBN EN 12697-5, Bituminous Mixtures—Test Methods for Hot Mix Asphalt—Part 5: Determination of the Maximum Density*; Bureau for Standardisation: Brussels, Belgium, 2010.
23. *NBN EN 12697-6, Bituminous Mixtures—Test Methods for Hot Mix Asphalt—Part 5: Determination of the Bulk Density of Bituminous Specimens*; Bureau for Standardisation: Brussels, Belgium, 2012.
24. Soenen, H.; Kara De Maeijer, P.; Blom, J.; Van den bergh, W. Peat as an example of a natural fiber in bitumen. Presented at RILEM 252-CMB: RILEM 252-CMB-Symposium, Braunschweig, Germany, 17 September 2018; Available online: https://www.researchgate.net/publication/325949164_Peat_as_an_Example_of_a_Natural_Fiber_in_Bitumen_Chemo-Mechanical_Characterization_of_Bituminous_Materials (accessed on 1 November 2018).
25. *CEN EN 16659, Bitumen and Bituminous Binders—Multiple Stress Creep and Recovery Test (MSCRT)*; Bureau for Standardisation: Brussels, Belgium, 2016.
26. Abiola, O.S.; Kupolati, W.K.; Sadiku, E.R.; Ndambuki, J.M. Utilisation of natural fiber as modifier in bituminous mixes: A review. *Constr. Build. Mater.* **2014**, *54*, 305–312. [CrossRef]

infrastructures

MDPI

Article

A Comparative Study of Laser Doppler Vibrometers for Vibration Measurements on Pavement Materials

Navid Hasheminejad [1,*], Cedric Vuye [1], Wim Van den bergh [1], Joris Dirckx [2] and Steve Vanlanduit [3]

1 EMIB Research Group, University of Antwerp, Groenenborgerlaan 171, 2020 Antwerp, Belgium; cedric.vuye@uantwerpen.be (C.V.); wim.vandenbergh@uantwerpen.be (W.V.d.b.)
2 BIMEF Research Group, University of Antwerp, Groenenborgerlaan 171, 2020 Antwerp, Belgium; joris.dirckx@uantwerpen.be
3 Op3Mech Research Group, University of Antwerp, Groenenborgerlaan 171, 2020 Antwerp, Belgium; steve.vanlanduit@uantwerpen.be
* Correspondence: navid.hasheminejad@uantwerpen.be; Tel.: +32-32658849

Received: 5 October 2018; Accepted: 30 October 2018; Published: 1 November 2018

Abstract: A laser Doppler vibrometer (LDV) is a noncontact optical measurement device to measure the vibration velocities of particular points on the surface of an object. Even though LDV has become more popular in road engineering in recent years, their signal-to-noise ratio (SNR) is strongly dependent on light scattering properties of the surface which, in some cases, needs to be properly conditioned. SNR is the main limitation in LDV instrumentation when measuring on low diffusive surfaces like pavements; therefore, an investigation on the SNR of different LDV devices on different surface conditions is of great importance. The objective of this research is to investigate the quality of two types of commercially available LDV systems—helium–neon (He–Ne)-based vibrometers and recently developed infrared vibrometers—on different surface conditions, i.e., retroreflective tape, white tape, black tape, and asphalt concrete. Both noise floor and modal analysis experiments are carried out on these surface conditions. It is shown that the noise floor of the He–Ne LDV is higher when dealing with a noncooperative dark surface, such as asphalt concrete, and it can be improved by improving the surface quality or by using an infrared LDV, which consequently improves the modal analysis experiments performed on pavement materials.

Keywords: laser Doppler vibrometer (LDV); pavements; vibration measurement; noise floor; modal analysis

1. Introduction

Nondestructive testing (NDT) is an important part of optimizing any pavement management system. Techniques such as falling weight deflectometer (FWD) using geophones [1] and rolling wheel deflectometer (RWD) using laser deflection system [2] are popular among researchers to find properties of the road. Moreover, accelerometers can be used to find the mechanical properties of asphalt concrete by applying a back calculation technique [3]. In recent years, laser Doppler vibrometer (LDV) has been introduced to conduct noncontact measurements in road engineering, and it is replacing the traditional vibration sensors [4,5].

LDV is an optical measurement system that is used to perform noncontact vibration measurements on a surface [6]. LDV devices were first introduced in the 1980s, but their limited sensitivity and low signal-to-noise ratio (SNR) allowed measurements only on very diffusive surfaces or by applying a retroreflective tape on the testing objects. It was only in the early 1990s that hardware and software developments increased instrumentation performances and applicability, leading to many researchers using LDV. LDV can significantly extend measurement capabilities compared to traditional vibration sensors, such as accelerometers, because the results will not be affected by errors due to mass loading of accelerometers. This is relevant for modal parameter estimation, especially when testing light or small structures or highly damped nonlinear materials [7]. LDV can also replace accelerometers for vibration measurement in cases where installing accelerometers in different measurement points is difficult [8]. One of the main applications of the LDV in road engineering is traffic speed deflectometer (TSD). TSD is an RWD that uses Doppler technology to measure pavement deflection while traveling at normal traffic speed. Using the measured deflection, bearing capacity indices can be derived, and pavement fatigue or residual life can be estimated [5]. Furthermore, scanning laser Doppler vibrometer (SLDV) has the ability to rapidly and precisely move the measurement point on the structure, allowing the analysis of a large surface with high spatial resolution. Using an SLDV, it is possible to perform modal analysis on targets and evaluate the natural frequencies, modal damping, and modal shapes of a structure [9,10]. This method can be used for pavement materials in order to conduct a modal analysis experiment and determine the mechanical properties of different types of asphalt concretes using a back-calculation technique [11].

For many years, the He–Ne laser was the leading technology used in commercial laser Doppler instruments. The desire for long-range measurements without reduction of the signal quality has seen the introduction of an instrument with a higher power infrared (invisible) fiber laser, which is used in conjunction with a green laser for sighting purposes. The infrared laser technology is now migrating into instrument designs for short-range applications on optically less cooperative surfaces, finally challenging the supremacy of the He–Ne laser [12]. This is an important improvement as the poor surface quality of the asphalt concretes can increase measurement uncertainties.

In data acquisition and signal processing, the noise floor is a measure of the summation of all the noise sources and unwanted signals generated by the entire data acquisition and signal processing system. In any measurement, the minimum resolvable signal level must be sufficiently larger than the noise content of the signal to obtain reliable measurements. To be able to use an LDV system for measurement, it is important to know the minimum detectable level. The noise floor can be established by examining the content of the spectrum of a signal measured by LDV where no external vibration is applied to the system. It is dependent on the optics, electronics, the software of the LDV, and the properties of the media and reflective target [13]. Speckle noise is one of the main sources of noise in LDV, especially in cases where there is relative motion between the test item and laser beam [7]. This particular type of noise has been investigated and modeled by several researchers [14–16].

In this research, noise floor measurements are firstly reported for two types of LDV on four different surface conditions. Then, a modal analysis experiment is designed to investigate the ability of both instruments to perform measurements on different types of pavements with both treated and untreated surface conditions.

This paper is divided into four sections. The first section includes the introduction and state of the art. In Section 2, an overview of the research methodology, experimental setup, and measurement procedure is given. In Section 3, the measurement results are discussed in detail. This entails a comparison of the noise floor of both instruments, the effect of the surface quality on the noise floor measurements, and modal analysis of three types of pavements. Finally, conclusions of the research are given in the last section.

2. Materials and Methods

Three measurement instruments were used in this research: a He–Ne SLDV (Polytec PSV-400), an infrared LDV (Polytec RSV-150) with two short-range and long-range lenses, and an infrared SLDV (Polytec PSV-500-3D Xtra). The LDV has the ability to carry out measurements at one point, and the SLDV has a computer-controlled mirror that can direct the laser to the desired measurement points so that measurements can be performed on a predefined grid on the surface of an object. The He–Ne SLDV has a class 2 laser with 633 nm wavelength and less than 1 mW power. The infrared LDV has a green targeting laser with 523 nm wavelength and a measurement laser with a wavelength of 1550 nm. The output power of the infrared LDV when both lasers are in operation is 10 mW class 2. The autofocus of both SLDV instruments is done automatically, but the infrared LDV has two long-range and short-range lenses and has to be manually focused. The short-range lens is for standoff distances between 1 to 5 m, and the long-range lens is used for a standoff distance larger than 5 m and up to 300 m.

Two sets of experiments were conducted in this research. The first experiment was to estimate the noise floor on targets with different surface conditions. The He–Ne SLDV and the infrared LDV were placed at the same standoff distance in front of the target, and measurements were conducted on one point on the target (Figure 1).

These instruments have different analog-to-digital converters (ADC); therefore, the sensitivity of the ADC is important for the experiments. To be able to compare the devices with each other, closest sensitivity values were chosen for both devices in a way that the less accurate device (He–Ne SLDV) had a lower sensitivity. Afterwards, to compare the noise floor of each device for different surfaces, the sensitivity of the He–Ne SLDV and infrared LDV were set to 20 and 122.5 mm/s/v, respectively. The four investigated targets were surfaces covered with a retroreflective tape, white tape, black tape, and an asphalt concrete.

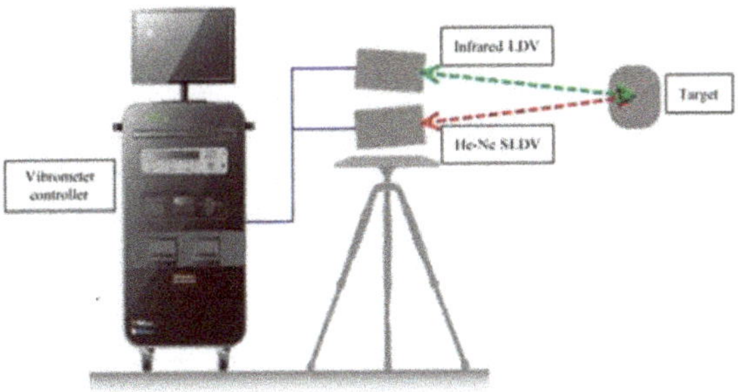

Figure 1. First experiment: noise floor measurement with two measuring systems on different targets.

The second experiment was designed to investigate the ability of different SLDVs to measure treated and untreated pavements. Modal analysis experiments were conducted to find the modal parameters of three different pavement slabs. First, a He–Ne SLDV was used for measurements on pavements with poor surface quality. To investigate the effect of surface quality, one side of the specimens were painted with a white spray paint (Ardrox® 9D1B aerosol), and the same modal analysis experiments were conducted on the painted side. Meanwhile, mode shapes of the specimens were predicted by a finite element model, and the modal assurance criterion (MAC) was calculated between the mode shapes acquired by SLDV and FEM. These type of experiments can be used to find mechanical properties of specimens with inverse method [11]. Then, the same modal analysis experiments on the same specimens were conducted with a 3D infrared SLDV.

The test items were three types of pavements. The first one was a thin asphalt layer (TAL) pavement with dimensions of 59 cm in length, 39 cm in width, and 2.6 cm in thickness. This type of asphalt is used as a top layer with an optimized fine texture in order to reduce tire vibrations and therefore the tire/road noise. The pavement used in the research project was the N19 in Kasterlee, Belgium [17,18] and Antwerpen, Belgium [19]. The second specimen was a 50*18*5.5 cm poroelastic road surface (PERS). PERS is a type of low-noise pavement with a higher elasticity than conventional road surfaces and a larger percentage of voids. The PERS has a porous structure composed of granular rubber made from recycled tires, aggregates, and polyurethane (PUR) resin as a binder [20,21]. The third test specimen was a 59*39*3.4 cm stone mastic asphalt (SMA). SMA has been used successfully in Europe for over 40 years to provide better rutting resistance and to resist studded tire wear [22]. As represented in Figure 2, in this part, the specimens were hung from a frame using two screw eyes and fishing lines to simulate the free-free condition. A Brüel & Kjær modal exciter type 4824 excited the specimens with a periodic chirp signal between the frequency range of 5 to 1000 Hz. Signals were generated using the Polytec onboard signal generator and amplified by a Brüel & Kjær power amplifier type 2732. A Brüel & Kjær force transducer type 8230-001 was placed between the tip of the shaker and the specimen to measure the exact force used for FRF calculations. Then, using an accurate modal parameter estimator called the Polymax estimator [23], modal parameters of the specimens were calculated from the spectrum of measured signals. An overview of all the experiments and their settings are presented in Tables 1 and 2.

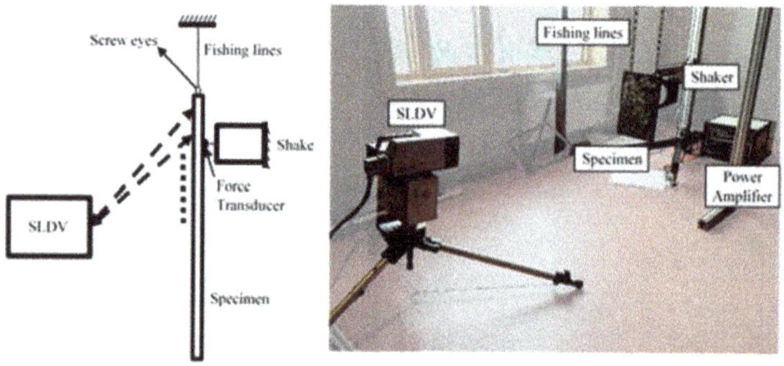

Figure 2. Second experiment: modal analysis on a pavement plate with two different scanning laser Doppler vibrometer (SLDV) systems.

Table 1. First experiment: noise floor measurement of two He–Ne and infrared laser Doppler vibrometers on four different surfaces.

Experiment	Test Number	Surface	Measurement System	Standoff Distance (m)
Noise Floor Measurements	1	Retroreflective tape	He–Ne SLDV and infrared LDV	9
	2	White tape	He–Ne SLDV and infrared LDV	9
	3	Black tape	He–Ne SLDV and infrared LDV	9
	4	Asphalt concrete	He–Ne SLDV and infrared LDV	1.7

Table 2. Second experiment: modal analysis of three different specimens with different surface conditions and different measurement systems.

Experiment	Test Number	Specimen	Surface Condition	Measurement System
Part 1	1	TAL	Unpainted	He–Ne SLDV
	2	TAL	Painted	He–Ne SLDV
	3	PERS	Unpainted	He–Ne SLDV
	4	PERS	Painted	He–Ne SLDV
	5	SMA	Unpainted	He–Ne SLDV
	6	SMA	Painted	He–Ne SLDV
Part 2	1	TAL	Unpainted	He–Ne SLDV
	2	TAL	Unpainted	3D infrared SLDV
	3	PERS	Unpainted	He–Ne SLDV
	4	PERS	Unpainted	3D infrared SLDV
	5	SMA	Unpainted	He–Ne SLDV
	6	SMA	Unpainted	3D infrared SLDV

3. Results

3.1. He–Ne vs. Infrared LDV

In this section, the noise floor measurements of a He–Ne SLDV and an infrared LDV on a retroreflective tape were investigated (see Table 1). The sensitivities of the ADCs of the devices were chosen as explained before. Both ADCs had the same low pass filters so that the noise floors could be compared until 10^5 Hz. Figure 3 shows that the noise floor of the He–Ne SLDV was higher than the noise floor of the infrared LDV in two different sensitivity settings, even though the sensitivity of the He–Ne SLDV was slightly lower in both cases compared to the sensitivity of the infrared LDV.

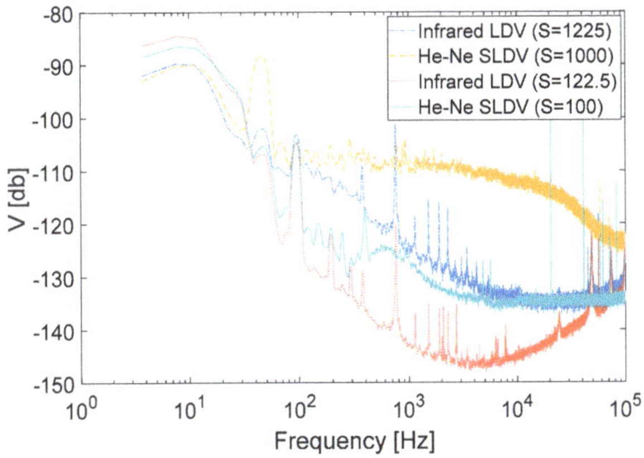

Figure 3. Noise floor comparison between He–Ne and infrared laser Doppler vibrometer (LDV); S = sensitivity (mm/s/v).

3.2. *He–Ne vs. Infrared LDV on Different Surfaces*

In this step, the same noise floor experiments were conducted on different surface conditions. Figure 4 shows that the noise floor of the He–Ne SLDV was higher than noise floor of the infrared LDV regardless of the surface conditions. The difference between noise floors of the instruments was at its lowest when retroreflective tape was used on the surface of the object, and it was more excessive for darker surfaces, especially at high frequencies. It could also be seen that the trend of the noise floor of each instrument was similar for all surfaces. For the He–Ne SLDV, the noise floor was almost the same with white tape and asphalt surface, and it was the highest for the black tape. It should be mentioned that the He–Ne SLDV is not able to autofocus the laser spot on black tape or the asphalt surface and therefore the laser spot was focused on the surface manually. Moreover, the reason for the sudden drop in the noise floor of the He–Ne LDV after 25 kHz was that the frequency range of the decoder of the He–Ne LDV was 0 to 25 kHz. The noise floor of the infrared LDV on different surfaces was almost the same on all surfaces, decreasing from more than −100 dB in low frequencies to around −140 dB at 1 kHz. After 1 kHz, it started rising for all surfaces and was highest for the black tape and lowest for the retroreflective tape.

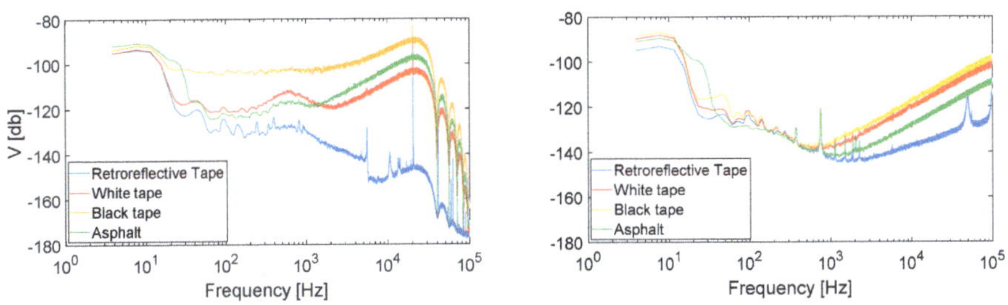

Figure 4. Noise floor measurement for four different surfaces by He–Ne SLDV (**left**) and infrared LDV (**right**). Stand-off distance is 1.7 m for asphalt and 9 m for all the other surfaces.

3.3. Modal Parameters with SLDV

In this section, a He–Ne SLDV was used to calculate the modal parameters of three types of pavement slabs (see Table 2). Table 3 lists the modal frequencies (*f*), damping ratios (*D*), and MAC calculated by the SLDV for both (unpainted and painted) sides of three specimens. More mode shapes of the specimens were acquired from the painted side of the specimens, which proved that the measurement on the painted side was more accurate. Furthermore, the measurement of each point was repeated eight times to calculate the coherence function. Figure 5 illustrates that the coherence function was much better on the painted side compared to the unpainted side.

Table 3. Modal parameters of three types of pavement slabs on their two sides (unpainted and painted). The gray rows are mode shapes that Polymax estimate was not able to detect.

Mode	TAL			PERS			SMA		
	f (Hz)	*D* (%)	MAC	*f* (Hz)	*D* (%)	MAC	*f* (Hz)	– (%)	MAC
				Unpainted side					
1				92.7	9.3	98.2	201.5	12.1	98.4
2	126.8	10.5	73.3	149.1	9.5	97.7	214.4	4.2 *	93.0
3	311.1	10.9	92.5	250.5	10.2	96.3			
4	337.2	9.0	90.7				493.1	8.6	94.1
5	378.5	8.8	86.4	454.1	9.3	90.0	567.7	12.1	93.3
6	457.9	8.3	90.1	497.9	10.3	92.6	663.2	8.9	92.2
7	561.5	8.7	95.1						
8	673.8	8.1	90.9						
				Painted side					
1				93.8	9.4	99.4	203.0	12.1	93.1
2	125.9	10.5	76.4	151.1	9.8	99.1	215.8	9.8	86.9
3	311.3	8.1	96.6	255.3	10.4	98.6	456.1	8.9	91.1
4	337.1	8.8	91.9	314.5	10.0	97.7	493.4	8.5	75.9
5	378.9	8.9	85.1	439.2	1.3 *	90.6	575.0	10.2	96.3
6	450.7	4.5	79.1	485.5	10.7	86.1	660.5	10.1	96.6
7	563.2	8.2	94.9	641.8	10.2	78.3			
8	682.9	8.5	95.0						

* In some cases, due to the heavy coupling between two mode shapes, the polymax estimator was not able to estimate the damping ratio of the mode shapes with a high accuracy.

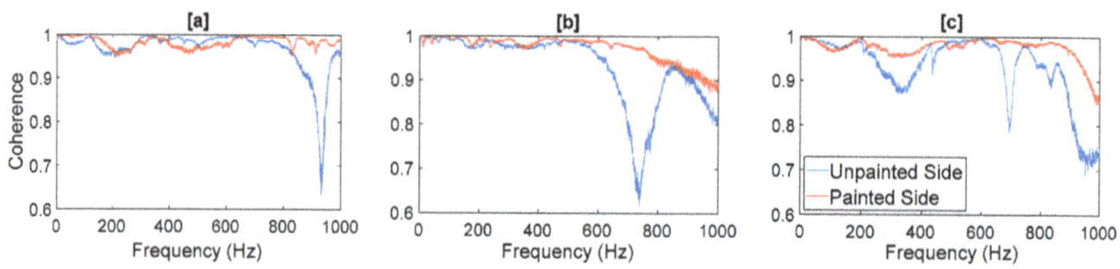

Figure 5. Coherence function of the specimens: (**a**) thin asphalt layer (TAL); (**b**) poroelastic road surface (PERS); (**c**) stone mastic asphalt (SMA).

3.4. He–Ne vs. Infrared SLDV for Modal Analysis on Pavement

In this step, a He–Ne SLDV and an infrared 3D SLDV were used to measure the modal parameters of three specimens. The specimens were the same pavement slabs used in the previous tests, hung in free-free condition with their unpainted side facing the SLDV. The natural frequencies and damping ratios of the pavement slabs, measured by the two instruments, are represented in Table 4. It was evident that using an infrared 3D SLDV led to finding more mode shapes of the specimens, especially in higher frequencies where the applied load by shaker was lower than that in the lower frequencies;

thus, the slightest noise could influence the results of the measurement. Therefore, as the infrared SLDV had a lower noise floor, it was able to conduct better measurements that led to finding more mode shapes of the specimen.

Table 4. Modal parameters of three specimens estimated from measurements conducted by two instruments: He–Ne SLDV and infrared 3D SLDV. The gray cells are the mode shapes that Polymax estimate was not able to detect.

Frequency (Hz)						Damping Ratio (%)					
He–Ne SLDV			Infrared 3D SLDV			He–Ne SLDV			Infrared 3D SLDV		
TAL	PERS	SMA	TAL	PERS	SMA	TAL	PERS	SMA	TAL	PERS	SMA
	92.7	201.5		90.5	188.7		9.2	12.1		9.5	12.9
126.8	149.1	214.4	128.9	144.8	207.0	10.5	9.5	4.2	10.3	10.2	10.7
311.1	250.5			246.5	445.9	10.9	10.2			13.3	12.3
337.2		493.1	337.6	302.3	484.3	9.0		8.6	8.5	9.7	9.9
378.5	454.4	567.7	382.2	450.6	589.9	8.8	9.3	12.1	9.4	11.0	11.6
457.9	497.9	663.2	459.1		651.7	8.3	10.3	8.9	9.1		7.6
561.5			574.2		830.2	8.7			7.6		9.7
673.8			676.0	641.5	964.4	8.1			8.0	11.1	10.6
			770.9						7.6		
			923.7						8.9		
			969.8						8.7		
			1044.9						10.2		

4. Conclusions

After 30 years of using He–Ne LDV as an accurate, noncontact measurement device, an infrared LDV with higher power compared to the conventional He–Ne LDV was developed to improve the quality of measurements in long-range applications. The infrared LDV is now becoming more popular, including in applications of optically low cooperative surfaces. In this paper, the noise floor of the two instruments (He–Ne and infrared LDV) were compared, and it was revealed that infrared LDV had lower noise level than He–Ne LDV in all surfaces, especially dark surfaces with low surface quality. Furthermore, it was shown that surface quality was more influential in measurements with He–Ne LDV. For instance, at some frequencies, there could be up to 60 dB difference between the noise floor measurements performed on the dark and retroreflective surfaces. Meanwhile, in an infrared LDV, surface quality was not important until 1000 Hz. For higher frequencies, retroreflective tapes could reduce the noise up to 20 dB. Therefore, in short-range measurements on materials with good surface quality, the difference of the noise between the instruments would not be significant. However, in cases where measurements are being conducted on materials with poor surface quality—like in road engineering where measurements are done on asphalt surface—using an infrared LDV could lead to better results (up to 30 dB reduction of noise floor in some frequencies).

Author Contributions: Data curation, N.H. and S.V.; Formal analysis, N.H.; Investigation, N.H.; Methodology, N.H., C.V., and S.V.; Resources, W.V.d.b.; Supervision, C.V., W.V.d.b., J.D., and S.V.; Writing—original draft, N.H.; Writing—review & editing, C.V. and S.V.

Funding: The authors would like to thank the research council of the Faculty of Applied Engineering for granting this project funded by the Everdepoel legacy.

Acknowledgments: Special thanks must be given to Polytec GmbH for the loan of Polytec LDV and 3D SLDV and for their technical assistance.

Conflicts of Interest: The authors declare that there is no conflict of interest regarding the publication of this paper.

References

1. Varma, S.; Emin Kutay, M. Backcalculation of viscoelastic and nonlinear flexible pavement layer properties from falling weight deflections. *Int. J. Pavement Eng.* **2015**, 1–15. [CrossRef]

2. Elseifi, M.A.; Abdel-Khalek, A.M.; Gaspard, K.; Zhang, Z.; Ismail, S. Evaluation of Continuous Deflection Testing Using the Rolling Wheel Deflectometer in Louisiana. *J. Transp. Eng.* **2012**, *138*, 414–422. [CrossRef]
3. Gudmarsson, A.; Ryden, N.; Di Benedetto, H.; Sauzeat, C. Complex modulus and complex Poisson's ratio from cyclic and dynamic modal testing of asphalt concrete. *Constr. Build. Mater.* **2015**, *88*, 20–31. [CrossRef]
4. He, L.; Lin, H.; Zou, Q.; Zhang, D. Accurate measurement of pavement deflection velocity under dynamic loads. *Autom. Constr.* **2017**, *83*, 149–162. [CrossRef]
5. Flintsch, G.W.; Ferne, B.; Diefenderfer, B.; Brayce, J. Evaluation of Traffic Speed Continuous Deflection Devices. *Transp. Res. Procedia* **2012**, *14*, 3031–3039.
6. Halliwell, N.A. Laser-Doppler measurement of vibrating surfaces: A portable instrument. *J. Sound Vib.* **1979**, *62*, 312–315. [CrossRef]
7. Castellini, P.; Martarelli, M.; Tomasini, E.P. Laser Doppler Vibrometry: Development of advanced solutions answering to technology's needs. *Mech. Syst. Signal Process.* **2006**, *20*, 1265–1285. [CrossRef]
8. Rossi, G.; Marsili, R.; Gusella, V.; Gioffrè, M. Comparison between accelerometer and laser vibrometer to measure traffic excited vibrations on bridges. *Shock Vib.* **2002**, *9*, 11–18. [CrossRef]
9. Stanbridge, A.B.; Ewins, D.J. Modal testing using a scanning laser Doppler vibrometer. *Mech. Syst. Signal Process.* **1999**, *13*, 255–270. [CrossRef]
10. Martatelli, M.; Revel, G.M.; Santolini, C. Automated Modal Analysis By Scanning Laser Vibrometry: Problems and Uncertainties Associated With the Scanning System Calibration. *Mech. Syst. Signal Process.* **2001**, *15*, 581–601. [CrossRef]
11. Hasheminejad, N.; Vuye, C.; Van den Bergh, W.; Dirckx, J.; Leysen, J.; Sels, S.; Vanlanduit, S. Identification of pavement material properties using a scanning laser Doppler vibrometer. In Proceedings of the 12th International A.I.VE.LA. Conference on Vibration Measurements by Laser and Noncontact Techniques, Ancona, Italy, 29 June–1 July 2016. [CrossRef]
12. Rothberg, S.J.; Allen, M.S.; Castellini, P.; Di Maio, D.; Dirckx, J.J.J.; Ewins, D.J.; Halkon, B.J.; Muyshondt, P.; Paone, N.; Ryan, T.; et al. An international review of laser Doppler vibrometry: Making light work of vibration measurement. *Opt. Lasers Eng.* **2016**. [CrossRef]
13. Harland, A.R.; Petzing, J.N.; Tyrer, J.R.; Bickley, C.J.; Robinson, S.P.; Preston, R.C. Application and assessment of laser Doppler velocimetry for underwater acoustic measurements. *J. Sound Vib.* **2003**, *265*, 627–645. [CrossRef]
14. Strean, R.F.; Mitchell, L.D.; Barker, A.J. Global noise characteristics of a laser Doppler vibrometer—I Theory. *Opt. Lasers Eng.* **1998**, *30*, 127–139. [CrossRef]
15. Martin, P.; Rothberg, S. Introducing speckle noise maps for Laser Vibrometry. *Opt. Lasers Eng.* **2009**, *47*, 431–442. [CrossRef]
16. Denman, M.; Halliwell, N.A.; Rothberg, S.J. Speckle noise reduction in laser vibrometry: Experimental and numerical optimisation. In Proceedings of the Second International Conference on Vibration Measurements by Laser Techniques: Advances and Applications, Ancona, Italy, 23–25 September 1996; Tomasini, E.P., Ed.; Volume 2868, pp. 12–21.
17. Vuye, C.; Devroye, G.; Stuer, W.; Van Geen, G. Van; Van den bergh, W.; Bergiers, A.; Goubert, L.; Vanhooreweder, B.; Buytaert, A. Acoustical Characteristics of Low-Noise Test Tracks in Flanders. In Proceedings of the 22nd International Congress on Sound and Vibration, Florence, Italy, 12–16 July 2015.
18. Vuye, C.; Bergiers, A.; Vanhooreweder, B. The Acoustical Durability of Thin Noise Reducing Asphalt Layers. *Coatings* **2016**, *6*, 21. [CrossRef]
19. Vuye, C.; Musovic, F.; Tyszka, L.; Van Den, W.; Kampen, J.; Bergiers, A.; Maeck, J. First experiences with thin noise reducing asphalt layers in an urban environment in Belgium. In Proceedings of the ISMA 2016 Noise and Vibration Engineering Conference, Leuven, Belgium, 19–21 September 2016.
20. Sandberg, U. Low noise road surfaces. A state-of-the-art review. *J. Acoust. Soc. Jpn.* **1999**, *20*, 1–17. [CrossRef]
21. Vuye, C.; Devroye, G.; Stuer, W.; Van Beveren, M. Acoustical and Mechanical Impedance Measurements on PoroElastic Road Surfaces. In Proceedings of the 10th European Congress and Exposition on Noise Control Engineering, Maastricht, The Netherlands, 31 May–3 June 2015; pp. 1205–1210.
22. Brown, E.R.; Haddock, J.E.; Mallick, R.; Lynn, T. A Development of a Mixture Design Procedure for Stone Matrix Asphalt. *Proc. Assoc. Asph. Paving Technol.* **1997**, *66*, 1–24.

23. Guillaume, P.; Verboven, P.; Vanlanduit, S. Frequency-Domain Maximum Likelihood Identification of Modal Parameters with Confidence Intervals. In Proceedings of the International Seminar on Modal Analysis Katholieke Universiteit, Leuven, Belgium, 16–18 September 1998; Volume 1, pp. 359–366.

MDPI
St. Alban-Anlage 66
4052 Basel
Switzerland
Tel. +41 61 683 77 34
Fax +41 61 302 89 18
www.mdpi.com

Infrastructures Editorial Office
E-mail: infrastructures@mdpi.com
www.mdpi.com/journal/infrastructures

www.ingramcontent.com/pod-product-compliance
Lightning Source LLC
Chambersburg PA
CBHW061458210926
51850CB00008B/1752
9783039363162